ENVIRONMENTAL IMPACT OF BIOFUELS

Edited by **Marco Aurélio dos Santos Bernardes**

Environmental Impact of Biofuels
Edited by Marco Aurélio dos Santos Bernardes

Published by InTech

Janeza Trdine 9, 51000 Rijeka, Croatia

Publishing Process Manager Petra Zobić

Technical Editor Teodora Smiljanic

Cover Designer InTech Design Team

Additional hard copies can be obtained from orders@intechopen.com

Environmental Impact of Biofuels, Edited by Marco Aurélio dos Santos Bernardes
p. cm.
ISBN 978-953-307-479-5

For EU product safety concerns:
IN TECH d.o.o., Prolaz Marije Krucifikse Kozulić 3, 51000 Rijeka, Croatia,
info@intechopen.com or visit our website at intechopen.com.

Meet the editor

Dr.-Ing. Marco Aurélio dos Santos Bernardes serves as a postdoc researcher at the Centre de Recherche Public Henri Tudor in Luxembourg. His expertise is in the area of energy analysis, life cycle assessment, renewable energy and biofuels. Dr.-Ing. Bernardes has had 10 papers published in journals such as Solar Energy, International Journal of Life Cycle Assessment, ASME Heat Transfer, a book and book chapters as well. Dr.-Ing. Bernardes' areas of interest include CFD, heat transfer modelling, Solar Chimney Power Plants, thermal processes, thermodynamics. He received his Ph.D. in Mechanical Engineering at Stuttgart University in Germany and conducted a postdoctoral research at the Stellenbosch University in South Africa. He was awarded with the UNEP/SETAC Life Cycle Assessment Award for LCA Projects in Development Countries. Dr.-Ing. Bernardes served as a full professor in the Department of Mechanical Engineering at CEFET-MG in Belo Horizonte for more than 13 years.

Contents

Preface

Over the past twenty years, there has been a substantial increase in research and development in the area of biofuels. Many researchers around the world have dealt with environmental, economic, policy and technical aspects relating to these studies. In a way, this book aspires to be a comprehensive summary of current biofuels issues and thereby contribute to the understanding of this important topic. Chapters include digests on the development efforts on biofuels, their implications for the food industry, current and future biofuels crops, the successful Brazilian ethanol program, insights of the first, second, third and fourth biofuel generations, advanced biofuel production techniques, related waste treatment, emissions and environmental impacts, water consumption, produced allergens and toxins.

Relating theoretical and experimental analyses with many important applied purposes of current relevance will make this book extremely useful for researchers, scientists, engineers and graduate students, who can make use of the experimental and theoretical investigations, assessment and enhancement techniques described in this multidisciplinary field. Additionally, the biofuel policy discussion is expected to be continuing in the foreseeable future, and the reading of the biofuel features dealt with in this book, are recommended for anyone interested in understanding this diverse and developing theme.

Marco Aurélio dos Santos Bernardes
Researcher Energy & Environment and LCA,
CRP Henri Tudor, CRTE
Luxembourg

1

Environmental Impacts of Production of Biodiesel and Its Use in Transportation Sector

Sippy K Chauhan and Anuradha Shukla
Traffic Planning & Environment Division,
Central Road Research Institute (CSIR), New Delhi,
India

1. Introduction

The world is presently confronted with the twin crises of fossil fuel depletion and environmental degradation. The search for alternative fuels, which promise a harmonious correlation with sustainable development, energy conservation, efficiency and environmental preservation, has become highly pronounced in the present context. The fuels of bio-origin can provide a feasible solution to this worldwide petroleum crisis. Gasoline and diesel-driven automobiles are the major sources of greenhouse gases (GHG) emission [3 - 5]. Scientists around the world have explored several alternative energy resources like biomass, biogas [6] primary alcohols, vegetable oils and biodiesel. These alternative energy resources are highly environment-friendly but need to be evaluated on case-to-case basis for their advantages, disadvantages and specific applications. Some of these fuels can be used directly, while some others need to be formulated to bring the relevant properties closer to conventional fuels.

Environmental concerns have increased significantly in the world over the past decade, particularly after the Earth Summit-92. Excessive use of fossil fuels has led to global environmental degradation effects such as greenhouse effect, acid rain, ozone depletion and climate change. So there is need to develop or find alternative ways to power the world's motor vehicles.

There are two global biorenewable liquid transportation fuels that might replace gasoline and diesel fuel. These are bioethanol and biodiesel. Bioethanol is good alternate fuel that is produced almost entirely from food crops. Biodiesel has become more attractive recently because of its environmental benefits.

Transport is one of the main energy consuming sectors. It is assumed that biodiesel is used as a fossil diesel replacement and that bioethanol is used as a gasoline replacement. Biomass based energy sources for heat; electricity and transportation fuels are potentially carbon dioxide neutral recycle the same carbon atoms. Due to its widespread availability, biorenewable fuel technology will potentially employ more people than fossil fuel based technology [7].

The term biofuel is referred to as solid, liquid or gaseous fuels that are predominantly produced from biorenewable or combustible renewable feedstocks [8]. Liquid biofuels are important for the future because they replace petroleum fuels. Biofuels are generally considered as offering many priorities, including sustainability, reduction of greenhouse gas emissions, regional developments, social structure and agriculture, security of supply [9].

The biggest difference between biofuels and petroleum feedstocks is oxygen content. Biofuels are non polluting, locally available, accessible, sustainable and are a reliable fuel obtained from renewable sources. Electricity generation from biofuel has been found to be a promising method in near future. The future of biomass electricity generation lies in biomass integration gasification / gas turbine technology, which offers high energy conversion efficiencies.

First generation biofuels refers to biofuels made from starch, sugar, vegetable oils or animal fats using conventional technology. The basic feedstocks for the production of first generation biofuels are often seeds or grains such as wheat, which yields starch that is fermented into bioethanol, or sunflower seeds, which are pressed to yield vegetable oil that can be used in biodiesel. Table 01 shows the classification of renewable biofuels based on their production technologies [10].

Generation	Feedstock	Example
First Generation biofuels	Sugar, Starch, vegetable oils, or animal fats	Bioalcohols, vegetable oil, biodiesel, biogas
Second Generation biofuels	Non food crops, wheat straw, corn, wood, solid waste, energy crops	Bioalcohols, bio – oil, bio – DMF, wood diesel
Third Generation biofuels	Algae	Vegetable oil, biodiesel
Fourth Generation biofuels	Vegetable oil, biodiesel	Biogasoline

Table 1. Classification of renewable biofuels based on their production technologies

Second and third generation bio fuels are also called advanced bio fuels. Second generation bio fuels are mainly made from non – food crops like wheat straw, corn, wood etc. On the other hand appearing the fourth generation is based in the conversion of vegoil and biodiesel into bio gasoline using the most advanced technology.

Renewable liquids bio fuels for transportation have recently attracted huge attention in different countries all over the world because of its renewability, sustainability, common availability, regional development, rural manufacturing, jobs, reduction of greenhouse gas emissions and its biodegradability. Table 02 shows the availability of modern transportation fuels. There are several reasons for biodiesel to be considered as relevant technologies by both developing and industrialized countries [7]. They include energy security reasons, environmental concerns, foreign exchange savings and socioeconomic issues related to rural sector.

Due to its environmental merits, the share of bio fuel in the automotive fuel market will grow fast in the next decade [11, 12]. The advantages of bio fuels are the following –

a. They are easily available from biomass sources
b. They represent a carbon dioxide cycle in combustion

c. They have a considerable environmentally friendly potential
d. They have many benefits for the environment, economy and consumer and
e. They are biodegradable and contribute to sustainability [13].

Fuel Type	Availability	
	Current	Future
Gasoline	Excellent	Moderate poor
Bioethanol	Moderate	Excellent
Biodiesel	Moderate	Excellent
Compressed natural gas (CNG)	Excellent	Moderate
Hydrogen for fuel cells	Poor	Excellent

Table 2. Availability of modern transportation fuels

Various scenarios have resulted in high estimates of bio fuels in the future energy system. The availability of resources is an important factor. The rationale is to facilitate the transition from the hydrocarbon economy to the carbohydrate economy by using biomass to produce bio ethanol and bio methanol as replacements for traditional oil based fuels and feed stocks.

The refining, transport and combustion of bio fuels can result in significant environmental costs, particularly on local water and air quality. Generally, these effects pale in comparison to those generated by the use of fossil fuels, where the main detrimental environmental effects originate from the vehicle exhaust pipe. Even so, these impacts could expand considerably as bio fuel production increases to meet rapidly rising global demand. However, more sustainable practices and new technologies offer the potential for environmental improvements.

This chapter elaborates the main environmental impacts associated with bio fuels processing, transport and use. In order to provide comparison, it first describes some of the environmental costs resulting from processing and the use of petroleum transport fuels.

2. Environmental costs of petroleum refining and use

While the use of oil has brought incalculable benefits to modern industrialised society, it has also extracted great costs, particularly to the local and global environments. Most of these things occur during oil refining and fuel consumption. Delucchi[14] estimated that in the US the costs of environmental externalities associated with oil and motor vehicle use totally between US $ 54 to $234 billion in 1991 alone. Human mortality and disease due to air pollution accounted for more than three quarters of these costs. In Germany it estimated that the quantifiable costs of air pollution and carbon dioxide emissions associated with the transport sector in 1998 totalled about US $ 14.5 billion.

3. Oil refining

Refining of petroleum is an energy intensive, water hungry and very highly polluting process. Everyday, average US refinery releases 41, 640 litres of oil and other chemicals into the air, soil and water [15]. Population lives around that location may feel higher incidences of respiratory problems, skin irritation, nausea, eye problems, headaches, birth defects, cancers etc.

Crude oil, chemical inputs and refined products leak from storage tank and spill during transfer points. Numerous toxins are likely to enter the groundwater, including benzene, toluene, ethyl benzene and xylene [16]. Other chemical may split into the air. Gases such as methane and slightly heavier hydrocarbons such as those in gasoline evaporate. Other chemicals enter in the air as combustion products; the most significant of these are sulphur dioxide (SO2), Nitrogen dioxide (NO2), carbon dioxide (CO2), carbon monoxide (CO), dioxins, hydrogen flouride, chlorine, benzene, large and small particulates and lead [17]. It is well established that oil refineries are the largest industrial source of volatile organic compounds and carbon dioxide, which leads to ozone and smog formation in tropospohere. The second known source of sulphur dioxide which contributes to particulate matter and acid rain and the third largest source of nitrogen oxide, all these are known as ozone precursors [18].

4. Oil transport

Most of the world's crude oil comes from field far from where it is refined and transported big distances from field to refinery and from refinery to fuel station. Large tanker vessels account for 68% of crude delivery to refineries covering an average of 6600 Km per trip. Oil pipelines, used mainly in places where deliveries can be land based, account for 30% while trucks and train transport the reminder [17].

Invariably oil spills occur along the journey. Although most tanker spillage is relatively minor, while during loading or unloading, even small amounts can damage ecosystems. Pipelines spills, although typically smaller, can also be ecologically disruptive, polluting soil and seeping into ground water. Such spills can be fairly common in region where pipelines are not maintained adequately [19].

Oil is shipped over distances to refineries and from refineries, gasoline and diesel fuels travel via pipelines and trucks to fuel depots. Upon leaving the refinery 59% of refined petroleum fuels enter pipelines before loading to trucks [17]. Gasoline and diesel are lighter hydrocarbons that tend to evaporate, participating in complex reactions that form ozone in the atmosphere. Benzene is another pollutant and is also evaporative and is well known for its carcinogenicity. The most significant hydrological pollutant is methyl tertiary - butyl ether (MTBE), fuel additive derived from petroleum that seeps quickly into nearby groundwater and is a likely carcinogen [20].

5. Combustion of petroleum fuels

Compared to bio fuels, petroleum contains a much wider variety of chemical molecules, including far more sulphur. Most of these have been sequestered in the earth for ten and even hundred years of millions years. The burning of gasoline and diesel fuels releases host

pollutants and heavy metals that affect local and regional air quality and these are well linked with global warming issues.
Transport related air pollution leads to reduce visibility, damage to vegetation and buildings and increased incidence of human illness and premature death [21]. Road transport is also growing contributor to air pollution in many developing countries / cities particularly where diesel remains the predominant fuel [22]. Table 03 is summarized below shows the main environmental and health impacts associated with the petroleum primary combustion products including CO2, CO, unburned hydrocarbons, NOx, SOx, particulates and in some countries lead [23].

Combustion product	Impacts
CO2	Contributes to global warming and climate change
CO	Results from incomplete combustion or burning. In the atmosphere, CO reacts with oxygen to form ozone, a highly reactive molecule that damages plant leaves and human and animal lungs
Benzene	The smallest aromatic hydrocarbon and a highly toxic carcinogen.
NO & NOx	Ozone precursor, they also react with atmospheric water and create acid rain
SO2 & SO3	Acid rain precursors,
Lead	Has been phased out from gasoline in most of the countries, but is still used as an octane enhancer
Particulate matter	Formed from SOx, NOx and hydrocarbons, particulates contribute to ozone formation and affect visibility and hence global warming.

Table 3. Environmental and health impacts of emissions from petroleum combustion

6. Environmental impacts of biofuel

Same as like petroleum fuels, bio fuels can have environmental impacts at all stages of their production and use. Relative to fossil fuels, however, the impacts resulting from refining, transporting and using bio fuels are generally significantly smaller. Moreover, there are ways to improve the resource efficiency and impacts of these activities.

7. Water use

Large quantities of water can be utilise for the processing the bio fuel feedstock into fuel. The primary uses of water for biodiesel refining are to wash plants and seeds for processing and then to remove the soap and catalysts from the oils before and final product is shipped out. A typical US Soybean crushing system requires just over 19 Kg water per tonne of oil produced [24]. For each tonne of soybeans that go into the refining process, 170 kg come out as crude de - gummed soybeans oil, 760 kg are soy meal and remaining 70 kg include air and solid and liquid waste [24]. The primary contaminant in wastewater is soybean oil [24]. Production of ethanol, in particular, requires a tremendous amount of water for processing and for evaporative cooling to keep fermentation temperatures at the required level [25]. But some feed stocks are more water intensive than other; each tonne of sugar cane in Brazil, for example, requires as much as 3900 litres for processing [26]. Ethanol processing also results in large volumes of nutrient rich wastewater that, if not cleaned and recycled can speed eutrophication of local rivers and streams by affecting the water's dissolved oxygen content [27]. In addition, sugar mills must be flushed every year, putting huge amount of organic matter into local waterways [28]. In Brazil, 1 litre of ethanol produces about 10 to 15 litres of vinasse which is very hot and corrosive, with a low pH and high mineral content [22]. Today, however, wastewater and vinasse are recycled and used for irrigation and fertilization of Brazil's sugarcane crops, with varying quantities of vinasse used under different conditions as regulated by law [29]. However, some experts caution that vinasse cannot be used where water tables are high, such as in India [30]. Also, if used excessively, vinasse can cause eutrophication of surface water due to the increased load [31]. Filter cake another waste stream from ethanol processing is also recycled as a fertilizer. As a result Brazil has been able to significantly reduce its use of petroleum fertilizers, saving money while creating value from waste products [22].

8. Air pollution

Among the pollutant that bio refineries emit in to the air are SOx, NO_X, VOCs and particulate matter. Emission from corn ethanol plants, for example include SOx, NOx, CO, mercury, particulates and CO [32]. Corn ethanol plants in low a have polluted both water and air, emitting cancer causing chemicals such as formaldehyde and toluene [33].

Biodiesel production require methanol, which has the same environmental cost as those associated with petroleum production. In addition, direct emission from biodiesel processing plants can include air, stream and hexane, which can be used to extract oil from plants and seeds. Hexane is air pollution, and through as much as possible is recovered and recycled, some is emitted into air as well. Sheehan et al (in 1998) estimate that the average US soybean crushing system releases just over 10 Kg of hexane per tonne of oil produced. Alternatives have been found so that hexane is no longer needed; but these options are more costly [34]. In addition, where renewable sources are not used to produce process energy, pollutants associated with the use for natural gas and the generation of steam and electricity are released into air. An estimated 3.6 Kilowatt hours of electricity are required per ton of soybeans entering in a soy biodiesel plant [35]. On the other hand, Fischer-Tropsch [F-T] biodiesel in gasification based and therefore has minimal local air pollution problem [36].

As plant size increases, concerns about pollution - including air emissions, odour releases during the drying of distillers grain in corn ethanol plants and waste water recharges- have risen as well [22]. However, with appropriate regulation and pollution control technologies, emission associated with bio fuels refining can be minimised significantly [37, 38]. For example NOx emission from boiler can be reduced by installing new NOx burner system [39]. VOCs emission, which results primarily from the blending of ethanol with gasoline, can be reduced by mixing fuels at locations where pollutants can be collected and treated [40]. In some cases, new and larger plants are incorporating such emission control system and are finding alternative options that are enable them to reduce such emissions [36].

In addition, much of the air pollution associated with bio fuels refining results from burning of fossils fuels for process heat and power - which in the US, Germany, China and many other countries is mainly coal. Thus, emission can be reduced through traditional power plant control technology or the use of renewably generated power [36].
In Brazil, today, mills and distilleries meet most if not all of their own energy needs with bagasse (a by product of sugar cane crushing), which can generate thermal, mechanical and electrical energy. Some plants even sell surplus electricity into the grid [22]. Elsewhere, agricultural and forestry residues can be used to produce required power and heat; however it is important to ensure that enough residues remain to maintain soil organic matter and nutrient levels [40].

9. Biofuel transport and storage

9.1 Water pollution

Pure ethanol and biodiesel fuels offer significant environmental benefits compared to petroleum fuels, making them highly suitable for marine and farm uses, among others. They result in dramatically reduced emission of VOCs and are less toxic to handle the petroleum fuels [41]. One other significant advantage relate specifically to water: both ethanol and biodiesel are biodegradable and break down readily, reducing their potential impact on soil and water [42].
Biodiesel is far more soluble than petroleum diesel, enabling marine animals survive in far higher concentrations of it than petroleum if fuel spills occur (due to lower risk of suffocation) [43]. Such benefits are helping to drive biofuel promotion policies in China, where vehicle have polluted water bodies and gasoline and diesel leakage pipelines has polluted ground water - affecting biodiversity, drinking water and soil resources [25].
At least one study has shown that biodiesel made with rapeseed oil can biodegrade in half the time required for petroleum diesel. Biodiesel also speeds the rate at which biodiesel petroleum blends can biodegrade, which is not the case of ethanol [43]. There is evidence that ethanol's rapid break down deplete the oxygen available in water and soil, actually slowing the breakdown of gasoline. This can increase gasoline's impact on the environment in two ways. First, the harmful chemicals in gasoline persist longer in environment than they otherwise would; benzene, in particular, can last 10 - 150 % longer when gasoline is blended with ethanol. Second, because gasoline breaks down more slowly, it can travel further (up to 2.5 times) in the marine environment, affecting a greater area [44].
Additionally, if ethanol is spilled, it can remobilize gasoline in previously contaminated soils, intensifying the impact of the initial spill. Since up to 85% of such spills occur at gasoline terminals. This is where such a problem is most likely to happen [45]. The transition of high label of ethanol needs to be planned with such impact in mind and should include regulations for the handing for fuels.

9.2 Air pollution

Most biomass is carried to processing plants by truck and most bio fuels are transported by truck as well, although some travel by train or, in Brazil, via pipelines. The environmental impacts associated with transport include the air emission and other pollutants associated with the life cycle of the fuel used - in most cases, petroleum diesel. As demand of bio fuel increases and as consumption exceeds, production in some countries, it is likely that a raising amount of feed stocks and bio fuel will be transported by ship. By shipping is a relatively energy-efficient means of transport, it is also a major source of pollution due primarily to a lack of regulations governing maritime emissions. Pollutants include NOx, SO_2, CO_2, particulate matter and a number of highly toxic substance, such a formaldehyde and poly aromatic hydrocarbons [46]. Emissions from diesel from marine engines represent an ever increasing share of air pollution, and most of these pollutants are released near coastlines, where they can easily be transported over land [46].

The over potential concern associated with bio fuels transport is possibility for spills and evaporation. Bio fuels can leak at the production facility, spill while being transported and leak from above and below ground tanks. They can also evaporate during fuelling and storage and from a vehicle's fuelling system.

In general, 'neat' bio fuels are distinctly less toxic than spills of petroleum fuels. For biodiesel, evaporative emissions are not a particular concern since biodiesel fuel does not have a higher vapour pressure. Neat ethanol has a low rigid vapour pressure (RVP), and when stored as a pure fuels (or even as an E-85 blends), it has a lower vapour pressure than gasoline and thus will have fewer evaporative emissions [36].

The primary concern regarding emission from bio fuel transport has to do with lower- level blends of ethanol in gasoline, which tend to raise vapour pressure of the base gasoline to which ethanol is added. When ethanol is blended up about 40% with gasoline, the two fuel combined have higher evaporative emission than either does on its own. The fuels are mixed via splash blending at the petroleum supply 'rack', so there is a potential for increased evaporative emission from these lower - level blends at the point in the distribution chain and 'downstream'- mainly during vehicles refuelling and from use in the vehicles. These evaporative emissions from a vehicle's fuelling system can increase ozone pollution.

Adding the first few per cent of ethanol generally causes the biggest increase in volatility so increasing the blend level to 2.5 or even 10 per cent will have similar results [47]. Evaporative emissions peak at the blend level between 5 to 10 percent and then start to decline. Once ethanol's share exceeds 40 percent, evaporative VOC emission from the blend are lower than those from gasoline alone [48].

Most international energy agencies (IEA) countries have emission standards requiring the VOC emissions and thus RVP, be controlled [47]. Emission resulting from higher vapour pressure can be controlled by requiring refiners to use base gasoline stock with a lower vapour pressure when blending with ethanol, although this increases cost and reduces production lavels. The US state of California and US federal reformulated gasoline programmes have set caps on vapour pressure that take effect during high ozone seasons in areas that do not meet ambient air quality standard of ozone. As a result, the addition of ethanol does not increase the vapour pressure of the gasoline available during summer months [49]. Emission from permeation are more difficult to control in the on-road fleet, although expert believes that most can be controlled in new vehicles that much meet stricter evaporative emission control standard (such as California LEV 2 and US Federal Tier II), with higher-quality tubes, hoses and other connectors [49].

10. Biofuel combustion

The level of exhaust emissions that results from the burning of ethanol and biodiesel depends upon the fuel (e.g. feedstock and blend), vehicles technology, vehicle tuning and driving cycles [22]. Most studies agree that using bio fuels can significantly reduced most pollutants compared to petroleum fuels, including reductions in controlled pollutant as well as toxic emissions [47]. NOx emissions have been found to increase slightly as blend level rise, although the levels of emissions differ from study to study.

10.1 Ethanol

Ethanol contains no sulphur, olefins, benzene and other aromatics [22]. All of which are component of gasoline that can affect air quality and threaten human health [22]. Benzene is carcinogen, while olefins and some other aromatics are precursor to ground-level Ozone (smog) [47]. Ethanol-gasoline blends also reduce toxic emissions of 1,3-butadiene, toluene, xylene, while few studies have a looked at the impacts on the pollution levels from high blends.

With ethanol fuel combustion, emission of the toxic air pollutants acetaldehyde, formaldehyde, peroxyacetyl nitrate (PAN) increase relative to straight gasoline [47]. Most is emitted as acetaldehyde, a less reactive and less toxic pollutant than formaldehyde. Neither pollutant present in fuel; they are created as by product of incomplete combustion. PAN is an eye irritant that is harmful to plants, is also formed as by product [47]. A US auto-oil industry study determined that combustion of E85 resulted in a slight increase in hydrocarbon emission relative to California reformulated gasoline. It is also found that toxic emission rose as much as two to three folds compared to conventional gasoline, due mainly to an increase in aldehyde emission.

There is concern that aldehydes might be carcinogenic; but the pollutant that are reduced by blending with ethanol (including benzene, 1,3 - butadiene, toluene and xylene) are considered more dangerous for human health. A study done in California determined that acetaldehyde and PAN concentrations increases only slightly with ethanol blends, and a Canadian study concluded the risks of increased aldehyde pollutants are negligible [90]. Because of reactivity of aldehydes, emissions can generally be managed with emission controls [36]. For example, three way catalysts can efficiently minimize aldehyde emissions [50].

Ethanol blended gasoline increase fuel oxygen content, making hydrocarbons in the fuel burn more completely in older vehicles, in particular, thus reducing emission of CO and hydrocarbon emissions [22]. Ethanol used an additive or oxygenate (e.g. 10 percent blend) has been found to achieve CO reductions of 25 percent or more in older vehicles [22]. In fact, one of the goals driving the use of ethanol in US during the 1990s was to reduce hydrocarbons and CO emissions particularly in winter when emission of these pollutants tend to be higher. Ethanol in higher blend will be positively affecting the efficiency of catalytic convertors because of the dilution of Sulphur [51]. Ethanol can be used to make ethyl tertiary butyl eather (ETBE), which is less volatile than ethanol and widely used in the European Union (EU) [51].

As a result of its national ethanol programme *proalcool*, Brazil was one of the first countries in the world to eliminate lead entirely from gasoline. According to Sao Poulo State Environmental Agencies (CETESB), ambient lead concentrations in the Sao Poulo metropolitan region declined from 1.4 gram per cubic meter in 1978 to less than 0.1 gram per cubic meter in 1991 [52]. Most of the countries however have been able to eliminate lead through other means, including a reduction in unnecessarily high octane grades and the development of the chapter refining alternatives (e.g. reforming and isomerisation) [22].

Ethanol use has resulted in significant reductions in other air pollutants as well. Emission of toxic hydrocarbons such as benzene has declined in Brazil, in addition to the emission of sulphur and CO. For Example, Brazil transport related CO emission declined from more than 50grams per kilometre in 1980 to less than 1gram per kilometre in 2000 due to ethanol use. CETESB estimates that urban air pollution in Brazil could be reduced in additional 20-40 percent if the entire vehicle fleet were fuelled by alcohol [22]. In 1998 Denver, Colorado, became the first US city to require blending of gasoline with ethanol; it is used in winter to improve fuel combustion and to reduce CO emissions. As a result, it is estimated that CO level have declined by 50 percent [53].

There is some evidence that emission reductions associated with using ethanol blends, compared to straight gasoline, are not as significant in the cleanest vehicles available today. Durbin et al (2006) tested vehicles that qualified as low emission and ultra low emission in California, and found that emission of non methane hydrocarbons increased as engine temperature rose and that benzene emission increased with higher concentrations of ethanol, while fuel efficiency declined. However, CO emission decreased somewhat with ethanol use [54]. Some of the findings were inconsistence with those of the studies, highlighting the need for further research [54].

As discussed earlier, ethanol used as an oxygenate can reduce emission of several pollutants particularly in older vehicles. However the use of oxygenates such as ethanol (and biodiesel), to alter the fuel to oxygen ratio will not necessarily have a positive effect on emission if a vehicle's air-to-fuel ratio is set low or if too much ethanol is added to gasoline in a vehicle with a fixed air-to-fuel ratio. If that is the case, oxygenate can increase NOx emissions and cause 'lean misfire' increasing hydrocarbon emissions [22]. In fact, Tyson et al (1993) argue that ethanol has no emission related advantages over reformulated gasoline other than the reduction of CO_2 [55].

Ethanol blended with diesel can provide substantial air quality benefits, Blends of 10-15 per cent ethanol (combined with performance additive) result is significantly lower emission compared with pure diesel fuel; exhaust emissions of PM, CO, and NOx decline. For high blends, the results are mixed. Some studies have found higher average CO and hydrocarbon emissions and other have seen reductions in these pollutants. However, all studies, to date, have seen significant decrease in both PM and NOx [47].

Flexible-fuel vehicles (FFV)- which can take virtually any ethanol- gasoline blend up to 85 percent in the US and up to 100 percent in Brazil- are widely used in the Brazil and are becoming increasingly available in the US. However, tests to date have found that the use of FFVs results in higher air emission than new gasoline vehicles [36]. Because it is not possible to tune the combustion controls of vehicles so that it is optimized for all conditions, controls are compromised somewhat to allow for different mixes [56,57]. It is possible that vehicles dedicated to specific blends, operated on those blends level, would achieve lower emissions than conventional vehicles.

10.2 Biodiesel

Biodiesel – whether pure or blended- results in lower emissions of most pollutants relative to diesel, including significantly lower emission of particulates, sulphur, hydrocarbons, CO, toxins [57]. Emissions vary with engine design, condition of vehicles and quality of fuel. In biodiesel- diesel blends, potential reductions of most pollutants increase almost linearly as the share of biodiesel increases, with the exception of NOx emission [47].

In one of the most comprehensive analyses to date, a US Environmental Protection Agency (EPA) study of biodiesel determined that the impacts on emissions vary depending upon type (feedstock) of biodiesel and the type of petroleum diesel that it is mixed with. Overall animal based biodiesel did better in the study than plant based biodiesel with regard to reducing emission of NOx, CO and particulates. On average, the EPA determined that B20 (made with soybeans) increase NOx emission the least, followed by rapeseed biodiesel and that soybean based biodiesel; the same relationship held true for CO reduction, as well. Reductions in particulate emissions were also greatest for animal based biodiesel [58].

The test carried out by the EPA showed that, when compared with conventional diesel, pure diesel (produced with Soybean oil) resulted an average reduction of particulate matter by 40 percent, CO by 44 per cent, unburned hydrocarbons by 68 percent, polycyclic hydrocarbons (PAHs) by 80 percent, carcinogenic nitrate by 90 percent, sulphate by 100 percent [59].

During 2000, biodiesel become the first alternative of fuel to successfully complete testing for tier 1 and 2 for health effect under the US Clean Air Act. Test determined that, with the exception of minor damage to the lung tissue at high level of exposure, animal observed in the study suffered non biological significant short term effect associated with biodiesel [22].

A 1999 Swedish study by Pedersen et al found that biodiesel (rapeseed methyl ester, or RME) led to an up to tenfold increase in emission of benzene and Ozone precursors compared with Swedish low sulphur diesel fuel, called MKI [60]. However, this study was conducted using a very small reactor; many US and European researches were sceptical about transferring results from this study to the real world for combustion in a diesel engine. Since then, other studies have produced results. For example, Krahl et al (2000) compared 100 percent RME to MKI, fossil diesel fuel and another low sulphur diesel fuel (with high aromatic compounds content and flatter boiling characteristics, known as DF05), using modern DaimlerChrysler diesel engine such as those generally installed in light duty transport vehicles. They concluded that RME lead to significant reduction in CO, hydrocarbons, (HCs), aromatics HCs (including Benzene) and aldehydes, ketones (which contribute to the formation of summer smog) compared with the other fuels [61].

11. Impact of NOx emissions

Most studies conclude that ethanol and biodiesel emit higher amounts of nitrogen oxides (NOx) than do conventional fuels, even as other emissions decline [47] there are exceptions, however. When ethanol is blended with diesel, NOx, emissions decline relative to pure diesel fuel; and some tropical oils are saturated enough- thus have a high enough cetane value - that they increase NOx less (and in the case of highly saturated oils such as coconut, actually decrease NOx) relative to diesel [62]. NOx are precursor to ground level ozone (smog). In addition, NOx emission increase acid rain and are precursor to fine particulate emissions; associated with health impact include lung tissue damage, reduction in lung function and premature health [63].

The level of NOx emissions found varies significantly from study to study. Some cities, particularly in the US state of California, have complained that ethanol has increased local problems with NOx and ozone [22]. California is using ethanol as an oxygenate meet requirements under US Clean Air Act because concern about water contamination led to the state to ban MTBE. More recently concern about evaporative VOCs emission and combustion emissions of NOx led California to sue the US EPA twice for a waiver; both times the waiver was denied [58]. But both the EPA and California Air Resources Board agreed during the process that ethanol increases NOx slightly in the on-road fleet [64].

Fulton et al (2004), on the other hand, report that the impact of bio fuels on NOx emissions level are relatively minor and can actually be higher or lower than conventional fuel, depending upon the conditions. In fact, there is evidence that NOx level from low ethanol blends range from a 10 percent decrease to a 5 percent increase relative to pure gasoline emission [47].

Studies by US National Renewable Energy Laboratory (NREL) show inconsistent results with regard to biodiesel and NOx, depending upon whether vehicle is driven on the road or in the laboratory. According to McCormick (2005), they have seen 'Nox reductions for testing of vehicles (chassis dyno) and Nox increases for testing of engines (engine dyno). The former, which involves driving an entire car on rollers rather than testing emissions directly from an engine removed from the vehicle, is considered more realistic than the latter [65].

NREL studies of in-use diesel buses have found a statistical significant reduction in NOx emissions with biodiesel. A US auto-oil industry six year collaborative study examined the impact of E85 on exhaust emissions and found that NOx emission were reduced by upto 50 percent relative to conventional gasoline [66]. But India's Central Pollution Control Board has determined that burning biodiesel is a conventional diesel engine increases NOx emissions by about 13 % [57].

Fortunately, newer vehicles designed to meet strict air standards, such as those in California, have very efficient catalyst system that can reduce VOC, NOx and CO emissions from ethanol-gasoline blends to very low levels [36]. With biodiesel, NOx increases can be minimized by optimizing the vehicle engine for the specific blend that will be used [47]. Emission can also be reduced with additives that enhance the cetane value or by using biodiesel made from feedstock with more saturated fats (e.g. tallow is better than canola, which is better than soy) [65].

It is possible to control diesel exhaust using catalysts and particulate filters. High efficiency Diesel Particulate Filter (DPF) remove particulate matter (PM) by filtering engine exhaust; such system can reduce PM emissions by 80 percent or more. However, because of concern about increased oil film dilution during the post- injections. German car manufactures do not accept neat biodiesel in DPF equipped vehicles [67]. There is also concern that the extra injection used to increase emission temperatures for regeneration of the particulate trap result in a dilution of engine oil when RME is used as a fuel, and this dilution can increase engine wear [68]. Rust particles filters, which are available in many new diesel automobiles and significantly reduce emissions of fine particulates, cannot operate with biodiesel [69].

According to some sources, biodiesel do not meet European air emissions standards that went into effect in January 2006 [69], although the Association of German Biofuel Industry noted that biodiesel can meet updated European standards for trucks and commercial vehicles.

Several groups are in the process of developing additives to address the issue of NOx, emissions associated with biodiesel blends, including NREL, the US National Biodiesel Board, the US Department of Agriculture and World Energy Alternatives [65].

12. Advanced technologies

In general, the air quality benefits of biofuel are greater in developing countries, where vehicle emission standards are non-existent or less stringent and where older more polluting cars are more common [70]. For example, the use of ethanol can effectively reduce emissions from CO and hydrocarbons in old technology vehicle today [22]. Less understood, however, are the impacts that biodiesel might have on exhaust emission from vehicles that are underpowered, over-fuelled, overloaded and not well maintained- vehicles that are most prevalent in the world's developing nation [22].

Advances in pollution control technologies for petroleum-fuelled vehicles will reduce. If not eliminate, the relative benefits of biofuels. Greene et al (2004) note that the main benefit of biofuels in such advanced vehicles may be to make it easier to comply with emission standard in the future, thus reducing the cost emission control technologies [36].

At the same time, new technologies are on the horizon. For example, Volkswagen and Daimler Chrysler have invested in biomass-to-liquid (BTL) technologies that convert lignocellulosic fibers into synthetic biodiesel. This process enables them to produce a cleaner burning biofuel. In the future, they hope to optimize fuels and vehicle engines in parallel.

13. Conclusion

The refining, transport and combustion of biofuels have environmental costs, particularly on local water and air quality, and these impacts could rise considerably as biofuel production increases to meet rapidly rising global demand. At the same time, more sustainable practices and new technologies offer the potential for environmental improvements.

Increasing efficiencies in water and energy use at refineries can help to reduce both air and water pollution. The UK-based biodiesel producer D1 Oils now recycles both water and methanol used in its refineries and uses biodiesel to run its facilities [71]. Standards and regulations are also needed to minimize pollutants. In addition, encouraging smaller scale distributed facilities will make it easier for communities to manage wastes, while possibly relying on local and more varied feed stocks for bio fuel production and thereby benefiting local economies and farmers.

The combustion of bio fuels- whether blended with conventional fuels or pure-generally results in far local emissions of CO, hydrocarbons, SO_2 and particulate matter (and, in some instances lead) than does the combustion of petroleum fuels. Thus, the use of bio fuels, particularly in order vehicles, can significantly reduce local and regional air pollution, acid deposition and associated health problems; such as asthma, heart and lung disease and cancer [72].

However, the air quality benefits of bio fuels relative to petroleum fuels will diminish as fuel standards and vehicle technologies continue to improve in the industrialised and developing worlds. Even today, the newest vehicle technologies continue to improve in the industrialised and developing worlds. Even today, the newest vehicles available for purchase largely eliminate the release of air pollutants (aside from CO_2) [73]. At the same time, concern about level of NOx and VOC emissions from bio fuels will probably diminish with improvements in vehicles and changes in fuel blends and additives. A combination of next generation bio fuels can make a major contribution to reducing air pollution in the transport sector.

In the developing world, ethanol should be used to replace lead, benzene and other harmful additives required for older cars and because of high blends or pure bio fuels pose minimal air emissions problems and are less harmful to water bodies than petroleum fuels, for all countries it is important to transition these high blends as rapidly as possible, particularly for road transport in highly polluted urban areas and few water transport, wherever feasible.

14. References

[1] The Earth has warmed by 0.6° C over the past 30 years and by 0.8° C over the past 100 years, according to US National Aeronautics and Space Administration (2006)

[2] Kevin A., Baumert, Timothy Herzog and Jonathan Pershing (2005); Working Group III (2001)

[3] Kesse D G (2000) Global warming - facts, assessment, countermeasures. J. Petrol Sci Eng 26: 141 - 149. Doi: 10, 1016/S0920 - 4105(00)00030 - 9

[4] Cao X (2003) Climate change and energy development: implications for developing countries. Resource Policy 29:61 - 67, doi:10.1016/j.resourpol.2004.05.001

[5] Johansson, Y. McCarthy S (1999) Global warming post Kyoto: continuing impasse or prospects for progress? Energy Dev Rep Energy, pp 69 - 71.

[6] Murphy J D, McCarthy K (2005) The optimal production of biogas for use as atransport fuel in Ireland. Renew Energy 30:2111 - 2127, doi:10.1016/j.renene.2005.02.004

[7] Demirbas A, (2006) Global biofuel strategies, Energy Edu Sci Technol, 17:27 - 63

[8] Demiras A, (2007) Gasoline and diesel fuel blends with alcohols, Energy Edu Sci Technol 19:87 - 92

[9] Reijnders L, (2006) Conditions for the sustainability of biomass based fuel use. Energy Policy 34:863 - 876.

[10] Demirbas, A , (2009) Biofuels, Green Energy & Technology

[11] Kim, S., Dale, B E (2005) Life cycle assessment of various cropping systems utilized for producing biofuels: Bioethanol and biodiesel, Biomass Bioenergy 29:426 - 439

[12] Demiras, M F, Balat M, (2006) Recent advances on the production and utilization trends of biofuels: A global perspective, Energy Convers Mgmt 47:2371 - 2381.

[13] Puppan, D., 2002, Environmental evaluation of biofuels, Periodica Polytechnica Ser Soc Man Sci, 2002; 10 : 95 - 116.

[14] 14] Delucchi, M A (1995), Summary of non monetary externalities of motor vehicle use, report 9 in series, The Annualized Social Cost of Motor Vehicle Use in US, based on 1990 - 1991 Data: Summary of Theory, Methods and Datadraft report prepared for Union of Concerned scientists (UCS), Davis, California, Institute of Transportation Studies.

[15] EC (2003) External Costs: Research Results on socio - environmental damage due to electricity and transport, Brussels, EC

[16] Sherertz, P C (1998), Petroleum products, Richmond, VA, Virginia Department of Health, 30 junewww.vdh.state.va.us/HHControl/petrofac.PDF

[17] Mann, M. And P. Spath (1997) Life cycle assessment of a biomass Gasification Combined cycle Gasification System, Golden, CO, National Renewable Energy Laboratorty

[18] Doniger, D. (2001) Oil Companies, making record profits, seek Environmental Rollbacks, New York, Natural resources Defense Council, 8 May.

[19] Bazilescu, I. and B Lyhus (1997) Russia Oil Spill, Trade and Environment Database case studies, Washington DC
[20] Pennsylvania Department of Environmental Protection (2005) Automobile Emissions: An overview, www.dep.state.pa.us/dep/subject/pubs/arr/aq/fs1829.pdf
[21] Abt associates Inc (2000) The Partculate Related Health Benefits of Reducing Power Plant Emissions, Bethesda, M D, Abt associates Inc, www.abtassociates.com/reports/particulat-related.pdf
[22] Kojima, M & T Johson (2005) Potential for Biofuels for Transport in Developing countries, Washington. DC , World Bank.
[23] McMillen, S et.al. (2005) Biodiesel: Fuel for Thought, Fuel for Connecticu's Futures, Stores, CT, Connecticut Centre for Economic Analysis, University of Connecticut.
[24] Sheehan, J et.al. (1998) Life Cycle Inventory of biodiesel and Petroleum Diesel for use in an Urban Bus, Golden, CO, National Renewable Energy Laboratory
[25] Gehua, W. et.al. (2005) Liquid Biofuels for transportation: Chinese Potential and Implications for sustainable Agriculture and Energy in the 21st Century, Report prepared for GTZ, Beijing
[26] Peplow, M (2005) Ethanol production harms environment, researchers claims, Nature, July
[27] Tampeir, M, et.al. (2004) Identifying Environmentally Preferable Uses for Biomass Resources, Stage 2 Report: Life cycle GHG Emissions Reduction Benefits of selected Feedstock - to - Product Threads, Prepared for Natural Resources Canada and National Research Council Canada, North Vancouver.
[28] Clay, J, (2004), World Agriculture and Environment, Washington, DC, Island press.
[29] Macedo, I C et.al. (2005), Sugar Cane's Energy: Twelve studies on Brazilian Sugar cane Agribusiness and its Sustainability, Sao Paulo.
[30] Hunt, S., and J. Sawin with P. Stair,(2006), Cultivative renewable alternatives to oil, in Worldwatch Institute (ed) State of the World 2006, New York.
[31] Sierra Club of Canada (2004) Mackenzie Pipeline to Fuel Americas' Gas Tank, Press release, Ottawa
[32] Environment News Services (2005) Illinois Ethanol Producer must install Air pollution controls, December
[33] Beeman P. (2005) 'Ethanol plants among Iowa's polluters', Des Moines Register, 11 September
[34] Core, J. (2005) 'new method simplifies biodiesel production', Agricultural Research, April
[35] Sheehan, j. et al (1998c) 'Life Cycle Inventory of Biodiesel and Petroleum Diesel for Use in an Urban Bus, Golden, CO, National Renewable Energy Laboratory, May
[36] Greenc ct al (2004) Growing Energy; How Biofuels Can Help End America's Oil Dependence, Washington, DC, Natural Resources Defence Council, December
[37] Beeman P. (2005) 'Ethanol plants among Iowa's polluters', Des Moines Register, 11 September; USEPA (2005b) 'Ethanol Plant Clean Air Act Enforcement Initiative, Washington, DC, updated 1 September 2005, www.epa.gov/compliance/resources/cases/civil/caa/ethanol ;
[38] Greene et al (2004) Growing Energy; How Biofuels Can Help End America's Oil Dependence, Washington, DC, Natural Resources Defence Council, December

[39] Doggett T. (2006) 'EPA seeks to ease US ethanol plants pollution rules', Reuters, 3 March

[40] Novozymes and BBI International (2005), Fuel Ethanol: A Technological Evolution, Grand Forks, ND, June

[41] TaTEDO (Tanzania Traditional Energy Development and Environmental Organization) (2005) Biofuels for transportation in Tanzania: Potential and Implications for Sustainable Agriculture and Energy in 21st Century, Prepared for the Deutsche Gesellschaft fur Technische Zusammenarbeit (GTZ) GmbH, Dar es salaam, September

[42] Perlack, R.D. et.al. (1995), Biomass fuel from woody crops for electric power generation, Oak Ridge, TN, http://bioenergy.ornl.gov/reports/fuelwoodtoc.html

[43] Cook J & J Beyea, An analysis of environmental impacts of energy crops in the USA: methodologies, conclusions & recommendations, Washington DC, (2005) www.panix.com/~jimcook/data/ec.workshop.html

[44] Northeast States of Coordinated Air Use Management (2001) Health, Environmental, Economic Impacts of Adding Ethanol to Gasoline in the Northeast State, vol 3; Water Resources and Associated Health impacts, Lowell, MA, New England Interstate Water Pollution Control Commission cited in Greene et al (2004) Growing Energy; How Biofuels Can Help End America's Oil Dependence, Washington, DC, Natural Resources Defence Council, December; TERI (2005) Liquid Biofuels for Transportation: India Country Study on Potential and Implications for Sustainable Agriculture and Energy, Report prepared for the Deutsche Gesellschaft fur Technische Zusammenarbeit (GTZ) GmbH, New Delhi, October www.gtz.de/de/dokumente/en-biofuels-for-transpotation-in-india-2005.pdf

[45] Von wedel, R. (1999) Technical Handbook for Marine Biodiesel in Recreational Boats, second addition, Point Richmond, CA, CytoCalture International, www.cytocalture.com/Biodiesel1%20Handbook.htm

[46] Northeast States of Coordinated Air Use Management (2001) Health, Environmental, Economic Impacts of Adding Ethanol to Gasoline in the Northeast State, vol 3; Water Resources and Associated Health impacts, Lowell, MA, New England Interstate Water Pollution Control Commission

[47] Clean Air Task Force et al (2005) Prevention of Air pollution from Ships: Reducing Shipping Emissions of Air pollution – Feasible and cost-effective Options submitted by Friends of the Earth International to the Marine Environment Protection Committee, International Maritime Organisation, 7 April.

[48] Fulton, L., Howes, T. and Hardy, J. (2004) Biofuel for transport: An International Perspective, Paris, France, International Energy Agency

[49] Northeast States of Coordinated Air Use Management (2001) Health, Environmental, Economic Impacts of Adding Ethanol to Gasoline in the Northeast State, vol 3; Water Resources and Associated

[50] Steven J. Brisby, Manager Fuels Section, Stationary, stationary Source Division, California Air Resources Board e-mail to Janet Sawin, world watch Institute, 1 March 2006.

[51] Prakash, C. (1998) Use of Higher than 10 volume Percent Ethanol/Gasoline Blend in Gasoline Powered Vehicles, Ottawa, Environment Canada, www.ec.gc.ca/cleanair-airpur/CAOL/transport/publications/ethgas/ethgastoc.htm

[52] Hammel-Smith, C. Et al (2002) Issues Associated with the Use of Higher Ethanol Blends (E17-E-24) Golden, CO, NREL, October
[53] CETESB (Sao Paulo State Environmental Agency) (2003) Relatorio de Qualidade do ar no Estado de Sao Paulo, CETESB, Sao Paulo
[54] F.O. Licht's World Ethanol and Biofuels Report (2005K) 'Denver winter ethanol to stay until 2008' 19December
[55] Durbin T. Et al (2006) Final Report: Effect of Ethanol and Volatility Parameter on Exhaust Emission, CRC Project No E-67, www.ec.gc.ca/cleanair-airpur/CAOL/transport/publications/ethgas/ethgastoc.htm
[56] Tyson, K.S., C. J. Riley and K. K. Humphreys (1993) Fuel Cycle Evaluations of Biomass Ethanol and Reformulated Gasoline, Golden, CO, NREL
[57] Earley J., T. Earley & M Straub (2005), Specific Environmental effects of trade liberalization: oil seeds, Washington, DC, International Policy council for food & Agriculture
[58] TERI (2005) Liquid Biofuels for Transportation: India Country Study on Potential and Implications for Sustainable Agriculture and Energy, Report prepared for the Deutsche Gesellschaft fur Technische Zusammenarbeit (GTZ) GmbH, New Delhi, October www.gtz.de/de/dokumente/en-biofuels-for-transpotation-in-india-2005.pdf
[59] USEPA (2005) EPA upholds Reformulated Gas Requirement in California, New York, and Connecticut, Press release, Washington, DC, 2 June
[60] Becker, K. and G. Francis (2005) Biodiesel from Jatropha Plantations on Degraded Land, Stuttgart, University of Hohenheim cited in TERI (2005) Liquid Biofuels for Transportation: India Country Study on Potential and Implications for Sustainable Agriculture and Energy, Report prepared for the Deutsche Gesellschaft fur Technische Zusammenarbeit (GTZ) GmbH, New Delhi, October www.gtz.de/de/dokumente/en-biofuels-for-transpotation-in-india-2005.pdf
[61] Perdersen, J. R., A. Ingemarsson and J. O. Olsson (1999) 'Oxidation of rapeseed oil, rapeseed methyl ester (RME) and diesel fuel studied with GC/MS', Chemosphere, vol 38, no 11, pp 2467-2474
[62] Krahl, J. et al (2001) Comparison of Biodiesel with Different Diesel Fuels Regarding Exhaust Gas Emissions and Health Effects, www.ufop.de/downloads/Biodiesel_Comperision.pdf ; Perdersen, J. R., A. Ingemarsson and J. O. Olsson (1999) 'Oxidation of rapeseed oil, rapeseed methyl ester (RME) and diesel fuel studied with GC/MS', Chemosphere, vol 38, no 11, pp 2467-2474
[63] Liezzel M. Pascual and Raymond R. Tan (2004)
[64] US EPA (no date) Health and Environmental Impacts of NOx, www.epa.gov/air/urbanair/nox/hlth.html
[65] Ronald Hwang, Vehicle policy director, Natural Resources Defense Council, San Francisco, CA, e-mail to Janet Sawin, Worldwatch Institute, 12 March 2006
[66] McCormick, R (2005) 'Effects of Biodiesel on NOx emissions', Presentation to ARB Biodiesel workgroup, National Renewable Energy Laboratory, Golden, CO, 8 June email to Peter Stair, Worldwatch Institute, 20 july 2005
[67] AQIRP (Auto/Oil Air /quality /improvement Research Program) (1997) Program Final Report, AQIRP, January

[68] Diesel Net News, May 2005 cited in AMFI Newsletter (Advanced Motor Fuels Information) (2005) October, http://virtual.vtt.fi/virtual/amf/pdf/amfinewsletter2005_4october.pdf
[69] Miljofordon Newsletter no 4 (2005), www.ufop.de, www.all4engineers.com, cited in AMFI Newsletter (Advanced Motor Fuels Information) (2005) October, http://virtual.vtt.fi/virtual/amf/pdf/amfinewsletter2005_4october.pdf
[70] Sinico, S. (2005) Fill it up with Natural, dw-world.de, 22 September, www.dw-world.de/dw/article/0,2144, 1717299,00.html ; Association of the German Biofuels Industry from Karin Retzlaff of the Association of the German Biofuels Industry, cited in Sinico, S. (2005) Fill it up with Natural, dw-world.de, 22 September, www.dw-world.de/dw/article/0,2144, 1717299,00.html
[71] Developing Countries in Fulton, L. (2004a) 'Driving ahead: Biofuels for transport around the world', Renewable Energy World, July-August, pp180-189
[72] www.d1plc.com
[73] Emissions and Reductions from Fulton, L., Howes, T. and Hardy, J. (2004) Biofuel for transport: An International Perspective, Paris, France, International Energy Agency
[74] Newest vehicles in Greene et al (2004) Growing Energy; How Biofuels Can Help End America's Oil Dependence, Washington, DC, Natural Resources Defence Council, December

2

The Impact of Oil Palm Expansion on Environmental Change: Putting Conservation Research in Context

Edgar C Turner[1,2], Jake L Snaddon[1,3], Robert M Ewers[2], Tom M Fayle[1,2] and William A Foster[1]

[1]*University Museum of Zoology, Cambridge, Cambridge,*
[2]*Imperial College London, Silwood Park Campus, Ascot, Berkshire,*
[3]*Biodiversity Institute, University of Oxford, Oxford,*
United Kingdom

1. Introduction

Agricultural expansion is one of the major drivers of tropical biodiversity loss worldwide (Foley et al., 2005; Green et al., 2005). Oil palm cultivation is among the main culprits, owing to its huge increase in cultivation in recent years (Food and Agriculture Organisation of the United Nations [FAO], 2011) and its centre of production being within the most biodiverse regions and habitats on the planet (Sodhi et al., 2010; Turner et al., 2008). Increasing demand for palm oil in food products and as a biofuel is likely to result in accelerating environmental change in the future (Koh & Ghazoul, 2008). Despite the importance of this crop and increasing global concern for environmental change, surprisingly little research has focussed on the actual impacts of conversion of forest to oil palm on biodiversity (Fitzherbert et al., 2008; Foster et al., 2011; Turner et al., 2008). In particular much still needs to be studied if we are to understand how human-modified landscapes can be managed to allow continued sustainable production of this globally important crop as well as maintenance of biodiversity. The development of more sustainable oil palm landscapes containing higher levels of biodiversity is not an alternative to conserving large areas of intact primary forest, as only these forested areas can provide a habitat for many rare and threatened species (Edwards et al., 2010). Rather it will allow preservation of a higher level of biodiversity within plantations, a greater connectivity and permeability for species to travel between reserve areas, and crucially the maintenance of important ecosystem functions within the agricultural landscape such as pollination, biological control, decomposition, maintenance of water quality, and environmental enrichment for people living in the vicinity of plantations. Central to the development of landscapes which support biodiversity and oil palm cultivation is increasing the dialogue between the oil palm industry, scientists and conservationists, as only this will allow new research findings to be applied to oil palm cultivation practices effectively.
In this chapter we will

- Describe in detail the change in palm oil production that has taken place over the last 30 years, the key regions where cultivation has taken place, and options for future conservation in the tropics

- Present an up-to-date review of the literature relating to the impacts on biodiversity of forest conversion to oil palm
- Assess how the focus of research relating to oil palm has changed in recent years
- Highlight gaps in existing knowledge and priorities for future research effort
- Assess the relationship between the oil palm industry, academic researchers and conservationists
- Highlight the importance of forging links between industry, science and conservation to understand and maintain functional tropical landscapes
- Introduce a new long-term large-scale collaborative research project between industry and science, the Stability of Altered Forest Ecosystems [SAFE] Project (Ewers et al., 2011; SAFE Project, 2010), which experimentally investigates landscape-scale biodiversity changes associated with the establishment of a new oil palm plantation in Sabah, Malaysia.

2. Global patterns of palm oil production

Agricultural ecosystems are now among the dominant habitat types on the planet (Foley et al., 2005). An expanding global population and a burgeoning demand for food have resulted in agricultural areas increasing dramatically in the tropics (Green et al., 2005), with 80% of the world's new agricultural land coming from the conversion of tropical forest (Gibbs et al., 2010). Conversion of natural ecosystems to agricultural landscapes has had a severe negative impact on global biodiversity (Sodhi et al., 2004, 2010), with losses of species already occurring and further regional and global extinctions predicted to occur. At the same time, global concerns for climate change have resulted in an accelerating demand for biofuel (Koh & Ghazoul, 2008), placing more pressure on remaining natural habitats.

Among the most important agricultural crops in the tropics is oil palm. Palm oil is used in a wide range of products, is a particularly important source of vegetable oil (Corley, 2009) and is increasingly used as a feedstock for biofuel production (Basiron, 2007; Henderson & Osborne, 2000; Koh, 2007). Globally, oil palm cultivation is centred in the tropics with the highest levels of production in Indonesia and Malaysia (Basiron, 2007). Both Indonesia and Malaysia are located in global biodiversity hotspots (Myers et al., 2000), so expansion in these areas is likely to have a large negative impact on biodiversity at the global scale (Sodhi et al., 2004).

Based on data from the Food and Agriculture Organisation of the United Nations [FAO] (FAO, 2011), we present trends in the global production of oil palm fruit over a 48-year period from 1961 to 2008 (Figure 1), as well as individual per country production for the top two palm oil producing nations in Southeast Asia, Africa and South America (Figure 2). In terms of quantity, these six nations are among the top ten oil palm producing countries worldwide (Figure 3). We present information on oil palm land area and yield per hectare. Where available, we also present trends in the producer prices for palm oil in each country. Global palm oil prices were estimated as the mean producer price from the 14 countries listed on the price domain of the FAOSTAT database (FAO, 2011).

Between 1961 and 2008 production of oil palm fruit has increased from 13 million tonnes to around 207 million tonnes worldwide (FAO, 2011). This rise has corresponded with substantial increases in land area under oil palm cultivation, with centres of oil palm production located throughout the tropics. Concerns for species losses as a result of palm oil expansion should therefore not be restricted to Southeast Asia, but rather to all tropical regions where forest is being converted (Wilcove & Koh, 2010). Although there have been increases in yield per unit area in most countries, this is not consistent and is very variable between nations and regions, with the well-developed oil palm industry in Malaysia and Indonesia showing the most marked increases in yield (Figures 2 & 3). Prices commanded for palm oil, although very variable, also continue to rise.

Between the 1960s and 1980s increases in global palm oil production were probably primarily obtained by increased yield per area. However since the 1980s this trend has shifted, with increased global production being driven instead by further conversion of land to oil palm cultivation (Murphy, 2009), threatening remaining forest habitats. The large difference in yield per area between different countries raises the possibility that, if yield can be increased in those regions at the lower end of the range, pressure on remaining forest habitats may be reduced. The recent development of higher-yielding seedling stock and more efficient processing technology (Donough et al., 2009; Mathews & Foong, 2010; Murphy, 2009) could enhance yield and productivity further, thereby also relaxing pressure to convert further natural habitats to oil palm cultivation. However, the rise in crop prices, which are closely linked to demand (Rudel et al., 2009), indicate that the market for palm oil is still expanding. This is probably owing to the continued high demand of palm oil as a source of edible oil and a biofuel feedstock (Corley, 2009; Koh, 2007), and diversification of its uses (Basiron, 2007; Henderson & Osborne, 2000). If further expansion of the area under oil palm cultivation is to be reduced, any rise in yield per area must therefore meet not only today's demand for palm oil, but also increased demand in the future.

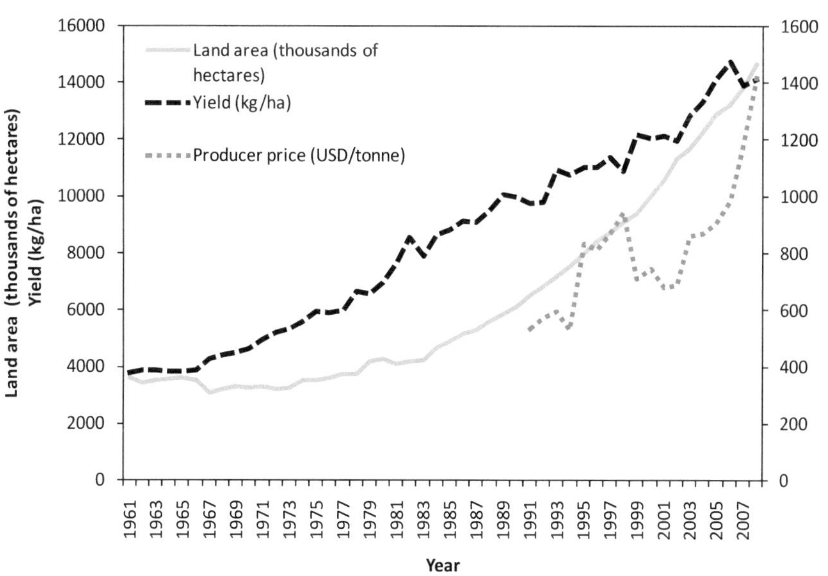

Fig. 1. Global oil palm land area under harvested cultivation, yield per unit area, and producer price of palm oil (in US Dollars per tonne produced). Land area under production has more than quadrupled since 1961, while yield and price have also increased substantially. Data from Food and Agriculture Organisation of the United Nations [FAO] (FAO, 2011)

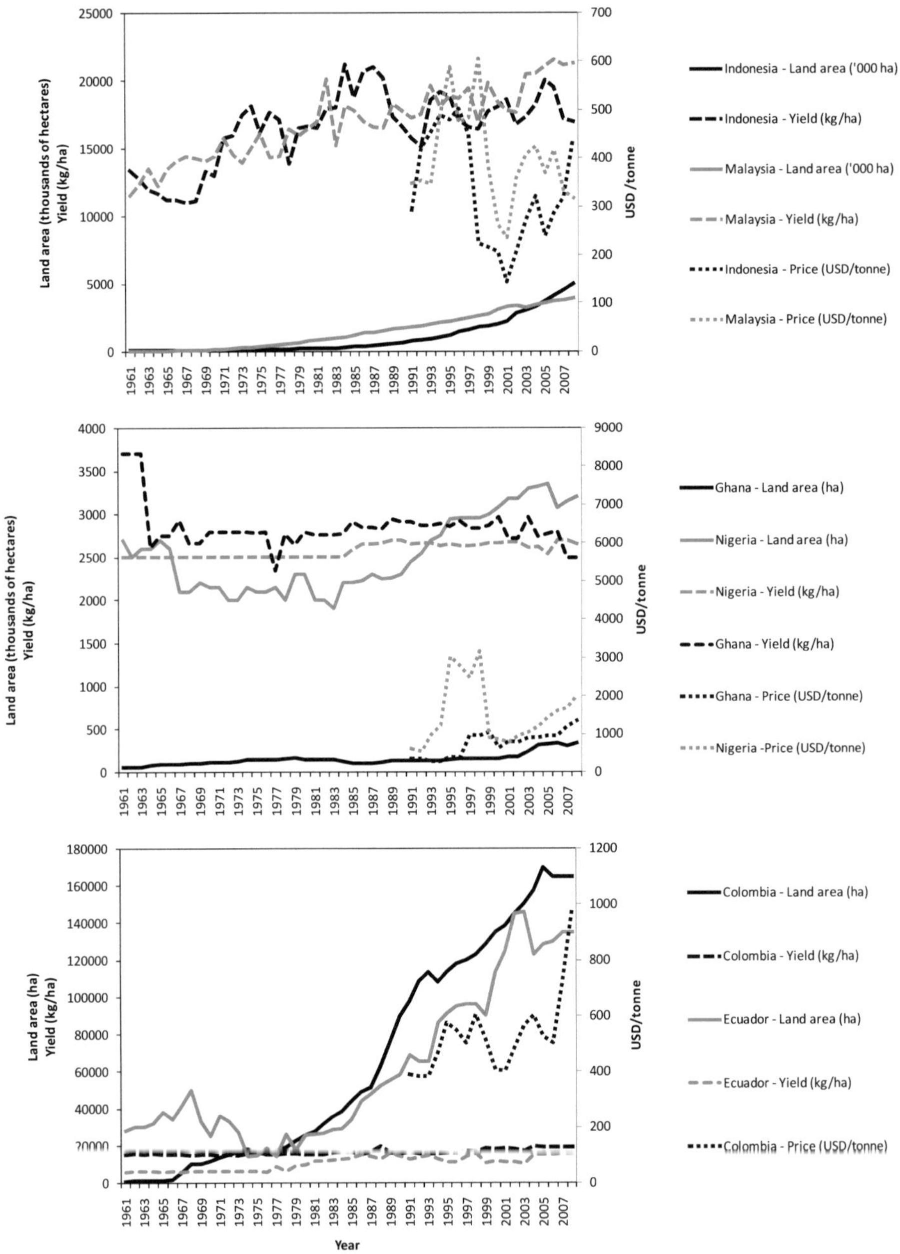

Fig. 2. Oil palm production, area under harvested cultivation, yield per unit area, and price of palm oil in US Dollars per tonne produced for the top two palm oil producing countries in each of the main tropical regions of production (SouthEast Asia, Africa and South America). Although production of palm oil has expanded in all countries, the level of productivity between regions varies widely, as does the price commanded by palm oil produced. Data from Food and Agriculture Organisation of the United Nations [FAO] (FAO, 2011). Note differing scales on the y-axes for different regions

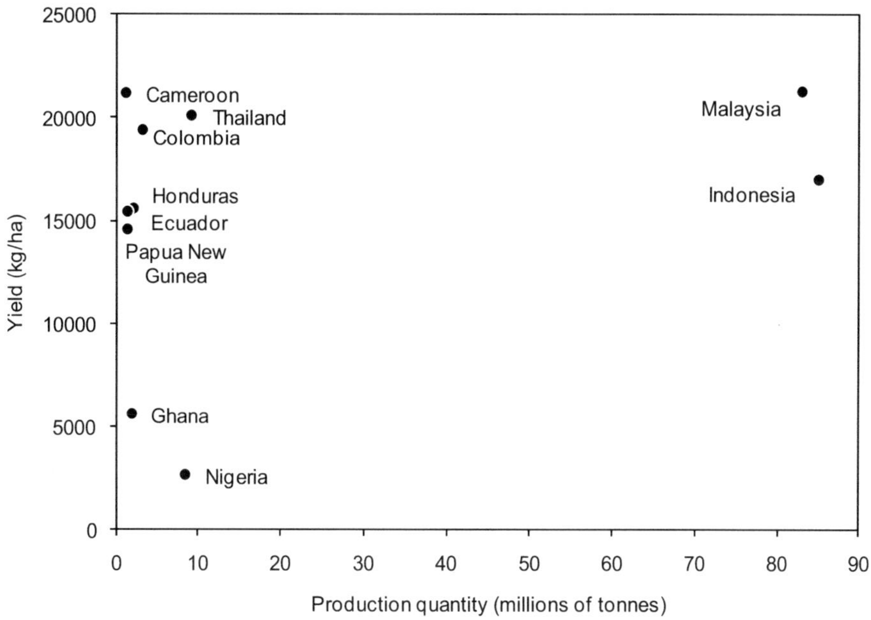

Fig. 3. Oil palm fruit production (in millions of tonnes) in relation to oil palm fruit yield per area (2008) for the top ten oil palm producing countries worldwide (FAO, 2011)

3. Oil palm impacts on biodiversity

Studies investigating the impacts of forest conversion to oil palm on biodiversity are surprisingly sparse (Foster et al., 2011; Turner et al., 2008). Despite this, there is now overwhelming evidence that conversion of natural or semi-natural habitats to oil palm has severe negative impacts on biodiversity (Fitzherbert et al., 2008; Foster et al., 2011)(Table 1). This is particularly the case if the land being converted is natural forest, but is also generally true if the land is under timber or another forest crop, which house higher levels of biodiversity than oil palm (Aratrakorn et al., 2006; Chung et al., 2000a, 2000b; Danielsen & Heegaard, 1995; Davis & Philips, 2005; Glor et al., 2001; Hassall et al., 2006; Peh et al., 2006; Room, 1975; Sheldon et al., 2010; Taylor, 1977). Studies have now been carried out on a diverse range of taxa including insects (ants, beetles, bees, butterflies and moths), other arthropods (woodlice), mammals (primates, tree shrews, squirrels and bats), birds, and lizards (Table 1). All of these taxa, with the exception of bees, show a decline in species richness from other habitats to oil palm, signalling a very high level of biodiversity loss as a result of oil palm expansion globally.

The majority of taxa also show a reduction in overall abundance in plantations compared to forest habitats, although this effect is more variable (Table 1). For example, in one study comparing arthropod abundance and biomass between forest habitats and oil palm plantations, some arthropod taxa showed the same levels of abundance and biomass in plantations, and others actually increased (despite arthropod numbers being reduced overall)(Turner & Foster, 2009). Similarly, in other studies, the total number of bats (Danielsen & Heegaard, 1995), dung beetles (Davis & Philips, 2005), woodlice (Hassall et al., 2006), and lizards (Glor et al., 2001) all increased in abundance as a result of habitat conversion. However, such increases are likely to be driven by an expansion in the populations of a few disturbance-tolerant species. These tend to be more wide-ranging "tramp" and invasive species and therefore have limited conservation value (e.g. Fayle et al., 2010). Despite this, disturbance-tolerant species may still be important in mediating ecosystem functioning in plantations and merit management to ensure their continued survival.

Group	Habitats compared to oil palm	Diversity	Abundance	Study location	Source
Arthropods					
All arthropods	Primary and secondary forest	-	↓	Sabah, Malaysia	Turner and Foster 2009
Ants	Primary forest	↓	-	Sabah, Malaysia	Brühl and Eltz 2010
Ants	Primary forest	↓↔	↓	Sabah, Malaysia	Fayle et al. 2010
Ants	Mangrove	↔	-	Peninsula Malaysia	Hashim et al. 2010
Ants	Primary forest, rubber and oil plantations, grassland, savanna, urban areas	↓	↓	Papua New Guinea	Room 1975
Ants	Primary/secondary forest and kola, cashew, coffee and plantain plantations	↓	-	Nigeria	Taylor 1977
Bees	Primary and secondary forest	↑	↓	Peninsula Malaysia and Singapore	Liow et al. 2001
Beetles	Primary and secondary forest and acacia	↓	↓	Sabah, Malaysia	Chung et al. 2000a
Rove beetles	Primary and secondary forest and acacia plantation	↓	-	Sabah, Malaysia	Chung et al. 2000b
Dung beetles	Primary and secondary forest and cacao plantation	↓	↑	Ghana	Davis and Philips 2005
Butterflies	Forest	↓	-	Sabah, Malaysia	Danielsen et al. 2008
Butterflies	Primary and secondary forest	↓	-	Peninsula Malaysia and Borneo	Koh and Wilcove 2008
Moths	Primary and secondary forest	↓	↓	Sabah, Malaysia	Chey VK 2006
Mosquitoes	Primary forest	↔	↓	Sarawak, Malaysia	Chang et al. 1997
Woodlice	Primary and secondary forest and fruit orchard	↓	↑	Sabah, Malaysia	Hassall et al. 2006

Group	Habitats compared to oil palm	Diversity	Abundance	Study location	Source
Mammals					
Primates	Primary forest and rubber plantation	↓	↓	Sumatra, Indonesia	Danielsen and Heegaard 1995
Squirrels	Primary forest and rubber plantation	↓	↓	Sumatra, Indonesia	Danielsen and Heegaard 1995
Tree shrews	Primary forest and rubber plantation	↓	↓	Sumatra, Indonesia	Danielsen and Heegaard 1995
Bats	Primary forest and rubber plantation	↓	↑	Sumatra, Indonesia	Danielsen and Heegaard 1995
Large mammals	Secondary forest and scrub	↓	-	Sumatra, Indonesia	Maddox et al. 2007
Small mammals	Primary forest and secondary forest	↓	↓	Sabah, Malaysia	Bernard et al. 2009
Small mammals	Forest	↓	-	Indonesia	Danielsen et al. 2008
Birds					
Birds	Primary forest and rubber plantation	↓	-	Thailand	Aratrakorn et al. 2006
Birds	Primary forest and rubber plantation	↓	-	Sumatra, Indonesia	Danielsen and Heegaard 1995
Birds	Primary forest and rubber plantation	↓	↓	Peninsula Malaysia	Peh et al. 2006
Birds	Secondary forest and acacia plantation	↓	-	Sabah, Malaysia	Sheldon et al. 2010
Reptiles					
Lizards	Secondary forest, cacao plantation, pasture, home gardens, undisturbed hilltops	↓	↑	Dominican Republic	Glor et al. 2001

Table 1. Species richness and abundance of various animal taxa compared between forest or plantation habitats and oil palm. – response not recorded; ↓ richness or abundance declines, ↔ richness or abundance is unchanged, ↑ richness or abundance increases

Biodiversity in most components of the forest ecosystem is likely to be negatively affected by habitat change. However, owing to varying levels of disturbance across the plantation landscape and differences in the environmental tolerances of species from different components of the forest ecosystem, some habitat components are more adversely affected than others. For example, one study comparing arthropods between forest and oil palm habitats, collected from the canopy, epiphytic bird's nest ferns and the forest floor, found that different sub-habitats exhibited differing levels of decline, with the forest floor arthropod community being the most severely affected and the epiphyte community the least affected (Turner & Foster, 2009). This was probably due to the high levels of disturbance that occur on the plantation floor and regular applications of herbicides at the base of individual palms. It is also likely that canopy species are comparatively less impacted by conversion of forest to plantation, as microclimatic conditions in the forest canopy are generally more similar to an oil palm plantation than the forest floor (Foster et al., 2011), and therefore canopy species may be better adapted to cope with habitat conversion. Epiphytes can also establish easily in oil palm plantations (Piggott, 1988), probably due to high light conditions and because the frond stumps, which are left on the trunks of the oil palms, trap organic matter and provide an attachment point for the plants. In fact some epiphytes, such as bird's nest ferns, can reach higher densities in plantations than in forests (Turner & Foster, 2009), although it is likely that only a subset of the forest species persist (Fayle et al., 2009). Epiphytes have also been found to modify the microclimatic conditions around them and therefore provide a more equitable temperature and humidity regime (Turner & Foster, 2006). It is therefore not surprising that epiphytes can house considerable densities of arthropods and act as an important habitat for species in plantations (Turner & Foster, 2009). The number of arthropod species living in plantation epiphytes can also be high. For example, the number of ant species in bird's nest ferns is the same in forest and oil palm plantation habitats (Fayle et al., 2010). However, the species found in plantation epiphytes are not the same as those in forests (Fayle et al., 2010). Therefore, although biodiversity as a whole was maintained in epiphytes, plantation communities were still fundamentally different from forest environments.

3.1 Drivers of biodiversity loss

Reasons for such a dramatic loss of species are almost certainly due to the simplification of the habitat that occurs when a forest is converted to oil palm (Foster et al., 2011). This includes the obvious loss of the diverse tree community that forms the basic structure of a forest (important in maintaining herbivore diversity for example (Novotny et al., 2006)), a reduction in above ground structural complexity, and a reduced canopy height. Partly due to this loss of canopy cover, microclimatic conditions are harsher for species in plantations with temperatures being on average hotter and humidity levels lower. Fluctuation in both temperature and humidity is also greater over 24 hours in plantations compared to forest habitats (Koh et al., 2009; Turner & Foster 2006). Direct disturbance effects, such as cutting and spraying of understory vegetation, and a higher proportion of invasive species probably also contributes to species' declines and extinctions.

3.2 Impacts of biodiversity loss on ecosystem functioning

The impact of reduced biodiversity on the healthy functioning of oil palm ecosystems has been little studied. However, there is considerable support from theoretical models and experimental systems that reductions in biodiversity can have significant negative impacts on ecosystem functioning (Schmid et al., 2009). Reliance on the function carried out by a single species or a few species is risky as if these species go extinct the function will fail. A higher diversity of species adds resilience to ecosystem processes and allows systems to adapt to future changes (Jackson et al., 2010). It is therefore likely that the documented losses in animal biodiversity associated with oil palm cultivation will have a detrimental effect, perhaps through a reduction in biological control of pest species or reduced pollination efficiency. For example, a wide and increasing range of species have been reported to attack oil palm (Corley, 2003; Mariau, 2001; Turner & Gillbanks, 2003), and it is clear that predators and parasitoids can have an important role in controlling their outbreaks. In oil palm management such species have long been included in Integrated Pest Management strategies (Wood, 2002), with examples including the use of the fungus *Metarhizium anisopliae* in the control of rhinoceros beetles, adult assassin bugs (Heteroptera) in the control of herbivorous insects, and barn owls (*Tyto alba*) in the control of rats (Turner & Gillbanks, 2003). The role of naturally occurring suites of predators, termed "Conservation Biological Control" (Jonsson et al., 2008; Tscharntke et al., 2007), in controlling pest species has been less studied. However, in one study where birds were excluded from young palms with netted cages, herbivory levels increased significantly, indicating that birds had an important effect in controlling herbivores (Koh, 2008b). Although the majority of oil palm pollination in Malaysia is said to be carried out by a single species of introduced weevil (*Elaiedobius kamerunicus* (Coleoptera: Curculionidae); Greathead, 1983), many other species of insects also visit oil palm flowers (Bulgarelli et al., 2002; Mariau & Genty, 1988; Mayfield, 2005; Syed et al., 1979) and may have a role in maintaining pollination (Caudwell et al., 2003). Taxa that show increases in abundance in oil palm systems might be important in maintaining ecosystem processes and have the potential to buffer functioning against losses of other species, even if they are of little direct conservation interest (e.g. are tramp or invasive species).

4. Strategies for conservation of global biodiversity

Since oil palm is widespread and its expansion is accelerating, the choice of tactics to mitigate the effects of oil palm cultivation on biodiversity is paramount. In recent years two alternative strategies for conservation in the tropics have emerged (Green et al., 2005). Generally referred to as "land sparing" and "land sharing", these competing ideas are that biodiversity can be best maintained by either setting aside (sparing) large areas of land in the tropics for reserves and intensifying production as much as possible elsewhere, or by developing agriculture over much larger areas but in a more wildlife-friendly way (sharing). A general consensus is now emerging in the conservation sector that the only way to conserve species of high conservation valuc in the tropics is by land sparing and the provision of large forest reserves (e.g. Edwards et al., 2010). However, it is important that these two approaches are not viewed as alternatives, but rather as opposite ends of a continuum of strategies that can be employed for different species and with different conservation outcomes in mind. There is no doubt that many species cannot be conserved in fragmented habitats and that intact forest reserves must therefore be maintained. However, the biodiversity still existing within plantation areas can be substantial, and a more biodiversity friendly environment can help to buffer and provide a foraging resource for species from forest reserves (e.g. Maddox et al., 2007). Most importantly as far as industry is concerned, biodiversity within plantation areas can provide important ecosystem functions and increase productivity within the crop area itself (Zhang et al., 2007). Finally the oil palm industry employs millions of workers and plantations are one of the commonest landscapes that people actually see or spend time in within the tropics (Koh & Wilcove, 2007). If popular engagement with conservation in oil palm producing countries is to be maintained, it is therefore vital that plantation diversity is not written off as unimportant. Koh et al. (2009) suggested that oil palm landscapes should be viewed more inclusively and could include both large reserves and also smaller forest patches within oil palm plantations. Such an approach paves the way to "designing" tropical landscapes with both agriculture and biodiversity in mind. However, these ideas have met with criticism by some conservationists, as funding and implementation of such research could divert resources away from land-sparing conservation projects (Struebig et al., 2009).

Decisions on optimal strategies for maintaining crop production while protecting global biodiversity will also depend on the level of demand for different commodities in the future. For example, central to the land sparing argument is the condition that if global demands for palm oil are met by intensified production in existing regions, then no more natural habitat need be converted. However, the price of oil palm is increasing rather than reaching an asymptote or declining as global production accelerates (Figures 1 & 2). Therefore demand is still rising and higher production in intensively farmed areas may not spare land in unconverted regions (Rudel et al., 2009). Indeed it would make sense economically for nations to clear more land and farm it intensively, as this yield would continue to command a high price on global markets.

4.1 Management strategies to reduce biodiversity loss in oil palm plantations

There has been little research effort to date focussing on methods that can be employed to maintain and enhance biodiversity in and around oil palm plantations. Increasing habitat complexity at both the local and regional scale can increase biodiversity within managed landscapes (Tscharntke et al., 2008). For example, leaving forest fragments in plantations (as is often done on steep slopes and riverine margins) can provide a habitat for non-plantation species (e.g. Maddox et al., 2007). Such areas may also provide source populations for species to "spill over" into the crop (e.g. Ricketts et al., 2004). Perhaps as a result of this, the level of forest cover surrounding oil palm areas has been shown to predict species richness of butterflies and birds (Koh, 2008a). The age structure of the oil palm could also be manipulated to increase landscape heterogeneity and therefore biodiversity. Oil palm is a long-lived crop and stands may exist for up to 30 years. Over its lifespan considerable biodiversity may therefore develop, with communities of animals and plants altering as a plantation ages (De Chenon & Susanto, 2006; Koh, 2008a; Mariau, 2001). Therefore management practices that maintain a diverse age structure (e.g. by clearing and replanting areas in rotation) could also increase plantation biodiversity.

Heterogeneity at the local scale may also be manipulated in long-lived agricultural ecosystems such as oil palm. Understory vegetation is usually cleared around individual palms, but if this is maintained it can be an important habitat for insect communities, as has been found for beetles (Chung et al., 2000a). This vegetation also produces more leaf litter, which itself may support a higher diversity and abundance of litter-dwelling arthropods. Finally, as has been mentioned before, epiphytes are numerous in plantations and can support diverse insect assemblages (Turner & Foster, 2009). Therefore, maintaining these plants in plantations rather than clearing them, as is sometimes done as part of management practices (Piggott, 1988), could also increase local biodiversity.

5. The changing focus of oil palm research

5.1 Oil palm research until 2007

In 2008 we used the scientific search engine ISI Web of Science (Web of Science [WoS], 2008) to assess the changing focus of oil palm research since 1970 (Turner et al., 2008). By entering the search term ""palm oil" or "oil palm"" we accessed over 3000 oil palm research papers published between 1970 and 2006. For each of these we recorded their main research focus as interpreted through their title, abstract, key words, journal title and subject classification. Based on this we classified each publication as belonging to one of the following categories:

1. Biodiversity and conservation
2. Environment
3. Social/human welfare
4. Diet and health
5. Pests, diseases and pollination
6. Industry improvements and oil palm biology
7. Chemistry, engineering and biotechnology
8. Biofuels
9. Alternative uses and by-products
10. Other

Based on analysis of these categories it was therefore possible to visualize how the focus of oil palm research had changed since 1970.

It was clear that there had been a dramatic increase in publications on oil palm over that time with a concurrent broadening in the scope of research. Surprisingly we found that less than 1% of publications related to biodiversity and species conservation, but that this number was increasing. There was also a marked increase in the number of publications on the subject of biofuel (Turner et al., 2008).

5.2 Oil palm research since 2007

Since 2007 there have been another 1722 new publications on oil palm featured in ISI Web of Science (WoS, 2011). Using the same methods as we employed before, we classified these new papers into the ten different research categories and examined those on the subject of biodiversity and conservation in greater detail. Since 2007 there has been a significant number of new publications on biodiversity and conservation (another 71 papers, 4% of the total), and biofuel (280 papers, 16% of the total) (Figure 4). There has also been a substantial increase in the number of publications investigating alternative uses of palm oil (153 publications, 9%). If these do indeed lead to more palm oil use in alternative industries, it will also result in increased demand for palm oil in the future.

The new studies have boosted our understanding of the impacts of oil palm expansion on biodiversity and have particularly provided information on a more diverse range of taxa, including arthropods (Turner & Foster, 2009), ants (Brühl & Eltz, 2010; Fayle et al., 2010; Hashim et al., 2010), butterflies (Danielsen et al., 2008; Koh & Wilcove, 2008), small mammals (Bernard et al., 2009; Danielsen et al., 2008), and birds (Sheldon et al., 2010). Results have illustrated unambiguously the severe threat that oil palm cultivation represents to global biodiversity. There have also been publications on the role of forest fragments in maintaining biodiversity in plantations, although this important subject is still little studied. These show that non-plantation species can be maintained in such areas (Struebig et al., 2008), although communities are markedly different from those in intact forest (Edwards et al., 2010) and genetic diversity may be reduced (Benedick et al., 2006; Bickel et al., 2006). Maintenance of large forest reserves is therefore essential for the conservation of tropical forest diversity.

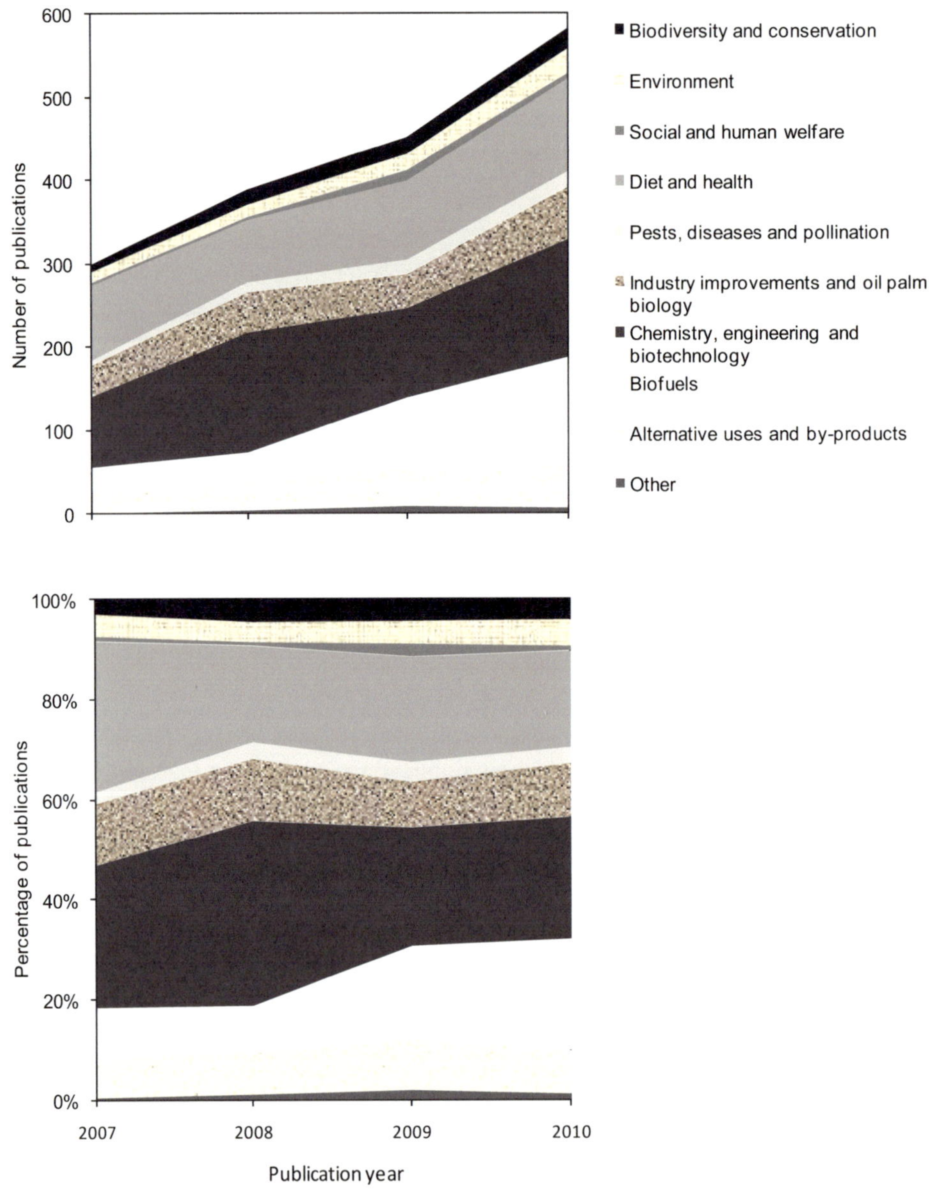

Fig. 4. Number and percentage of publications on oil palm in different research areas published since 2007. Papers were accessed using the scientific search engine, ISI Web of Science (WoS, 2008), by entering the search term ""palm oil" or "oil palm"" and assigned to categories based on their title, abstract, key words, journal title and subject classification

5.3 Gaps in existing knowledge and future research priorities

Despite this increase in knowledge, substantial gaps still exist in our understanding of the impacts of oil palm expansion and the functioning of oil palm ecosystems. In particular, it is still not clear how communities and ecosystem functions fare as plantations develop, although limited research indicates that communities change as oil palm matures (De Chenon & Susanto, 2006). Indeed only half of the publications directly comparing biodiversity between oil palm and forest habitats actually report the age of oil palm in which the study was carried out. There are still no publications linking habitat management to biodiversity and ecosystem functioning in plantations, although this is a crucial topic for the continued sustainable production of palm oil. Finally, few studies in oil palm have investigated the role of non-crop habitats in maintaining biodiversity, although again these areas can be important in maintaining biodiversity in oil palm plantations, may provide corridors for species to move between forest reserves, and could act as reservoirs for species which may spill out into plantations to perform important ecosystem functions (Ricketts et al., 2004) or indeed act as pests.

6. Links between the oil palm industry, scientists and conservationists

It is clear that much more research must be carried out if we are to move towards the development of tropical landscapes which sustainably produce palm oil and have minimal detrimental impacts on tropical biodiversity. Central to this goal is forging more links between the oil palm industry, conservation and science bodies; an aim which has proved difficult to achieve in the past, owing to widely diverging philosophies and knowledge bases (Koh et al., 2009; Struebig, 2010). Only by working closely together, can these different stakeholders ensure that their respective goals are met. For example, conservation scientists must be free to work in plantations if they are to understand how these ecosystems function and this requires industry collaboration and involvement. Similarly, industry stakeholders must be free to advise conservation researchers on existing management practices and the economic realities of oil palm cultivation if unrealistic and uneconomic policy advice is to be avoided. Finally, more links must be made between the conservation and industry-grounded research that is taking place, to ensure a free exchange of ideas and to avoid duplication of research effort. Indeed many additional studies on oil palm ecosystems probably exist within the grey literature that are not cited in this chapter owing to difficulties in locating such material (Anderson, 2008). The impact of such work outside of the industry on conservation and international policy is therefore limited. Similarly, many of the publications in academic journals are not readily available to industry workers, reducing their impact on policy implementation. Such a disparity in the circulation of literature can increase differences in the knowledge-bases between industry, conservation and science, exacerbating misunderstandings and direct conflicts between stakeholders (Koh et al., 2009). Implementation of new policies, informed by novel research also requires close engagement with the industry as well as with industry regulating bodies. The Round Table on Sustainable Palm Oil [RSPO] (RSPO, 2011), with a mission statement to promote the growth of sustainable palm oil, is already actively engaged with major oil palm producers and consumers, and can provide a platform for the launch of such new policies.

Closer collaboration between industry, research and conservation can also have potential mutual benefits as far as funding is concerned. The oil palm industry makes considerable profits each year and it has been suggested that these could be used in part to fund conservation practices at the plantation scale (Koh & Wilcove, 2007). Such work, if properly implemented, could also help plantations achieve sustainability criteria and therefore command a higher price for their products. Collaboration of this kind can also provide access to international funding designed to minimize further conversion of forest: these include identifying and protecting High Conversion Value forest, Reducing Emissions from Deforestation and Forest Degradation (REDD), and biodiversity banking (Yaap et al., 2010).

6.1 Analysis of the relationship between conservation and industry research

Despite the potential benefits of closer collaboration, there is still a wide divide between conservation and industry in the oil palm sector. To determine the level of engagement between the oil palm industry and conservation science, we examined the top 10 most cited research papers on the subject of biodiversity and conservation that we found during a Web of Science search with the search terms ""oil palm" or "palm oil"" and "biodiversity" and "conservation". For each publication, we recorded which papers had cited it and assigned each of these to biodiversity and conservation or industry sectors, based on the focus of the journal the paper was in and the home institution of the first author (Figure 5).

We found that a quarter of the citations were from the industry sector, indicating a fairly high level of engagement of industry with conservation research. This also indicates that conservation research results are being disseminated successfully to the oil palm industry, hopefully signalling a greater level of understanding between these sectors in the future. More now needs to be done to increase collaboration between conservation and industry to increase the transfer of ideas and results. Central to this is a greater awareness of industry grey literature by conservation scientists.

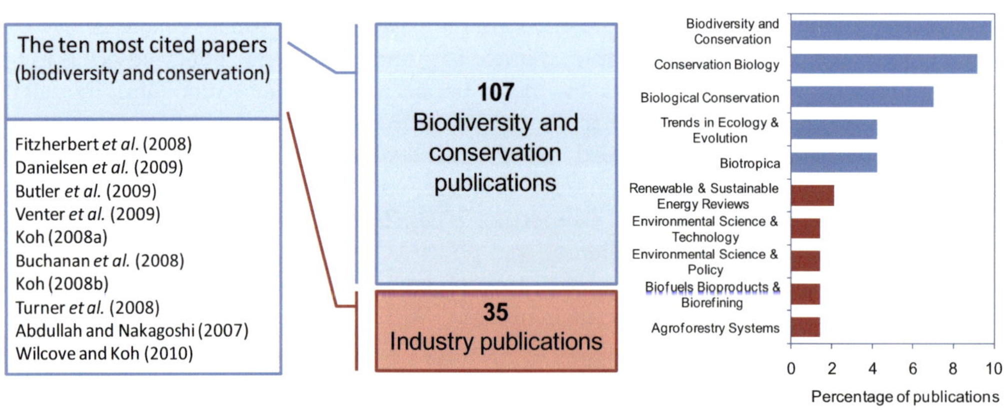

Fig. 5. Citation map showing the links between the top ten most cited biodiversity and conservation publications on the subject of oil palm accessed using the Web of Science search engine (WoS, 2011) (see reference list for full reference details). Between them, the ten papers were cited 142 times, with one quarter of citations being in industry publications. The histogram on the right shows percentage of citations by the different conservation and industry journals. Although there is overlap between conservation and industry research, there is clearly scope for more collaboration

7. The SAFE Project

The Stability of Altered Forest Ecosystems [SAFE] Project (SAFE Project, 2011; Ewers et al., 2011) has recently been set up in Sabah, Malaysia to investigate the impacts of tropical habitat change on biodiversity and ecosystem functioning in tropical ecosystems - with a particular focus on forest fragmentation and conversion to oil palm plantation. The success of this project relies on a close working relationship between the oil palm industry, academic research institutions, and the Malaysian Government and provides a template for collaboration between oil palm stakeholders. Development of such large-scale, long-term projects is crucial in developing scientific understanding of the impacts of forests to environmental change (Clark et al., 2001).

The project itself is based within a concession area managed by the Sabah Foundation (a state government body charged with the socio-economic development of the Malaysian state of Sabah (Yayasan Sabah, 2011)), and includes areas of logged forest and oil palm plantation managed by Benta Wawasan and Sabah Softwoods (subsidiary companies of the Sabah Foundation). Funding for the project has been guaranteed for ten years by the Sime Darby Foundation (Sime Darby Foundation, 2011), with in kind contributions from Benta Wawasan. Academically, the project is led by Imperial College London in collaboration with the Royal Society South East Asia Rainforest Research Programme [SEARRP] (SEARRP, 2011). Finally, the research itself is carried out by an international team of scientists, with the help of a team of 15 full-time Malaysian research assistants. The majority of these researchers come from independent institutions: to date more than 150 scientists from over 50 different institutions in 13 countries have worked on or expressed an interest in working on the project. In addition to these independent researchers, the project funds both Malaysian and international Ph.D. students and post-doctoral researchers.

Research plots for the project range from pristine primary rainforest around Maliau Basin Studies Centre (an area of over 58,840 hectares of unlogged forest), logged forest and areas of established oil palm. In addition to logged forest areas which will remain under forest, research plots are also located in a 7200 ha area of the Benta Wawasan forestry estate that has been earmarked for conversion to oil palm plantation in 2011. Working closely with Benta Wawasan, the SAFE Project has designed a landscape in which 800ha of forest will be spared clearance, and will be maintained in an arrangement of circular fragments of 100ha, 10ha and 1ha (42 experimental fragments in total). This design allows the comparison of biodiversity and ecosystem functioning across a range of disturbances, as well as direct experimental tests of the impacts of tropical forest fragmentation and conversion. Within this major topic the project has a wide remit, including research on biodiversity, carbon and nutrient dynamics, ecosystem services within plantations, and disease transfer. The project also encompasses research on a very wide range of taxa including plants (trees, epiphytes and vines), insects (particularly beetles, termites and ants), birds, mammals and amphibians. By setting up an experimentally-designed landscape, which includes forest fragments within the oil palm matrix, the project will directly investigate the importance of habitat heterogeneity in maintaining biodiversity and ecosystem functioning in human-managed landscapes. This will provide answers to key research questions for conservationists and agronomists alike. As well as representing an important step forward in collaboration between stakeholders, this project is on a scale that would not be possible without industry involvement, and will directly facilitate knowledge transfer between science and industry. We hope that collaborative research projects such as this and others (for example the

Zoological Society of London's [ZSL] Biodiversity and Oil Palm Project (ZSL, 2011)) will become more common in the future, facilitating conservation in the tropics, as well as spearheading sustainable development projects.

8. Conclusion

The rapid expansion of agriculture in the tropics poses a huge threat to tropical and therefore to global biodiversity. However, it also presents opportunities for conservation and research through closer collaboration between industry players and conservationists. Until now there has been only a limited transfer of ideas and knowledge between different oil palm stakeholders. It is vital that this situation changes to ensure that landscapes can be designed to fulfil the functions of production and conservation. This is not only important for biodiversity conservation within and outside of reserves, but also represents the best opportunity for palm oil to be produced sustainably.

9. Acknowledgement

We would like to thank Sime Darby, Yayasan Sabah and Benta Wawasan, the Sabah Forestry Department and the Royal Society South East Asia Rainforest Research Programme for supporting the SAFE Project. Researchers interested in collaborating on the SAFE Project should visit the SAFE Project website (www.safeproject.net) for more information.

10. References

Anderson, JM. (2008). Eco-Friendly Approaches to Sustainable Palm Oil Production. *Journal of Oil Palm Research,* Special Issue, (October 2008), pp.127-142, ISSN 1511-2780

Aratrakorn, S.; Thunhikorn, S. & Donald, PF. (2006). Changes in bird communities following conversion of lowland forest to oil palm and rubber plantations in southern Thailand. *Bird Conservation International,* Vol.16, No.1, (March 2006), pp. 71-81, ISSN 0959-2709

Basiron, Y. (2007). Palm oil production through sustainable plantations. *European Journal of Lipid Science Technology,* Vol. 109, No.4, (Apil 2007), pp. 289-295, ISSN 1438-7697

Benedick, S.; Hill, JK.; Mustaffa, N.; Chey, VK.; Maryati, M.; Searle, JB.; Schilthuizen, M. & Hamer, KC. (2006). Impacts of rain forest fragmentation on butterflies in northern Borneo: species richness, turnover and the value of small fragments. *Journal of Applied Ecology,* Vol.43, No.5, (October 2006), pp. 967-977, ISSN 0021-8901

Benedick, S.; White, TA.; Searle, JB.; Hamer, KC.; Mustaffa, N.; Chey, VK.; Mohamed, M.; Schilthuizen, M. & Hill, JK. (2007). Impacts of habitat fragmentation on genetic diversity in a tropical forest butterfly on Borneo. *Journal of Tropical Ecology,* Vol.23, No.6, (November 2007), pp.623-634, ISSN 0266-4674

Bernard, H.; Fjeldså, J. & Mohamed, M. (2009). A case study on the effects of disturbance and conversion of tropical lowland rain forest on the non-volant small mammals in north Borneo: Management implications. *Mammal Study,* Vol34, No.2, (June 2009), pp.85-96, ISSN 1343-4152

Bickel, TO.; Brühl, CA.; Gadau, JR.; Hölldobler, B. & Linsenmair, KE. (2006). Influence of Habitat Fragmentation on the Genetic Variability in Leaf Litter Ant Populations in

Tropical Rainforests of Sabah, Borneo. *Biodiversity and Conservation*, Vol.15, No.1, (January 2006), pp.157-175, ISSN 0960-3115

Brühl, CA. & Eltz, T. (2010). Fuelling the biodiversity crisis: species loss of ground-dwelling forest ants in oil palm plantations in Sabah, Malaysia (Borneo). *Biodiversity and Conservation*, Vol.19, No.2, (February 2010), pp.519-529, ISSN 0960-3115

Bulgarelli, J.; Chinchilla, C. & Rodríguez, R. (2002). Male inflorescences, population of *Elaeidobious kamerunicus* and pollination in a young commercial oil palm plantation in a dry area of Costa Rica. *ASD Oil Palm Papers*, Vol.24, pp.32-37

Caudwell, RW.; Hunt, D.; Reid, A.; Mensah, BA. & Chinchilla, C. (2003). Insect pollination of oil palm - a comparison of the long term viability and sustainability of *Elaeidobious kamerunicus* in Papua New Guinea, Indonesia, Costa Rica, and Ghana. *ASD Oil Palm Papers*, Vol.25, pp.1-16

Chang, MS.; Hii, J.; Buttner, I. & Mansoor, F. (1997). Changes in abundance and behaviour of vector mosquitoes induced by land use during the development of an oil palm plantation in Sarawak. *Transactions of the Royal Society of Tropical Medicine and Hygiene*, Vol.91, pp.382-386

Chey, VK. (2006). Impacts of forest conversion on biodiversity as indicated by moths. *Malayan Nature Journal*, Vol.57, pp.383-418

Chung, AYC.; Eggleton, P.; Speight, MR.; Hammond, PM. & Chey, VK. (2000a). The diversity of beetle assemblages in different habitat types in Sabah, Malaysia. *Bulletin of Entomological Research*, Vol.90, No.6, (December 2000), pp.475-496, ISSN 0007-4853

Chung, AYC.; Hammond, PM.; Eggleton, P.; Speight, MR. & Chey, VK. (2000b). A general survey of the Staphylinidae (Insecta: Coleoptera) assemblage in Sabah, Malaysia . *Malayan Nature Journal*, Vol.54, pp.355–368

Clark, JS.; Carpenter, SR.; Barber, M.; Collins, S.; Dobson, A.; Foley, JA.; Lodge, DM.; Pascual, M.; Pielke, R.; Pizer, W.; Pringle, C.; Reid, WV.; Rose, KA.; Sala, O.; Schlesinger, WH.; Wall, DH, &Wear, D. (2001). Ecological forecasts: an emerging imperative. *Science*, Vol.293, No.5530, (July 2001), pp.657-660, ISSN 0036-8075

Corley, RHV. (2009). How much palm oil do we need? *Environmental Science & Policy*, Vol.12, pp.134-139, ISSN: 1462-9011

Corley, RHV. & Tinker, PBH. (2003). *The Oil Palm*, Blackwells, ISBN 0632052120. Oxford, UK

Danielsen, F. & Heegaard, M. (1995). Impact of logging and plantation development on species diversity: a case study from Sumatra, In *Management of tropical forests: towards an integrated perspective*, O. Sandbukt, (Ed.), Centre for Development and the Environment, ISBN 8290391250, Oslo, Norway.

Danielsen, F.; Beukema, H.; Burgess, ND.; Parish, F.; Brühl, CA.; Donald, PF.; Murdiyarso, D.; Phalan, B.; Reijinders, L.; Struebig, M. & Fitzherbert, EB. (2008). Biofuel Plantations on Forested Lands: Double Jeopardy for Biodiversity and Climate. *Conservation Biology*, Vol.23, No.2, (April 2009), pp.348-358, ISSN 0888-8892

Davis, ALV. & Philips, TK. (2005). Effect of Deforestation on a Southwest Ghana Dung Beetle Assemblage (Coleoptera: Scarabaeidae) at the Periphery of Ankasa Conservation Area. *Environmental Entomology*, Vol.34, No.5, (October 2005), pp.1081-1088, ISSN 0046-225X

De Chenon, RD. & Susanto, A. (2006). Ecological observations on diurnal birds in Indonesian oil palm plantations. *Journal of oil palm research*. Special Issue, pp.122-143

Donough, CD.; Witt, C.; Fairhurst, TH. (2009) Yield intensification in oil palm plantations through best management practice. Better Crops, Vol.93, No.1, pp.12-14, ISSN 0006-0089

Edwards, DP.; Hodgson, JA.; Hamer, KC.; Mitchell, SL.; Ahmad, AH.; Cornell, SJ. & Wilcove, DS. (2010). Wildlife-friendly oil palm plantations fail to protect biodiversity effectively. *Conservation Letters,* Vol.3, No.4, (August 2010), pp.236-242, ISSN 1755-263X

Ewers, RM.; Didham, RK.; Fahrig ,L.; Ferraz, G.; Hector, A.; Holt, RD.; Kapos, V.; Reynolds, G.; Sinun, W.; Snaddon, JL. & Turner, EC. (2011). A large-scale forest fragmentation experiment: the Stability of Altered Forest Ecosystems Project. *Philosophical Transactions of the Royal Society of London B,* in press

FAO (2011). In: *FAOSTAT Online Statistical Service. Food and Agriculture Organization of the United Nations,* Accessed March 2011, Available from: http://faostat.fao.org

Fayle, TM.; Chung, AYC; Dumbrell, AJ.; Eggleton, P. & Foster, WA. (2009). The effect of rain forest canopy architecture on the distribution of epiphytic ferns (*Asplenium* spp.) in Sabah Malaysia. *Biotropica,* Vol.41, No.6, (November 2009), pp.676-681, ISSN 0006-3606

Fayle, TM.; Turner, EC.; Snaddon, JL.; Chey, VK.; Chung, AYC.; Eggleton, P. & Foster, WA. (2010). Oil palm expansion into rain forest greatly reduces ant biodiversity in canopy, epiphytes and leaf-litter. *Basic and Applied Ecology,* Vol.11, No.4, (2010), pp.337-345, ISSN 1439-1791

Fitzherbert, EB.; Struebig, MJ.; Morel, A.; Danielsen, F.; Brühl, CA.; Donald, PF. & Phalan, B. (2008). How will oil palm expansion affect biodiversity? *Trends in Ecology & Evolution,* Vol.23, No.10, (October 2008), pp.538-545, ISSN 0169-5347

Foley, JA.; DeFries, R.; Asner, GP.; Barford, C.; Bonan. G.; Carpenter, SR.; Chapin, FS.; Coe, MT.; Daily, GC.; Gibbs, HK.; Helkowski, JH.; Holloway, T.; Howard, EA.; Kucharik, CJ.; Monfreda, C.; Patz, JA.; Prentice, IC.; Ramankutty, N. & Snyder, PK. (2005). Global consequences of land use. *Science,* Vol.309, No.5734, (July 2005), pp.570-574, ISSN 0036-8075

Foster, WA.; Snaddon, JL.; Turner, EC.; Fayle, TM.; Cockerill, TD.; Ellwood, MDF.; Broad, GR.; Chung, AYC.; Eggleton, P.; Chey, VK.; & M. Yusah, K. (2011). Establishing the evidence base for maintaining biodiversity and ecosystem function in the oil palm landscapes of South East Asia *Philosophical Transactions of the Royal Society of London B,* in press

Gibbs, HK.; Ruesch, AS.; Achard, F.; Clayton, M.K.; Holmgren, P.; Ramankutty, N. & Foley, JA. (2010). Tropical forests were the primary sources of new agricultural land in the 1980s and 1990s. *Proceedings of the National Academy of Sciences,* Vol.107, No.38, (September 2010), pp.16732-16737, ISSN 0027-8424

Glor, RE.; Flecker, AS.; Benard, MF. & Power, AJ. (2001). Lizard diversity and agricultural disturbance in a Caribbean forest landscape. *Biodiversity and Conservation,* Vol.10, No.5, (May 2001), pp.711-723, ISSN 0960-3115

Greathead, DJ. (1983) The multi-million dollar weevil that pollinates oil palms. *Antenna,* Vol.7, pp.105-107

Green, RE.; Cornell, SJ.; Scharlemann, JPW. & Balmford, A. (2005). Farming and the fate of wild nature. *Science,* 307, No.5709, (January 2005), pp.550-555, ISSN 0036-8075

Hashim, NR.; Akmal, WF.; Jusoh, W. & Nasir, MNSM. (2010). Ant diversity in a Peninsular Malaysian mangrove forest and oil palm plantation. *Asian Myrmecology,* Vol.3, No.3, (2010), pp:5-8, ISSN 1985-1944

Hassall, M.; Jones, DT.; Taiti, S.; Latipi, Z.; Sutton, SL. & Mohammed, M. (2006). Biodiversity and abundance of terrestrial isopods along a gradient of disturbance in Sabah, East Malaysia. *European Journal of Soil Biology,* Vol.42, No.1, (November 2006), pp.S197-S207, ISSN 1164-5563

Henderson, J. & Osborne, DJ. (2000). The oil palm in our lives: how this came about. *Endeavour,* Vol.24, No.2, (2000), pp. 63-, ISSN 0160-9327

Jackson, L.; van Noordwijk, M.; Bengtsson, J.; Foster, W.; Lipper, L.; Pulleman, M.; Said, M.; Snaddon, J. & Vodouhe, R. (2010). Biodiversity and agricultural sustainagility: from assessment to adaptive management. *Current Opinion in Environmental Sustainability,* Vol.2, No.1-2, (May 2010), pp.80-87, ISSN 1877-3435

Jonsson, M.; Wratten, SD.; Landis, DA. & Gurr, GM. (2008). Recent advances in conservation biological control of arthropods by arthropods. *Biological Control,* Vol.45, No.2, (May 2008), pp.172-175, ISSN 1049-9644

Koh, LP. (2007). Potential habitat and biodiversity losses from intensified biodiesel feedstock production. *Conservation Biology,* Vol.21, pp.1373–1375, ISSN 1523-173

Koh, LP. & Wilcove, DS. (2007). Cashing in palm oil for conservation. *Nature,* Vol.448, No.7157, (August 2007), pp.993-994, ISSN 0028-0836

Koh, LP. & Wilcove, DS. (2008). Is oil palm agriculture really destroying tropical biodiversity? *Conservation Letters,* Vol.1, No.2, (June 2008), pp.60-64, ISSN 1755-263X

Koh, LP. & Ghazoul, J. (2008). Biofuels, biodiversity, and people: understanding the conflicts and finding opportunities. *Biological Conservation,* Vol.141, No.10, (October 2008), pp.2450 - 2460, ISSN 0006-3207

Koh, LP. (2008a). Can oil palm plantations be made more hospitable for forest butterflies and birds? *Journal of Applied Ecology,* Vol.45, No.4, (2008), pp.1002-1009, ISSN 0021-8901

Koh, LP. (2008b) Birds defend oil palms from herbivorous insects. *Ecological Applications,* Vol.18, No.4, (June 2008), pp.821- 825, ISSN 1051-0761

Koh, LP.; Levang, P. & Ghazoul, J. (2009) Designer landscapes for sustainable biofuels. *Trends in Ecology and Evolution,* Vol.24, No.8, (August 2009), pp.431-438, ISSN 0169-5347

Koh, LP.; Ghazoul, J.; Butler, RA.; Laurance, WF.; Sodhi, NJ.; Mateo-Vega, J. & Bradshaw, CJA. (2010). Wash and spin cycle threats to tropical biodiversity. *Biotropica,* Vol.42, No.1, (January 2010), pp.67-71, ISSN 0006-3606

Liow, LH.; Sodhi, NS. & Elmqvist, T. (2001). Bee Diversity along a Disturbance Gradient in Tropical Lowland Forests of South-East Asia. *Journal of Applied Ecology,* Vol.38, No.1, (February 2001), pp.180-192, ISSN 0021-8901

Maddox, T.; Priatna, D.; Gemita, E.; & Salampessy, A. (2007). The conservation of tigers and other wildlife in oil palm plantations. *ZSL Conservation Report,* No.7

Mariau, D. & Genty, P. (1988). IRHO contribution to the study of oil palm insect pollinators in Africa, South America and Indonesia. *Oleagineux,* Vol.43, No.6, (June 1988), pp.233-240, ISSN 0030-2082

Mariau, D. (2001) *The fauna of oil palm and coconut: insect and mite pests and their natural enemies.* CIRAD, ISBN 2-87614-478-6, Montpellier, France

Mathews, J. & Foong, LC. (2010). Yield and harvesting potentials. *The Planter*, Vol.86, No.1015, pp.699-709, ISSN 0126-575x

Mayfield, MM. (2005) The importance of nearby forest to known and potential pollinators of oil palm (*Elaeis guineensis* Jacq.; Areceaceae) in southern Costa Rica. *Economic Botany*, Vol.59, No.2, (Summer 2005), pp.190- 196, ISSN 0013-0001

Murphy, DJ. (2009). Oil palm: future prospects for yield and quality improvements. *Lipid Technology*, Vol.21, No.11-12, (November/December 2009), pp.257–260, ISSN 1863-5377

Myers, N.; Mittermeier, RA.; Mittermeier, CG.; de Fonseca, GAB. & Kent, J. (2000). Biodiversity hotspots for conservation priorities. *Nature*, Vol.403, No.6772, (February 2000), pp.853-858, ISSN 0028-0836

Novotny, V.; Drozd, P.; Miller, SE.; Kulfan, M.; Janda, M.; Basset, Y.& Weiblen, GD. (2006). Why are there so many species of herbivorous insects in tropical rainforests? *Science*, Vol.313, No.5790, (August 2006), pp.1115-1118, ISSN 0036-8075

Peh, KS-H.; Sodhi, NS.; Jong, JD.; Sekercioglu, CH.; Yap, CA-M. & Lim SL-H. (2006). Conservation value of degraded habitats for forest birds in southern Peninsular Malaysia. *Diversity and Distributions*, Vol.12, No.5, (September 2006), pp.572-581, ISSN 1366-9516

Piggott, AG. (1988). *Ferns of Malaysia in Colour*. Tropical Press, ISBN 9677300296, Kuala Lumpur, Malaysia

Ricketts, TH.; Daily, GC.; Ehrlich, PR. & Michener CD. (2004). Economic value of tropical forest to coffee production. *Proceedings of the National Academy of Sciences of the United States of America*, Vol.101, No.34, (August 2004), pp.12579-12582, ISSN 0027-8424

Room, P M. (1975). Diversity and organization of the ground foraging ant faunas of forest, grassland and tree crops in Papua New Guinea. *Australian Journal of Zoology*, Vol.23, No.1, (1975), pp.71-89, ISSN 0004-959X

Round Table on Sustainable Palm Oil (RSPO) (2011). Accessed March 2011, Available from www.rspo.org

Royal Society South East Asia Rainforest Research Programme (SEARRP) (2011). Accessed March 2011, Available from www.searrp.org

Rudel, TK.; Schneider, L.; Uriarte, M.; Turner, BL.; Defries, R.; Lawrence, D.; Geoghegan, J.; Hecht, S.; Ickowitz, A.; Lambin, EF.; Birkenholtz, T.; Baptista, S. & Grau, R. (2009). Agricultural intensification and changes in cultivated a2, 1970-2005. *Proceedings of the National Academy of Sciences*, Vol. 106, No.49, pp.20675-20680, ISSN-0027-8424

Sheldon, FH.; Styring, A. & Hosner, PA. (2010) Bird species richness in a Bornean exotic tree plantation: A long-term perspective. *Biological Conservation*, Vol.143, No.2, (February 2010), pp.399-407, ISSN 0006-3207

Schmid, B.; Balvanera, P.; Cardinale, BJ.; Godbold, J.; Pfisterer, AB.; Raffaelli, D.; Solan, M. & Srivastava, DS. (2009). Consequences of species loss for ecosystem functioning: meta-analyses of data from biodiversity experiments, In: *Biodiversity, Ecosystem Functioning, and Human Wellbeing an Ecological and Economic Perspective*, S. Naeem; DE. Bunker; A. Hector; M. Loreau & C. Perrings (Ed.), 14-29, Oxford University Press, ISBN 0199547955, Oxford, UK.

Sime Darby Foundation (2011). Accessed March 2011, Available from www.yayasansimedarby.com

Sodhi, NS.; Koh, LP.; Brook, BW. & Ng, PKL. (2004). Southeast Asian biodiversity: an impending disaster. *Trends in Ecology and Evolution,* Vol.19, No.12, (December 2004), pp.654-660, ISSN 0169-5347

Sodhi, NS.; Posa, MRC.; Lee, TM.; Bickford, D.; Koh, LP. & Brook, BW. (2010). The state and conservation of Southeast Asian biodiversity. *Biodiversity and Conservation,* Vol.19, No.2, (February 2010), pp.317-328, ISSN 0960-3115

Stability of Altered Forest Ecosystems (SAFE) Project (2011). Accessed March 2011, Available from www.safeproject.net

Struebig, M.; Kingston, T.; Zubaid, A.; Adnan, A. & Rossiter, A. (2008). Conservation value of forest fragments to Palaeotropical bats. *Biological Conservation,* Vol.141, No.8, (August 2008), pp.2112-2126, ISSN 0006-3207

Struebig, MJ.; Paoli, G. & Meijaard, E. (2009). A reality check for designer landscapes. *Trends in Ecology and Evolution,* Vol.25, No.1, (January 2010), pp.7-8, ISSN 0169-5347

Struebig, MJ. (2010). Reassessing the "real scenario" regarding the environmental sustainability of palm oil. *Renewable and Sustainable Energy Reviews,* Vol.14, No.8 (October 2010), pp.2443-2444, ISSN 1364-0321

Syed, RA. (1979) Studies on oil palm pollination by insects. *Bulletin of Entomological Research,* Vol.69, No.15, (1979), pp.213-224, ISSN 0007-4853

Taylor, B. (1977). The ant mosaic on cocoa and other tree crops in Western Nigeria. *Ecological Entomology,* Vol.2, No.3, (1977), pp.245-255, ISSN 0307-6946

Tscharntke, T.; Bommarco, R.; Clough, Y.; Crist, TO.; Kleijn, D.; Rand, TA.; Tylianakis, JM.; van Nouhuys, S & Vidal, S. (2007). Conservation biological control and enemy diversity on a landscape scale. *Biological Control,* Vol.43, No.3, (December 2007), pp.294-309, ISSN 1049-9644

Tscharntke, T.; Sekercioglu, CH.; Dietsch, TV.; Sodhi, NS.; Hoehn, P. & Tylianakis, JM. (2008). Landscape constraints on functional diversity of birds and insects in tropical agroecosystems. *Ecology,* Vol.89, No.6, (June 2008), pp.944-951, ISSN 0012-9658

Turner, EC. & Foster, WA. (2006). Assessing the influence of bird's nest ferns (*Asplenium* spp.) on the local microclimate across a range of habitat disturbances in Sabah, Malysia. *Selbyana,* Vol.27, No. 2, (2006), pp.195-200, ISSN 0362-185X

Turner, EC.; Snaddon, JL.; Fayle, TM. & Foster, WA. (2008). Oil palm research in context: identifying the need for biodiversity assessment. *PLoS ONE,* Vol.3, No.2, (Feb 2008), ppe2579, ISSN 1932-6203

Turner, EC. & Foster, WA. (2009). The impact of forest conversion to oil palm on arthropod abundance and biomass in Sabah, Malaysia. *Journal of Tropical Ecology,* Vol.25, No.1, (January 2009), pp.23-30, ISSN 0266-4674

Turner, PD. & Gillbanks, RA. (2003). *Oil palm cultivation and management.* The Incorporated Society of Planters, ISBN 0-632-05212-0, Kuala Lumpur, Malaysia.

Web of Science (2011). ISI Web of Science for UK Education, Thomson Scientific Product, Accessed September 2007, January 2008 and March 2011, Available from http://wos.momas.ac.uk/

Wilcove, DS. & Koh, LP. (2010). Addressing the threats to biodiversity from oil-palm agriculture. *Biodiversity and Conservation,* Vol.19, pp.999-1007

Wood, BJ. (2002). Pest control in Malaysia's perennial crops: A half century perspective tracking the pathway to integrated pest management. *Integrated Pest Management Reviews,* Vol.7, pp.173-190

Yaap, B.; Struebig, MJ.; Paoli, G. & Koh, LP. (2010). Mitigating the biodiversity impacts of oil palm development. *CAB Reviews,* Vol.5, pp.1-11

Yayasan Sabah (2010). Accessed March 2011, Available from www.ysnet.org.my

Zhang, W.; Ricketts, TH.; Kremen, C.; Carney, K. & Swinton, SM. (2007). Ecosystem services and dis-services to agriculture. *Ecological Economics,* Vol.64, No.2, (December 2007), pp.253-260, ISSN 0921-8009

Zoological Society of London (ZSL) (2011). Biodiversity and Oil Palm Project. Accessed March 2011, Available from http://oilpalm-biodiversity.info/?page_id=38

3

Allergens and Toxins from Oleaginous Plants: Problems and Solutions

Natália Deus de Oliveira and Olga Lima Tavares Machado
Universidade Estadual do Norte Fluminense (UENF)
Brazil

1. Introduction

Population growth, industrialisation worldwide and the consequent increase in the use of fossil fuels such as petroleum contribute to the advancement of environmental damage. Burning fossil fuels releases carbon dioxide (CO_2) into the atmosphere, which contributes to an increase in the earth's greenhouse effect and climatic change (Balat & Balat, 2010). Factors such as increased demand for energy, price hikes in crude oil, global warming due to emission of green house gases, environmental pollution, and a fast diminishing supply of fossil fuels contribute to the search for alternative sources of energy (Atadashi et al., 2010).
Petroleum is a finite fuel resource that is rapidly becoming scarcer and more expensive. Petroleum-based products are one of the main causes of anthropogenic carbon dioxide (CO_2) emissions to the atmosphere (Balat & Balat, 2010). The current scenario of the world, confronted with the twin crises of fossil fuel depletion and environmental degradation, encourage research programs to reduce reliance on fossil fuels by the use of alternative and sustainable fuel sources, such as solar energy, wind energy, geothermal energy, tidal energy, ocean thermal energy, hydropower, biofuels and others (Atadashi et al., 2010; Sharma & Singh, 2009).
Biodiesel are one a biofuel that can be compound for fatty acid alkyl esters (methyl esters) that are produced from renewable natural sources such as vegetable oils, animal fats and microalgal oil by a new technology, the transesterification reaction (Atadashi et al., 2010). Biodiesel is considered a biodegradable, sustainable and clean energy because the oleaginous plants used to produce the biofuel absorb carbon dioxide during growth to a greater extent than that which is released to the atmosphere when used as a fuel in diesel engines (Sharma & Singh, 2009).

1.1 Sources of biodiesel

Various raw materials and technologies have been used for biodiesel production; however, to be profitable, biofuels need provide a net gain of energy, be environmentally sustainable, be cost-competitive and be produced in sufficient quantities without reducing the food supply (Nass et al., 2007). Biofuels are produced from renewable natural sources such as vegetable oils, animal fats and microalgal oil, and at present, many natural sources have been researched as prospective renewable fuels. With advances regarding the search for new sources of energy show, there are well-established raw materials for the processing and synthesis of biofuels.

Among the oilseeds used for biodiesel production are soybean (*Glycine max* L.), sunflower (*Helianthus annus* L.), cottonseed (Gossypium spp.), rapeseed (*Brassica napus* L.), castor bean (*Ricinus communis* L.), physic nut (*Jatropha curcas* L.) and other plants (Singh & Singh, 2010).

The source of biodiesel usually depends on the crops amenable to the regional climate. Countries such as the U.S.A. and those belonging to the European community are self-dependent in the production of edible oils and even have a surplus amount to export (Sharma & Singh, 2009).

Different countries are looking for different types of vegetable oils as substitutes for diesel fuels, depending upon the climate and soil conditions. Biodiesel has been in use in countries such as the U.S.A., Malaysia, Indonesia, Brazil, Germany, France, Italy and other European nations. However, the potential for its production and application is much more. Malaysia is far ahead of the rest in production terms. The feedstock available for the development of biodiesel in these nations is 28% for soybean oil, 22% for palm oil, 20% for animal fats and 11% for coconut oil, while rapeseed, sunflower and olive oils constitute 5% each (Sharma & Singh, 2009).

The source material for biodiesel production in Brazil varies widely among regions. Soybean, *Helianthus annuus* (sunfl ower), *Gossypium hirsutum* (cotton), *Ricinus communis* (castor bean), and *Brassica* spp. (colza) are grown in the south, southeast, and central regions; *Elaeis guineensis* (African palm), *Attalea speciosa* (babassu), soybean, and castor bean are found in the northeast and north regions (Nass et al., 2007).

This review discusses the major products obtained from oil plants that are used in biodiesel synthesis and their allergenic and toxic by-product compounds and describes the research already carried out with castor bean (*Ricinus communis* L.) oleagionous widely used for biodiesel production and other oilseeds used in the synthesis of biodiesel such as physic nut (*Jatropha curcas* L) species under investigation in the world and rapeseed (*Brassica napus* L.) widely used culture in European countries.

2. Products obtained after biodiesel synthesis

During biodiesel production, several residues such as press cakes, husks and glycerol are generated. The integral utilisation, according to the biorefinery concept, of all the fractions generated in biodiesel production is a requirement for the economy and the sustainability of the process, and for the rational exploitation of the raw materials.

2.1 Press cakes

Press cakes, the residues remaining after mechanical or solvent extraction of oils from seed kernels, can be utilised as raw materials in different bioprocesses for the production of chemicals and value-added products such as amino acids, enzymes, vitamins, antibiotics and biopesticides (Martín et al., 2010). However, those uses are restricted to edible oil cakes, which are recognised to have a high nutritional value, due to their high protein content. Non-edible oil cakes have been less investigated, and their uses are limited to organic fertilisers and biogas production (Ramachandran et al., 2007).

Possible non-food applications of cake proteins are in the field of adhesives, coatings, chemicals, fertiliser, such as seed press cake fertiliser and amino acid chelated micronutrient fertiliser. Protein research is of the high interest to many research groups with various industrial applications (Lestari et al., 2010).

Many oilseeds are used for the synthesis of biodiesel and with the growing worldwide interest in biodiesel production, is expected that the planting of oilseeds will grow exponentially. Among the candidates for oil biofuel synthesis are the castor bean, the physic nut and the rapeseed.

2.1.1 Castor bean cake

The castor bean plant (*Ricinus communis* L.) originated in Ethiopia and gradually dispersed towards South Africa, the Mediterranean region and warm areas of Asia, until finally establishing itself as a natural species in the majority of warm climate regions of the world. This plant is distributed throughout the tropics and subtropics, and is well adapted to temperate regions (Garcia-Gonzalez et al., 1999). This seed contains 45-48% oil and is important as a source of vegetable and medicinal oil with numerous benefits to humanity. As for industrial uses, dehydrated castor oil is used in the paint and varnish industry, in the manufacture of a wide range of sophisticated products like nylon fibers, jet engine lubricants, hydraulic fluids, plastics, artificial leather, fibre optics, bulletproof glass and bone prostheses and as an antifreeze for fuels and lubricants utilised in aircraft and space rockets (Ogunniyi, 2006; Conceição et al., 2005).

After oil extraction by pressing the seeds of *R. communis* L., organic matter known as castor cake is retained in the filters (Gandhi et al., 1994). The castor cake, once considered a byproduct of oil extraction, is today a product of castor bean that arouses considerable economic interest (Morais & Silva, 2008). This organic mass has constituents similar to those found in the endosperm of the seeds, such as proteins, tannins, etc. Many of these constituents are toxic or have allergenic activity (Felix et al., 2008).

Castor cake has a high protein content (~43%) and is often used as an organic fertiliser as an excellent nitrogen source and presenting insecticide and nematicide properties (Directorate of Oilseeds Research, 2004). As constituents of the high protein content of castor cake, 60% of the proteins are globulins (only soluble in salt solutions), 20% are glutelins (soluble in dilute acids and alkalis), 16% are albumins (soluble in water and dilute neutral pH buffer) and 4% are proteases (Silva Jr. et al., 1996). The protein content is not recommended for use as an animal feed because it is toxic due to the presence of the proteins ricin (toxoalbumin) and ricinin and the allergen complex, CB-1A (castor bean allergen) that is a mixture of proteins of low molecular weight (Felix et al., 2008; Silva JR. et al., 1996). Martín et al. (2010) have proposed that the high protein and carbohydrate content in castor press cake can be used as a potential feedstock following some fermentation processes.

2.1.2 Physic nut cake

Jatropha curcas L. is a tropical plant belonging also to the family of Euphorbiaceae,. It is cultivated mainly as a hedge in many Latin American, Asian and African countries and it is an oilseed crop, grown mainly for oil production. Besides oil, the jatropha seed kernel contains approximately 25–30% protein (Openshaw, 2000). After oil removal, the proteins remain in the jatropha cake. Jatropha seed protein may have similarities with other well-known oilseed proteins such as soy, canola or sunflower protein. In contrast to soy and sunflower, jatropha seed contains toxic compounds such as curcin (Lin et al., 2003) and phorbol esters (Devappa et al., 2010; Li et al., 2010; Martinez-Herrera et al., 2006) which make jatropha protein unsuitable for food applications. In addition to the several toxic or antinutritional compounds previously cited, trypsin inhibitors, lectins, saponins and phytate also might cause or at least aggravate the adverse effects, but the short-term toxicity of the kernels has been ascribed mainly to phorbol esters (Makkar et al., 2009).

2.1.3 Rapeseed cake

Rapeseed *(Brassica napus L.)* is mainly produced for its high oil content (45-50%). It is the most commonly grown oilseed crop in Europe. *Brassica napus* (rape) has as the main components of it seeds lipids (about 35% of the dry weight of the seed) and proteins (about 20–25%) (Pantoja-Uceda et al., 2004; Schmidt et al., 2004). The main storage proteins of *Brassica napus* (oilseed rape) are the 2S albumins (napins) and the 12S globulin cruciferin (Barciszewski et al., 2000).

Rapeseed cake is a high-protein product (30-40%) from industrial oil extraction, obtained from the mechanical pressing of seeds (Swiatkiewicz et al., 2010). Originally, its use was limited to animal feed because of the presence of undesirable substances (glucosinolates, erucic acid) (Swiatkiewicz et al., 2010; Schmidt et al., 2004).

Rapeseed cake contains a considerable amount of protein, rich in sulphur amino acids, and, because of its higher crude fat level and low fibre content, rapeseed cake is a richer source of metabolisable energy for monogastric animals as compared to solvent-extracted rapeseed meal (Swiatkiewicz et al., 2010).

In the processing of rapeseed oil seeds for biodiesel production, 65% of the feedstock is converted into a lignocellulosic cake residue. This product, which is rich in hemicelluloses and has a high content in hydroxyl groups, is currently used as cattle feed or for energy production. Nevertheless, the upgrading of this byproduct through its conversion to low-cost polyols by oxypropylation and their incorporation into polymer formulations could entail a considerable valorization of the residue and, thus, economic and environmental improvements for the process.

2.2 Husks

Husks, generated during dehusking of the seeds for obtaining the kernels, generally are of low economic value, and they are mainly disposed of or burnt. In some cases, the husks are used as solid fuel or as raw materials for activated charcoal production (Martín et al., 2010).

Singh et al. (2008) have proposed that all parts of the *J. curcas* fruit can be utilised efficiently for energy purposes. That paper showed how a holistic approach was been taken to utilise all the components, including the husks, that can be used for gasification. Jatropha seed husk could be used successfully as feedstock for an open core down-draft gasifier, either as a feedstock or in briquetted form.

Pollution of the environment by heavy metal ions is a serious problem because of their toxic effects on humans and other living organisms. The use of hazelnut husks for the removal of copper and lead ions from aqueous solution has been described by Imamoglu and Tekir (2008). Ngah and Hanafiah (2008) have presented a review that describes the use of husks for the removal heavy metal ions from wastewater by the use of chemically modified plant wastes as adsorbents. In this paper, they described a number of plant wastes as adsorbents, including rice husks.

The investigation carried out by Martín et al. (2010) revealed that the husks of neem (*Azadirachta indica*) and moringa (*Moringa oleifera*) can be considered potential substrates for ethanol production due to their high cellulose content (approximately 30%).

Today, a higher level of utilisation of all parts of a raw material is shown as a promising economic alternative. The production of biofuels generates many products which may have high value and be used in various industrial applications.

2.3 Glycerol

Increased biodiesel production has been driven by rapidly depleting fossil fuels, plus increasing concerns about global warming and the environment. For each gallon of biodiesel produced, 1 lb of glycerol is also produced as a by-product. One mole of glycerol is produced for every 3 mol of methyl esters, which is equivalent to approximately 10 wt.% of the total product (Karinen & Krause, 2006). This increase in glycerol production has depressed the price of refined glycerol.

Glycerol is a trivalent alcohol widely used in the pharmaceutical, food, cosmetic and chemical industries. It is produced from a diversity of procedures, among them the transesterification of vegetable oils and animal fats. During biodiesel production from vegetable oil and animal fats, two phases are produced after transesterification and distillation of the excess alcohol: one upper ester phase (EP) that contains the main product, biodiesel and the lower glycerol phase (GP) that consists of glycerol and many other chemical substances such as water, organic and inorganic salts, a small amount of esters and alcohols, traces of glycerides and vegetable colours (Hájek & Skopal, 2010).

3. Toxic and allergenic compounds

Many oilseed plant candidates and those currently used for the synthesis of biodiesel present toxic or allergenic compounds that are constituents of the seeds, which, as a consequence, can also be found in some products obtained after extracting the oil. Other problem is that some of these compounds are also found in others parts of the plant such as the 2S albumin from *R. communis* (an allergen) present in the pollen of this oilseed. The presence of these compounds limits the economic applications of the press cake and is a risk to the workers and the population living nearby.

3.1 Toxins

3.1.1 *Ricinus communis*

Castor bean is an oleaginous candidate for oil production (Singh & Singh, 2010), which will contribute to enhancing the cultivation of this plant. Castor bean seeds, however, contain a strong toxin (ricin), a toxic volatile alkaloid ricinine (1,2-dihydro-4-methoxy-1-methyl-2-oxo-3-piridinocarbonitrila-$C_8H_8N_2O_2$) and an allergenic protein fraction (CB-1A or 2S albumin isoforms), which severely limits the usefulness of the castor meal after oil extraction (Godoy et al., 2009; Audi et al., 2005; Garcia-Gonzalez et al., 1999; Thorpe et al., 1988).

The castor bean is an oilseed member to the Euphorbiaceae family and yields an oil that is used for biodiesel production. Furthermore, the residual cake is very useful for fertilisation and it is rich in proteins, opening the possibility of its use as animal feed. However, this second application addresses the problem of the presence of ricin, an extremely toxic protein.

Ricin is a protein found exclusively in the endosperm of castor bean seeds and has not been detected in other plant parts such as the roots, leaves or stems. It represents 1.5 to 2% of the total weight of the seed (Anandan et al. 2005; Cook et al., 2006). It is primarily responsible for the toxicity of castor oil and is among the most toxic proteins known to man (Moskin, 1986). The ricin toxin is a 62-66 kDa protein produced by castor beans (*Ricinus communis*). This holotoxin consists of two polypeptide chains, approximately 32 kDa and 34 kDa in size, linked by a disulphide bond (Figure 1). The A chain (RTA) is a potent ribotoxin, inhibiting protein synthesis in mammalian cells at doses as low as a single RTA molecule per cell. The B chain (RTB) is a lectin, which binds to galactose residues on the cell surface. (Audi et al., 2005; Rao et al., 2005; Brandt et al., 2005).

Sehgal et al., 2010 demonstrated the presence of three isoforms of ricin in castor seeds. The isoforms were sub fractionated into ricin I, II and III by chromatography. Their molecular weights lie between 60-65 kDa. Ricin I, II and III were highly cytotoxic against Vero cell line with IC50 values of 60, 30 and 8 ng/ml respectively. Difference in cytotoxicity of isoforms was confirmed through hemagglutination assay and ricin III caused higher degree of hemolysis.

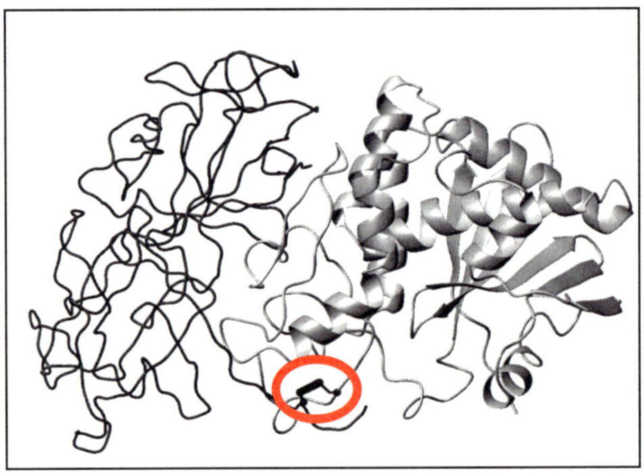

Fig. 1. Structure of the ricin molecule. The B chain is located on the left and the chain A is on the right. The red circle indicates the disulfide bridge linking the A and B chains (Rutenber & Robertus, 1991)

Ricin is a potent toxin that kills eukaryotic cells by inhibiting protein synthesis. Therefore, it is a protein of the class of toxins known as ribosome inactivating proteins, RIPs (Cook et al., 2006; Olsnes et al., 1999).

RIPs can be either type 1 (monomer) or type 2 (dimeric) (Stirpe & Bartelli, 2006). Type 1 RIPs present only the A chain, which is a glycosidase that removes an adenine residue from 28S ribosomal RNA. The RNA, after depurination, is susceptible to hydrolysis in alkaline pH and to acids in the presence of aniline. The region of the modified rRNA is essential for elongation factor binding and modified ribosomes cannot support protein synthesis (Olsnes, 2004).

The B chain is required for binding to the target cell and intracellular direction of the A chain (Olsnes, 2004; Day et al., 1996). When there are A and B chains, the toxin is classified as a type 2 RIP, which is the case for ricin (Cook et al., 2006). The ricin A chain is very efficient inside the cell, since a single molecule inactivates thousands of ribosomes per minute. Thus, one molecule can inactivate ribosomes faster than the cells can synthesize new ribosomes and, therefore, only one molecule kills the cell (Olsnes & Kozlov, 2001). The value of the oral LD50 for rats and mice is between 20 and 30 mg/kg body weight, while in humans the toxic oral dose is 1-20 mg/kg of body weight (Alexander et al., 2008; Audi et al., 2005; Rao et al., 2005).

Despite its high toxicity, it is possible to develop immunity against ricin, as demonstrated in the studies of Tokarnia and Döbereiner (1997) in which cattle that received small doses of ricin (by ingestion) developed some immunity and later supported a higher dose with symptoms of intoxication, but stayed alive, while animals that received the higher dose directly were not resistant.

In the medical area, ricin has been prominent among a group of toxic proteins that have been used as immunotoxins and therapeutic agents used in the treatment of cancer and autoimmune diseases (Brandt et al. 2005). This toxin has also drawn attention due to its criminal use in the murder of Bulgarian journalist Georgi Markov in 1978 in London (Olsnes, 2004).

- Solutions:

Ricin is a major impediment to the use of castor cake for animal food (Na et al., 2004). The transformation of castor cake into a non-toxic product that can be used for animal feed already has long drawn the interest of many researchers around the world, and some satisfactory results have been obtained. A number of methods have been employed to detoxify castor oil seed meal, some of which appear to be more effective than others (Puttaraj et al., 1994).

In recent years, several methodologies have emerged to detoxify castor bean cake and use it as animal feed. Anandan et al. (2005) reported that physical processes based on heat (boiling, autoclaving, hot air oven) and alkali-based chemical processes (sodium hydroxide, calcium hydroxide and ammonia) could detoxify castor cake. The efficacy of the treatments was assessed based on the qualitative and quantitative changes in ricin content. Of all the methods employed, autoclaving (15 psi., 60 min) and lime treatment (40 g/kg) completely destroyed the toxin as observed by electrophoresis, however, toxicologic assays were not done.

Godoy et al. in 2009 used solid-state fermentation (SSF) of castor bean waste to achieve ricin detoxification, to reduce allergenic potential and to stimulate lipase production. The fungus, *Penicillium simplicissimum*, an excellent lipase producer, was able to grow and produce the lipase enzyme in castor bean waste. The biodetoxification process described could extend the use of fermented castor bean waste and potentially be used as an animal feed or fertiliser, without causing damage to the environment.

The SSF processes used by Godoy et al. (2009) permitted the total detoxification as observed by electrophoresis and toxicological analysis. This process offers potential advantages in bioremediation and biological detoxification of toxic compounds second Pandey et al., 2000.

3.1.2 *Jatropha curcas*

Jatropha curcas is another member of the Euphorbiaceae family and is known for its toxicity. It is grown in Central America, South America, Southeast Asia, India and Africa. The kernels have about 50% oil and the seeds contain curcin, a toxic glycoprotein with a 54% homology with the ricin A chain and with a similar mode of action (Alexander et al., 2008; Kumar & Sharma, 2008), as well as phorbol esters, which are polycyclic compounds

(Devappa et al., 2010; Martinez-Herrera et al., 2006) that can induce skin tumours when administered to mice (Chen et al., 1988).

Curcin, a kind of type I RIP, was first isolated from the seeds of *Jatropha curcas* by Stirpe et al. (1976). It was found to inhibit the growth of some tumour cells (Lin et al., 2003). Curcin is a similar protein to ricin, a toxic protein isolated from castor beans (*Ricinus communis*), which has two chains, one a functional lectin and the other capable of inhibiting protein synthesis (Rakshit et al., 2008; Stirpe et al., 1976). The absence of a lectin portion of this protein prevents binding to cells and impairs internalisation, thus becoming much less toxic than the type II RIPs such as ricin present in the seeds of castor bean. Recently, Lin et al, 2010 have purified a curcin molecule that was a glycoprotein with 4,91% of the total neutral-surge content. It strongly inhibits the protein synthesis of rabbit reticulocyte lysate, with an IC50 of 0.42 nM. The isolated curcin had a hemagglutinating activity, when its concentration was more than 7.8 mg=L. The secondary structure of curcin was analyzed by Circular Dichroism (CD) spectrum. The results of acute toxicity in mice show that mice oral Semi-lethal dose LD(50) was 104.737 +/- 29.447 mg=kg; mice parenteral semi-lethal dose LD(50) was 67.20 +/- 10.445 mg=kg.

Due to the toxic compounds found in physic nut seeds, the press cake cannot be used for animal feed, despite its high protein content. Experiments have shown the toxicity of the seeds of *J. curcas* in mice, rats, sheep, calves and chicks (El-Badwi et al., 1995). In contrast to this, Panigrahi et al. (1984) found no dramatic effects of poisoning in mice and rats fed on seeds of Mexican origin (edible varieties) that naturally occur in Mexico (King et al., 2009).

Beyond the concern about the presence of curcin in physic nut cake, there is another concern to be addressed: the presence of phorbol esters. The term phorbol ester is used today to describe a naturally occurring family of compounds widely distributed in plant species of the Euphorbiaceae and Thymelaeceae families (Rakshit et al., 2008). They are defined as polycyclic compounds in which two hydroxyl groups on neighbouring carbons are esterified into fatty acids. These compounds are present in many plants, including the physic nut. The structure of phorbol esters is dependent on a tetracyclic diterpene carbon skeleton known as tigliano, the main portion of alcohol in the phorbol esters (Goel et al., 2007).

Phorbol esters and their various derivatives are said to promote tumours. In addition to this effect, they induce significant biological effects, even at low concentrations. The primary action of phorbol esters occurs in biological membranes. This toxin tends to bind to receptors of membrane phospholipids (Weinstein et al., 1979). The phorbol esters are analogues of diacylglycerol, an activator of many isoforms of protein kinase C (PKC). The most investigated activity of these esters is their binding and activation of protein kinase C (PKC), which plays a critical role in signal transduction pathways and regulates cell growth and differentiation (Goel et al., 2007). Contradictory to their tumour-promoting ability, there are reports on the pro-apoptosis capacity of phorbol ester on tumour cells (Brodie & Blumberg, 2003; Gonzalez-Guerrico & Kazanitez, 2005). Some phorbol esters are inhibitors of HIV replication and have antileukemic activity (Goel et al., 2007).

The phorbol esters are acutely toxic, and oils containing phorbol esters are known purgatives (Gandhi et al., 1995). Adoption of varieties lacking phorbol esters, in addition to providing a potential source of income from animal feed, would also eliminate any potential risks associated with prolonged exposure to phorbol esters.

- Solutions:

A range of methods have been used to try detoxify defatted seed meal. Hass and Mittelbach (2000) suggest a method for detoxification of the seed oil using traditional oil refining processes to examine the effect processing on the content of phorbol esters. That paper, almost no effect could be observed with degumming and deodorisation, whereas the steps of deacidification and bleaching could reduce the content of phorbol esters by up to 55%.

Extraction with polar organic solvents and combined heat/$NaHCO_3$ treatments using a combination of both solvent extraction and heat/$NaHCO_3$ treatment, have been shown to promote a 48-fold reduction in phorbol ester content in the seed meal of the physic nut (Martinez-Herrera et al., 2006).

Heat treatments, such as autoclaving for example, usually inactivate the curcin, allowing the use of this as food for ruminants. It is known that heat treatment alone is not able to decrease the concentration of phorbol esters. Then, in 2008, Rakshit et al. described satisfactory results in toxicity studies with rats using alkali (2% NaOH or 2% $Ca(OH)_2$) and heat treatments (autoclaved at 121°C) to deactivate phorbol esters as well as the lectin content of the physic nut meal. After these treatments, the phorbol ester content was reduced up to 89% in whole and dehulled seed meal. The rats fed with treated meals exhibited delayed mortality compared to untreated meal-fed rats.

The phorbol ester content was analysed in fractions obtained at different stages of oil pre-treatment and biodiesel production from the physic nut by Makkar et al. (2009). Makkar et al. observed that the phorbol esters were destroyed by the stripping process during biodiesel production. In physical refining (degumming, silica/bleaching, deodorisation/stripping at 240–260°C and under vacuum) the deodorisation conditions were much more severe, leading to phorbol ester degradation.

3.1.3 Brassica napus

Rapeseed (*Brassica napus* L.) is an important crop for the production of vegetable oil for human consumption, and more recently for the biodiesel. *B. napus* is member of the Brassicaceae family and the crop is mainly grown for its biodegradable oils which can be used for the production of cooking oil, machine oil, diesel substitutes and as a base oil for the plastics industry. The high protein seed residue following oil extraction provides a good source of animal feed (Welch et al., 2000).

After oil extraction, a residue with high protein content is obtained that can be used as a valuable animal feed. However, anti-nutritive factors, such as the glucosinolates or erucic acid in rapeseed may cause various specific physiological effects in humans and in animals (Fahey et al., 2001). Glucosinolates are considered anti-nutritive factors for animal production but, on the other hand, they have an important role in plant protection against pests, diseases and also weeds (Rahmanpour et al., 2010; Haramoto and Gallandt, 2004).

Originally, the rapeseed cake uses for animal feed were limited because of the presence of undesirable substances (glucosinolates, erucic acid) (Swiatkiewicz et al., 2010; Schmidt et al., 2004). Major deleterious effects of glucosinolate ingestion in animals are reduced palatability, decreased growth and production. Ruminants are less sensitive to dietary glucosinolates. Among the monogastric animals, pigs are more severely affected by dietary glucosinolate compared to rabbits, poultry and fish (Tripathi & Mishra, 2007).

- Solutions:

The oil meal of Brassica origin is a good source of protein for animal feed but the glucosinolate content limits its efficient utilisation. Various processing techniques have been applied to remove glucosinolates in order to minimise their deleterious effects on animals. Tripathi and Mishra (2007) presented in their review some techniques, described by other authors, to remove glucosinolates; water extraction, heat and $CuSO_4$ treatments were found to be suitable for rapeseed meal quality improvement.
The work presented by Petisco et al. in 2010 measured the quality parameters of intact seeds of Brassica species using visible and near-infrared spectroscopy (NIRS). Petisco t al, 2010 demonstrated that NIRS technology is viable for the quantification of oil, protein and total glucosinolates in seed samples of *B. napus* and/or *B. carinata* without sample preparation. The accurate predictions provided by the NIR equations confirmed that NIR technology could be very useful for the rapid quality evaluation of intact rapeseeds, thus avoiding the need for grinding and thereby saving time. The speed of analysis and the non-destruction of the seed make this technique well-adapted for breeding purposes as well as for quality control in oil factories and in feed manufacturing. The problem of erucic acid has been solved by conventional breeding technology of rapeseed. The term canola (**CAN**adian **O**il **L**ow **A**cid) refers to strains of *B. napus* and *B. campestris* containing less than 2% of total fatty acids as erucic acid.
Despite the problems regarding the presence of glucosinolates in rapeseed cake for use in animal feed, it can also be used as a biopesticide. The utilisation of the meal as a biopesticide requires seed meal storage prior to field application. Morra and Borek (2010) studied the effect of a storage period to maintain glucosinolate stability in *B. napus, B. juncea* and *S. alba* seed meals. Glucosinolate concentrations measured every six months using HPLC-MS decreased only in meal samples stored at 4°C, and to the greatest extent in samples stored within paper bags. This procedure can be used for maintaining glucosinolate stability and facilitating the utilisation of rapeseed cake as a biopesticide.

3.2 Allergens

The term allergen is used to identify substances that have the ability to promote two or three distinct molecular properties: i) the property to raise awareness (i.e., induce the production of antibodies of high affinity, particularly IgE, by the immune system), ii) ability to bind to IgE antibodies and also iii) the property to enable an allergic reaction (i.e., trigger allergic symptoms in a sensitised person) (Aalbers, 2000).
IgE-mediated reactions are believed to be responsible for most induced allergic reactions of the immediate hypersensitivity type (type 1), and the diagnosis relies on specific biological and clinical features. Such allergic reactions involve activation of effector cells, mainly mast cells and basophils, leading to an inflammatory response and specific clinical manifestations (Aalbers, 2000).
The pathogenesis in allergy has two phases: (i) usually, the primary contact with an allergen involves awareness of the naïve immune system to produce an IgE response and (ii) later repetitive exposure to the same allergen results in elicitation of an allergic reaction and the clinical manifestations (Moreno, 2007). The body's cells, having been previously sensitised, upon contact with the allergen are attracted to the place of antigen inoculation, and then try to orchestrate cellular mechanisms to eliminate and/or protect the body from further

damage, thus helping to exacerbate symptoms in allergic individuals (Sichere & Leung, 2008) such as sneezing, difficulty breathing, cramps, hives, itching, etc.

3.2.1 Plant allergens

The plant seed is not only an organ of propagation and dispersal but also the major plant tissue harvested by humankind, a major source of dietary protein. Although the vast majority of the individual proteins present in mature seeds have either metabolic or structural roles, a seeds also contain one or more groups of proteins that are present in high amounts and that serve to provide a store of amino acids for use during germination and seedling growth (Shewry et al., 1995).

Because of their abundance and economic importance, proteins are characterized into groups on the basis of their extraction and solubility in water (albumins), dilute saline (globulins), alcohol/water mixtures (prolamins) and dilute acid or alkali (glutelins). The major seed storage proteins include albumins, globulins and prolamins (Breiteneder & Radauer, 2004; Shewry et al., 1995).

Plant allergens can be also proteins that act as defense, enabling the plant to defend itself against biotic and abiotic stresses. Many plant tissues which are consumed by humans contain thousands of these allergenic proteins. Approximately 0.5% of the U.S. population is affected by various stages of IgE-mediated food allergy (El-Agamy, 2007, Breiteneder & Radauer, 2004).

Plant allergens are classified into families and superfamilies, based on their structure and function. Proteins are clustered into families if they have residue identities of 30% or greater or if they have lower sequence identities but their functions and structures are very similar. Families whose members have low sequence identities but whose structures and functional features are placed together in superfamilies (Breiteneder & Radauer, 2004). All storage protein fractions are mixtures of components that exhibit polymorphism both within single genotypes and among genotypes of the same species. This polymorphism arises from the presence of multigene families and, in some cases, proteolytic processing and glycosylation (Shewry et al., 1995). Additionally, it has become evident that the level of exposure and the properties of the allergen itself are important factors for determining its allergenic potential (Breiteneder & Mills, 2005).

Plant food allergens belong to the most abundant cupin and prolamin superfamilies and the protein families of the plant defence system. In the cupin superfamily are grouped the 7S and 11S seed storage proteins and transfer proteins [nsLTPs], α-amylase/trypsin inhibitors; the prolamin storage proteins of cereals are grouped into the prolamin superfamily (Breiteneder & Radauer, 2004).

According to many studies, it is known that the oilseeds described in this review (castor bean, physic nut and rapeseed) have allergens from the 2S albumin family. So, the study of the structure/function of these allergens will be addressed.

3.2.2 2S albumin

The 2S albumins are a major group of storage proteins present in many dicotyledonous plant species (Shewry et al., 1995). These proteins belong to the prolamin superfamily that includes the nsLTPs and cereal seed inhibitors of α-amylase/trypsin or both. All the proteins of this superfamily are of low molecular weight, are rich in cysteine and present similar three-dimensional structures rich in α-helix (Breiteneder & Radauer, 2004).

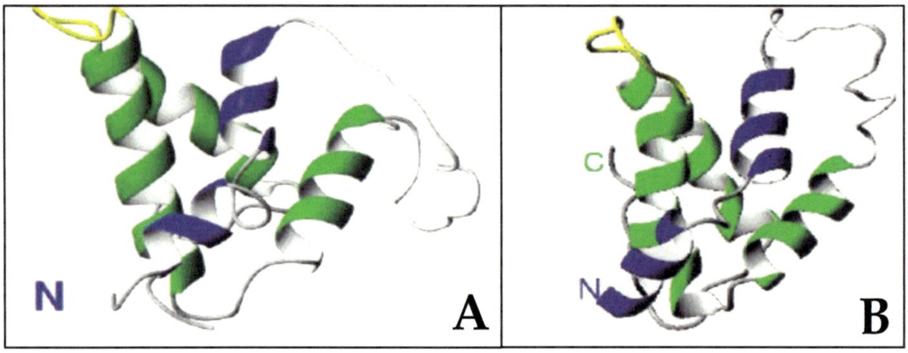

Fig. 2. (A) Ribbon representation of the solution structure of rRicC3, showing helices (blue and green) and loops (gray, but the hypervariable loop in yellow) (Pantoja-Uceda et al., 2003). (B) Ribbon representation of the solution structure of rproBnIb (Pantoja-Uceda et al., 2004)

The Figure 2 shows two 2S albumins; in (A) the three dimensional structure of recombinant RicC3 determined by NMR methods (Pantoja-Uceda et al., 2003) and in (B) the structure of the precursor form of the recombinant napin BnIb, rproBnIb (Pantoja-Uceda et al., 2004). Both 2S albumins show similar three-dimensional structures rich in α-helix.

- Ric c 1 and Ric c 3:

The 2S albumins from castor bean are synthesized at specific times during seed development and deposited within vacuoles (corpuscle protein) during seed development, then can be degraded during germination, supporting the growth of the seed (Ahn & Chen, 2007; Regente & La Canal, 2001). They are synthesized in the endoplasmic reticulum as a precursor protein of high molecular weight, Figure 3. Later, this precursor is proteolytically cleaved, generating a peptide ligand and other small peptides (Jolliffe et al. 2004; Shewry et al., 1995). Glycosylation of proteins may occur during protein synthesis when carbohydrates are incorporated, mostly mannose and glucosamine (Jolliffe et al. 2004; Bewley & Black, 1994).

It was believed that the 2S albumins were metabolically inactive, but currently, due to their ability to inhibit proteinases, alpha amylase (Nascimento, 2011) as well as their allergenic (Machado & Silva, 1992) and antifungal (Aggizio et al. 2003) properties, it is believed they are involved in defence functions in plants (Regente& La Canal, 2001). The allergenic properties of 2S albumins are resistant to thermal and chemical denaturation, possibly even detoxification treatment, and the allergy may be triggered by contact and inhalation (Machado & Silva, 1992; Silva Jr. et al., 1996). The 2S albumins are also able to reach the gut immune system intact so as to induce sensitization and elicitation of allergic reactions at the gut mucosa (Pantoja-Uceda et al., 2004).

Historically, in 1943, Spies and Coulson described one protein fraction of low molecular weight, heat stable protein from castor bean seeds, which was designated CB-1A (Castor Bean allergen). In 1947, hypersensitivity triggered by castor bean was first described, and in 1977, Li and co-workers isolated and characterized a protein from the seeds of *Ricinus communis* L. with low molecular weight and high glutamine content, which showed properties similar to those proteins previously isolated from castor beans. Later, in 1978,

Youle and Huang showed that CB-1A was the same storage protein characterized by Li et al. in 1977. In 1982, Sharief and Li isolated and sequenced a protein from the seeds of *Ricinus communis* L. (Ric c 1), with coefficient 2S sedimentation, consisting of two subunits linked by sulphur bridges. The smallest contained 34 amino acids (Ric c 1 small chain) with an apparent molecular mass of 4 kDa and the larger subunit contained 61 amino acids (Ric c 1 large chain) with a molecular mass of 7 kDa.

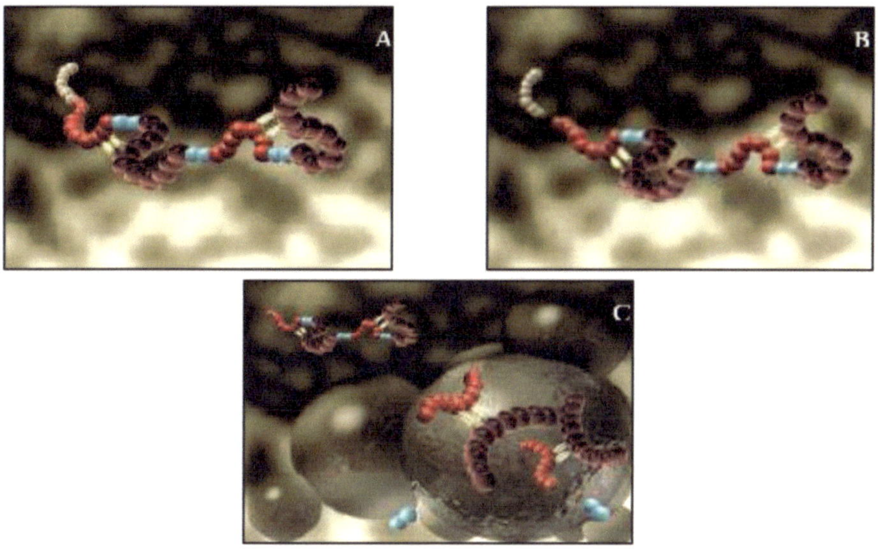

Fig. 3. Schematic of the processing of the precursor isoforms Ric c 3 and Ric c 1. A) Precursor signal peptide intact with beige, yellow sulphur bridges, Ric c 3 and Ric c 1 respectively in red (light chain) and brown (heavy chains), peptide binding in blue, B) Loss of signal peptide, C) loss of peptide connection with subsequent separation of the two isoforms

In 1992, Machado and Silva isolated and sequenced one second allergen of the castor bean seeds, named Ric c 3, with molecular weight around 11 kDa, present in the same precursor of Ric c 1 with 29 kDa. The primary structure of the allergen was fully elucidated in 1996. Since 2003, many other allergenic proteins belonging to the 2S albumin class have been identified in castor bean seeds by Machado and co-workers (Felix et al. 2008).

Currently, it is known that the allergen complex CB-1A represents about 12.5% by weight of the cake, as determined by the precipitation test with the antigen diluted. This complex consists of approximately 20 isoforms, with molecular mass between 10 and 14 kDa (Machado et al, 2003, Machado & Silva, 1992).

It is known that allergic diseases have increased in recent years and that over 30% of the population suffers from allergic diseases. The main causative agents are pollen, fungal spores, dust mites, animal epithelia. (Prueksakorn & Gheewala, 2008; Robotham et al., 2002). Medical problems such as conjunctivitis, rhinitis and urticaria have been associated with castor bean seeds, as well as the pollen (Garcia-Gonzalez et al., 1999).

The allergy triggered by the 2S albumin of castor bean is mainly caused by the inhalation of cake dust, representing a problem for the workers in extraction plants and for the population that inhabits the area around of these extraction plants (Garcia-Gonzalez, et al.,

1999). Another factor to be considered is the risk of allergic reactions of field workers using the castor cake as a fertilizer and who are subject to the dust.

There are few reports regarding the role of allergens in their pollen. In India, a study conducted by Singh and co-workers in 1992 demonstrated that there is variation in the protein profile of extracts of castor bean pollen in different years and places in this country. In 1997, the cross-reaction and the presence of common epitopes between seed and pollen extracts of castor beans were confirmed (Singh et al., 1997). That same year, some studies demonstrated a cross-reaction of castor bean pollen with pollen from other plant species, *Mercurialis annua* (Vallverdú et al., 1997) and *Putranjiva roxburghii* (Singh et al., 1997). In 1999, studies performed by Garcia-Gonzalez et al. demonstrated that the castor bean pollen causes symptoms of respiratory allergy. Accordingly, Paru and co-workers in 1999 proposed a new approach for identification and partial characterisation of allergenic proteins from the pollen of *Ricinus communis* L. In 2002, Palosuo et al. demonstrated the cross-reactivity between allergens from castor beans and other vegetables of the Euphorbiaceae family, confirming the importance of studies of cross-reactivity in diagnostic research.

Singh & Kumar in 2003 demonstrated, quantitatively and qualitatively, the prevalence of pollens in the region of India, noting that, among other aeroallergens, there is a significant distribution of castor bean pollen in this area. Knowing also that air pollution has been described as an important factor for the recent increase in the incidence of respiratory diseases and that the air carries many grains of pollen, the work done by Bist et al. in 2004 observed a variability of castor bean pollen protein before and after exposure to air pollutants.

- Jat c 1:

Seeds and pollen in general present allergenic proteins with additional defense properties such as proteases, amylase inhibitors or antifungal factors. Though protective for the plant, these antinutritional and toxic factors may have deleterious effects or even be toxic to animals and humans. Nothing was known about the presence of allergens in *J. curcas* seeds until the work of Maciel et al. (2009) which provided further information on the presence of allergenic proteins in this oilseed.

Maciel and co-workers, in 2009, described the presence of an allergenic 2S albumin (12 kDa), called a Jat c 1 (Figure 4), isolated from seeds of *Jatropha curcas* L. These N-terminal sequences presented similarities with 2S albumin from *Ricinus communis, Cucurbita maxima, Sesamum indicum, Solanum lycopersicum* and *Helianthus annus*. Sequence analysis revealed an important common feature: the conservation of four cysteine residues that are important for 2S albumin folding.

Jat c 1

(Small chain): VRDK**C**GEEAERRTLXG**C**ENYISQRR

(Large chain): PREQVPRQ**CC**NQALE

Fig. 4. Partial sequence data of *Jatropha curcas* 2S albumin. Data sequencing was performed by Edman degradation (Maciel et al., 2009)

Maciel et al. in 2009 also demonstrated the ability of this allergenic protein binding to IgE attached to rat mast cells, inducing histamine release from these cells. Its allergenic properties were demonstrated by the PCA test, a type I allergic reaction in vivo. Another feature shown by Maciel was that 2S albumin isolated from physic nut also showed strong crossreactivity with the major allergens from castor bean, Ric c 1 and Ric c 3. These data indicated that an individual sensitized to allergens from the castor bean (Ric c 1 and Ric c 3) could become sensitive to 2S albumin from *J. curcas* (Jat c 1) and that the inverse condition may also be possible, suggesting that Jat c 1 has potential intrinsic allergenicity.
Since allergy to oleaginous seeds has emerged as an important clinical condition following an increase in the use of biodiesel, and given the risk due to cross-reactive allergens (as observed for allergens from *J. curcas* and *R. communis*), advances in the identification and characterization of common aeroallergens and allergens from oleaginous seeds are necessary for the establishment of a specific therapy.

- Napins:

The oilseed rape (*B. napus*) ranks as the most commonly grown oilseed crop in Europe (Krzyzaniak, et al., 1998). Rapeseed (*Brassica napus* L.) is mainly produced due its high oil content (45-50%). After oil extraction, a meal is obtained containing most of the proteins (30-40%) (Boucher et al., 2007; Pantoja-Uceda et al., 2004).
Rapeseed protein meal contains two predominant classes of seed storage proteins: 12S globulin (cruciferin) which represents 25–65% of its protein content (Raab et al., 1992) and 2S albumin (napin). Napins belong to the 2S albumin class of proteins and hence are water soluble, stable at high temperature (up to 88±C) (Krzyzaniak, et al., 1998) and represent 15-45% of the total rape seed protein content depending on the variety (Raab et al., 1992). These proteins belong to the albumin storage proteins; in the seeds of recent varieties, they are present in lower quantities than cruciferins.
Various forms of napins (2S albumin) are also found in seeds of other Brassicaceae. They can be classified into three classes according to molecular weight 12.5, 14.5 and 15 kDa (Monsalve & Rodrigues, 1990).
Mature napins exhibit molecular weights between 12,500 and 14,500 Da (Raab et al., 1992). They are encoded by a multigenic family, initially synthesized as a precursor which is proteolytically cleaved to generate mature napin chains. Napins are expressed during seed development as precursors of 21 kDa. They comprise two polypeptide chains held together by two disulphide bonds: a small (4500 Da) and a large one (10,000 Da) (Krzyzaniak et al., 1998). The large chain includes two additional intrachain disulphide bonds, which reinforce the stability of the proteins (Byczynska & Barciszewski, 1999; Monsalve & Rodriguez, 1990). Napins are characterised by their strong basicity (isoelectric point, pI ~ 11) mainly due to a high amidation of amino acids (Raab et al., 1992).
Napins are polymorphic proteins due to their origin from multigene families. As a result, their isolation from the seeds renders a microheterogeneous material unsuitable for three-dimensional structure determination, by either X-ray diffraction or NMR (Rico et al., 1996).
Many isoforms of napin exist because of the large number of napin genes and differences in proteolytic cleavages. Five isoforms were first identified according to their molecular weights (Monsalve et al., 1991). One of them (isoform BnIb, called 2SSI-_BRANA in the Swiss-prot databank nomenclature) has been totally sequenced and its three-dimensional structure determined by NMR (Pantoja-Uceda et al., 2004; Rico et al., 1996). BnIb (12.7 kDa)

is a representative member of a distinct group of rapeseed 2S albumins, referred to as "low molecular weight napins" (LMW-napins) to distinguish them from the more common and abundant group of "high molecular weight napins" (14.0-14.7 kDa) (Monsalve et al., 1991).
The 2S albumin class of proteins constitutes the major seed storage protein group in *Brassica napus*, representing about 20% of the total protein content in mature rape seeds. 2S Albumins from several species such as mustard, castor bean, Brazil nut, English walnut, sunflower and peanut have been shown to be type I allergy inducers of remarkable incidence, suggesting that this family of storage proteins is intrinsically allergenic (Pantoja-Uceda et al., 2004).
Coincidental with the expansion of rapeseed cultivation, there have been increases in the number of reported cases of asthma and other conditions related to allergenicity and irritancy, but it is not clear evidence that rapeseed has adverse effects on human health (Murphy, 1999). The work conducted by Murphy (1999) described that the allergens present in rapeseed pollen have only a minimal impact on public health.
The distinction between oilseed rape and grass pollen was described by Welch and co-workers in 2000. They showed that these pollens are immunologically distinct and there is no evidence of cross-reactivity between them. Individuals allergic to grass pollen will not necessarily develop a specific nasal or airway response to inhaled oilseed rape pollens.
Chardin et al. 2008 aimed to characterize the IgE specificity of various patients suffering from pollen polysensitization to identify both peptidic and carbohydrate cross-reactive determinants. They showed the rapeseed, grass and Arabidopsis proteins were separated by isoelectric focusing, followed by SDS-PAGE, and transferred to a nitrocellulose sheet. They showed that multiple pollen sensitizations could result from multiple sensitizations to specific proteins or from a cross-sensitization to a wide range of glycoproteins. That paper also allowed for improving the diagnosis of allergy and its medical treatment.
Knowing that the oilseed rape production is widespread in cereal growing areas and that many patients who attend the clinic (district general hospital, UK) for seasonal allergies claim that they are allergic to it, the aim of the work in development by Trinidade et al. (2010) is to determine the prevalence of oilseed rape allergy in this population. They observed that oilseed rape hypersensitivity was relatively uncommon, comprising only 2% of the population tested (n = 28). Oilseed rape does not cause significant allergy, even in areas of high production. It is likely that those patients exhibiting oilseed rape allergy may in fact be symptomatic due to the effect of other allergens, acting synergistically with the oilseed rape allergen (Trinidade et al., 2010).

3.2.3 Solutions

Several methodological solutions for reducing or eliminating allergens can be used to obtain positive results. Heat processing induces, in most cases, irreversible denaturation of proteins, leading to aggregation, and such structural changes do not always correlate with decreased allergenicity. Depending on the system, heating may have no effect or it may decrease or increase allergenicity. This occurs because of the existence of sequential and/or conformational epitopes in allergen structure.
The knowledge of the protein's primary structure is essential for initial strategies for protein modification of its epitopes. Many studies have shown positive results with various experiments performed with unmodified and chemically modified proteins. In 2002, Cai and

co-workers identified the amino acid residues of allergenic proteins (trichosanthin, a Chinese herb) with an important role in the IgE response. Using an assay with these proteins mutated at their residues important for IgE binding, they showed that the protein specifically lost its binding activity and exhibited reduced IgE induction in the immunized mice. Kamal et al. (2005) described that the tryptophan residue is essential for immunoreactivity of a diagnostically relevant peptide epitope of *A. fumigatus.* The loss of specific IgG and IgE antibody binding of the modified protein by ELISA confirmed the critical role of tryptophan (Trp17) in the immunoreactivity of this protein. With the same objective, allergen modification and a better understanding of the functional role of castor bean allergens is fundamental to preventing allergy induced by *R. communis* (Ric c 1 and Ric c 3). Accordingly, Felix and co-workers (2008) showed the mapping of IgE binding epitopes of Ric c 1 and Ric c 3, the allergens from castor bean, by a mast cell degranulation assay. They identified four continuous epitopes in Ric c 3 and two in Ric c 1. This knowledge may allow the induction of protective antibody responses to antagonise the IgE recognition. All the data showed that the IgE epitope of these proteins were determined and shown to play a critical role in induction of IgE, and modification of the IgE epitope may be a useful strategy to reduce the allergenicity of an allergen. Deus-de-Oliveira evaluated the possibility of use of compounds of calcium in order to inactivate allergenicity of isolated 2Salbumin and castor cake. The samples were incubated with a solution of calcium hydroxide, calcium carbonate or calcium oxide, 4 and 8% in the ratio of 1:1 (v/v), during 12 hours, at the room temperature. The calcium treatments modified the allergen of castor bean and all they are effectives as was valued by reducing the allergenicity as observed by quantification of mast cells degranulation. Simultaneously, castor meal detoxification was also obtained using treatments with $CaCO_3$, $Ca(OH)_2$ and CaO. The results obtained in by Deus-de- Oliveira contribute to get of a safer product for manipulation of the workers and with the possibility of expanding the economical applicability, for example, in animal feed.

4. Conclusion

Oilseeds are renewable sources of oil, protein and carbohydrate for edible and industrial applications. Traditionally, the commodity value for oilseeds has been the meal (or cake) produced after mechanical pressing or solvent extraction oil from the seed. The press cake obtained after oil production could be used for animal feed but each of these cakes may have in its constitution toxic or allergenic compounds (Thelen, 2009).

The study of these structures, allergens and toxins allows better choices on the oilseed crop being planted extensively in order to allow better worker and population health. In addition, an understanding of the allergens and/or toxic compounds present in oilseeds allows us to propose methodological strategies to eliminate or reduce such compounds.

The challenge is huge in this direction because there is a large expansion in the application of other oilseeds for biofuel synthesis, and new allergens and toxic compounds need to be unravelled.

5. References

Aalberse, R.C. (2000). Structural biology of allergens. *Journal of Allergy and Clinical Immunology*. Vol.106, No.2, pp. 228-238, ISSN: 0091-6749

Aggizio, A.P.; Carvalho, A.O.; Ribeiro, S.F.F.; Machado, O.L.T.; Aves, E.W.; Bloch Jr, C.; Okorokov, L.A.; Samarao, S.S.; Prates, M.V. & Gomes, V.M. (2003). A 2S albumin homologous protein from passion fruit seeds inhibits the fungal growth and acidification of the médium by *Fusarium oxisporum*. *Archives of Biochemistry and Biophysics*. Vol.416, Issue2, pp. 188-195, ISSN: 0003-9861

Ahn, Y-J & Chen, G. Q. (2007). Temporal and Spatial Expression of 2S Albumin in Castor (Ricinus communis L.) *Journal Agricultural and Food Chemistry*. Vol.55, No.24, pp. 10043–10049, ISSN:1520-5118

Alexander, J.; Benford, D.; Cockburn, A.; Cravedi, J.P.; Dogliotti, E.; Di Domenico, A.; Férnandez-Cruz, M.L.; Fürst, P.; Fink-Gremmels, J.; Galli, C.L.; Grandjean, P.; Gzyl, J.; Heinemeyer, G.; Johansson, N.; Mutti, A.; Schlatter, J.; Leeuwen, R. V.; Peteghem, C. V. & Verger, P. (2008). Ricin (from *Ricinus communis*) as undesirable substances in animal feed. *European Food Safety Authority Journal*. Vol.726, pp. 1-38

Anandan, S.; Anil Kumar, G. K.; Ghosh, J. & Ramachandra, K. S. (2005). Effect of different physical and chemical treatments on detoxification of ricin in castor cake. *Animal Feed Science and Technology*. Vol. 120, pp. 159–168, ISSN: 0377-8401

Atadashi, I.M.; Aroua, M.K. & Abdul Aziz, A. (2010). Biodiesel separation and purification: A review. *Renewable Energy*. Vol.36, Issue 2, pp. 437 - 443, ISSN: 0960-1481

Audi, J.; Belson, M.; Patel, M.; Shier, J. & Osterloh, J. (2005). Ricin poisoning—a comprehensive review. *Jama*. Vol.294, No.18, pp. 2342–2351, ISSN: 00987484

Balat, M. & Balat, H. (2010). Progress in biodiesel processing. *Applied Energy*. Vol.87, pp. 1815 - 1835, ISSN: 0306-2619

Barciszewski, J.; Szymanski, M. & Haertlé, T. (2000). Minireview: Analysis of Rape Seed Napin Structure and Potential Roles of the Storage Protein. *Journal of Protein Chemistry*. Vol.19, No.4, pp. 249- 254, ISSN: 0277-8033

Bewley, J.D. & Black, M. (1994). Seeds: *Physiology of development and germination*. New York: Plenum Press. Second edition. p. 68, ISBN13: 9780306447471, ISBN10: 0306447479

Bist, A.; Pandit, T.; Bhatnagar, A.K. & Singh, A.B. (2004). Variability in protein content of pollen of Castor bean (Ricinus communis) before and after exposure to the air pollutants SO_2 and NO_2. *Grana*. Vol.43, Issue2, pp. 94-100, ISSN: 0017-3134

Boucher, J.; Steiner, L. & Marison, I.W. (2007). Bio-sorption of atrazine in the press-cake from oilseeds. *Water Research*. Vol.41, Issue15, pp. 3209 - 3216, ISSN: 0043-1354

Brandt, N. N.; Chikishev, A. Yu.; Sotnikov, A. I.; Savochkina, Yu. A.; Agapov, I. I. & Tonevitsky, A. G. (2005). Ricin, ricin agglutinin, and the ricin binding subunit structural comparison by Raman spectroscopy. *Journal of Molecular Structure*. Vol.735-736, pp. 293–298, ISSN: 0022-2860

Breguet, V.; Boucher, J.; Pesquet, F.; Vojinovic, V.; Von Stockar, U. & Marison, I.W. (2008). Immobilization of rapeseed press-cake in an alginate matrix for the sorption of atrazine. *Water Research*. Vol.42, Issue 6-7, pp. 1606 - 1612, ISSN: 0043-1354

Breiteneder, H. & Mills, C.E.N. (2005). Plant food allergens - structural and functional aspects of allergenicity. *Biotechnology Advances*. Vol.23, Issue 6, pp. 395-399, ISSN: 0734-9750

Breiteneder, H. & Radauer, C. (2004). A classification of plant food allergens. *Journal of allergy and clinical immunology*. Vol.113, No.5, pp. 821- 830, ISSN: 0091-6749.

Brodie, C. & Blumberg, P. M. 2003. Regulation of cell apoptosis by protein kinase C. *Apoptosis*. Vol.8, pp. 19–27, ISSN: 1360-8185

Byczynska, A. & Barciszewski, J. (1999). The biosynthesis, structure and properties of napins the storage protein from rape seeds. *Journal of Plant Physiology*. Vol.154, pp. 417-425, ISSN: 1677-0420

Cai, X.; Yao, G.; Xu, G.; Yang, C.; Xu, H.; Lin, Y.; Yu, J.; & Suna, B. (2002). Identification of the amino acid residues in trichosanthin crucial for IgE response. *Biochemical and Biophysical Research Communications*. Vol.297, pp. 510–516, ISSN: 0006-291X

Chardin, H.; Sénéchalz, H.; Wal, J.M.; Desvauxz, F.X.; Godfrin, D. & Peltre, G. (2008). Characterization of peptidic and carbohydrate cross-reactive determinants in pollen polysensitization. *Clinical and Experimental Allergy*. Vol.38, No.4, pp. 680–685, ISSN: 0954-7894

Chen, M.J.; Hou, L.L. & Zhang, G.W. (1988). The diterpenoids from *Jatropha curcas* L. *Acta Botanica Sinica*. Vol.30, Issue 3, pp. 308-311, ISSN: 0577-7496

Chi, Z.; Pyle, D.; Wen, Z; Frear, C. & Chen, S. (2007). A laboratory study of producing docosahexaenoic acid from biodiesel-waste glycerol by microalgal fermentation. *Process Biochemistry*. Vol.42, pp. 1537–1545, ISSN: 1359-5113

Conceição, M.M.; Candeia, R.A.; Dantas, H.J.; Soledade, L.E.B.; Fernandes, V.J. & Souza, A.G. (2005). Rheological behavior of castor oil biodiesel. *Energy & Fuel*. Vol.19, No.5, pp. 2185–2188, ISSN:0887-0624

Cook, D. L.; David, J. & Griffiths, G. D. (2006). Retrospective identification of ricin in animal tissues following administration by pulmonary and oral routes. *Toxicology*. Vol.223, Issue 1-2, pp. 61-70, ISSN:0300-483X .

Culcuoglu, E.; Unay, E.; Karaosmanoglu, F. (2002). Rapeseed cake as a biomass source. *Energy Sources*, Vol. 24, Issue 4, pp. 329-336, ISSN: 0090-8312.

Day, P. J.; Ernst, S. R.; Frankel, A. E.; Monzingo, A. F.; Pascal, J. M. & Robertus, J. D. (1996). Structure and activity of an active site substitution of ricin A chain. *Biochemistry*. Vol.35, pp. 11098-11103, ISSN:0006-2960

Deus-de-Oliveira, N.; Felix, S.P.; Carrielo-Gama, C.; Fernandes, K.V.; DaMatta, R.A.; Machado, O.L.T. (2011). Identification of critical amino acids in the IgE epitopes of Ric c 1 and Ric c 3 and the application of glutamic acid as an IgE blocker. *PLoS ONE*, Vol.6, Issue.6, e21455, ISSN: 1932-6203.

Devappa, R. K.; Makkar, H.P.S. & Becker, K. (2010). Optimization of conditions for the extraction of phorbol esters from Jatropha oil. *Biomass and Bioenergy*. Vol.34, pp. 1125 - 1133, ISSN: 0961-9534

Directorate of Oilseeds Research. (2004). Diversifiel uses of Castor. In: International Seminar On Castor Seed, Castor Oil And Its Value Added Products. *Proceedings*... Ahmedabad: The Solvent Extractors Association of India, pp.50-57

El-Agamy, E. I. (2007). The challenge of cow milk protein allergy. *Small Ruminant Research*. Vol.68, Issue 1-2, pp. 64–72, ISSN: 0921-4488

El-Badwi, S. M. A.; Adam, S. E. I. & Hapke, H. J. (1995). Comparative toxicity of *Ricinus communis* and *Jatropha curcas* in Brown Hisex chicks. *Deutsche Tierarztliche Wochenschrift*. Vol.102, No.2, pp. 75-77, ISSN: 0341-6593.

Fahey, J.W.; Zalcmann, A.T. & Talalay, P. (2001). The chemical diversity and distribution of glucosinolates and isothiocyanates among plants. *Phytochemistry*. Vol.56, No. 1, pp. 5-51, ISSN: 0031-9422

Felix, S.P.; Mayerhoffer, R.O.; Damatta, R.A.; Verícimo, M.A.; Nascimento, V.V. & Machado, O.L.T. (2008). Mapping IgE-binding epitopes of Ric c 1 and Ric c 3, allergens from Ricinus communis, by mast cell degranulation assay. *Peptides*. Vol.29, Issue 4, pp. 497-504, ISSN: 0196-9781

Fernando, S.; Adhikari, S.; Kota, K. & Bandi, R. (2007). Glycerol based automotive fuels from future biorefineries. *Fuel*. Vol.86, Issues 17-18, pp. 2806–2809, ISSN: 0016-2361

Forrest, A. K.; Sierra, R. & Holtzapple, M. T. (2010). Effect of biodiesel glycerol type and fermentor configuration on mixed-acid fermentations. *Bioresource Technology*. Vol.101, Issue 23, pp. 9185–9189, ISSN: 0960-8524

Gandhi, V.M.; Cherian, K. M.; Mulky, M.J. (1994). Detoxification of cartor seed meal by interaction with sal seed meal. *Journal of the American Oil Chemists Society*. Vol. 71, pp. 827- 831, ISSN: 0003-021X

Gandhi, V.M.; Cherian, K.M. & Mulky, M.J. (1995). Toxicological studies on ratanjyot oil. *Food and Chemical Toxicology*. Vol.33, Issue 1, pp. 39–42, ISSN: 0278-6915

Garcia-Gonzalez, J. J.; Bartolomé -Zavala, B.; Del Mar Trigo-Pérez, M.; Barceló–Muñoz, J. M.; Fernández-Meléndez, S.; Negro-Carrasco, M. A.; Carmona-Bueno, M. J.; Vega-Chicote, J. M.; Muñoz-Román, C.; Palacios-Peláez, R.; Cabezudo-Artero, B. & Martínez-Quesada, J. (1999). Pollinosis to *Ricinus communis* (castor bean): an aerobiological, clinical and immunochemical study. *Clinical and Experimental Allergy*. Vol.29, Issue 9, pp. 1265-1275, ISSN: 0954-7894

Giannakopoulou, K.; Lukas, M.; Vasiliev, A.; Brunner, C. & Schnitzer, H. (2010). Conversion of rapeseed cake into bio-fuel in a batch reactor: Effect of catalytic vapor upgrading. *Microporous and Mesoporous Materials*. Vol.128, Issue 1-3, pp. 126–135, ISSN: 1387-1811

Godoy, M.G.; Gutarra, M.L.E.; Maciel, F.M.; Felix, S.P.; Bevilaqua, J.V.; Machado, O.L.T. & Freire, D.M.G. (2009). Use of a low-cost methodology for biodetoxification of castor bean waste and lipase production. *Enzyme and Microbial Technology*. Vol.44, Issue 5, pp. 317–322, ISSN: 0141-0229

Goel, G.; Makkar, H. P. S.; Francis, G. & Becker, K. (2007). Phorbol Esters: Structure, Biological Activity, and Toxicity in Animals. *International Journal of Toxicology*. Vol.26, pp. 279–288, ISSN: 1091-5818

Gonzalez-Guerrico, A. M. & Kazanietz, M. G. (2005). Phorbol ester-induced apoptosis in prostate cancer cells via autocrine activation of the extrinsic apoptotic cascade a key role for protein kinase C . *The Journal of Biological Chemistry*. Vol.280, pp. 38982–38991, ISSN: 0021-9258

Haas, W. & Mittelbach, M. (2000). Detoxification experiments with the seed oil from *Jatropha curcas* L. *Industrial Crops and Products*. Vol.12, Issue 2, pp. 111–118, ISSN: 0926-6690

Hájek, M. & Skopal, F. (2010). Treatment of glycerol phase formed by biodiesel production. *Bioresource Technology*. Vol.101, pp. 3242–3245, ISSN: 0960-8524

Haramoto, E.R. & Gallandt, E.R. (2004). Brassica cover cropping for weed management: a review. *Renewable Agriculture and Food Systems*. Vol.19, pp. 187–198, ISSN:1742-1705

Imamoglu, M. & Tekir, O. (2008). Removal of copper (II) and lead (II) ions from aqueous solutions by adsorption on activated carbon from a new precursor hazelnut husks. *Desalination*. Vol.228, pp. 108–113, ISSN: 0011-9164

Jain, S. & Sharma, M.P. (2010). Biodiesel production from Jatropha curcas oil. *Renewable and Sustainable Energy Reviews*. Vol.14, pp. 3140–3147, ISSN: 1364-0321

Jolliffe, N. A.; Brown, J. C.; Neumann, U.; Vicré, M.; Bachi, A.; Hawes, C.; Ceriotti, A.; Roberts, L. M. & Frigerio, L. (2004). Transport of ricin and 2S albumin precursors to the storage vacuoles of *Ricinus communis* endosperm involves the Golgi and VSR-like receptors. *The Plant Journal*. Vol.39, pp.821–833, ISSN: 1365-313X

Kamal, N.; Chowdhury, S.; Madan, T.; Sharma, D.; Attreyi, M.; Haq, W.; Katti, S.B.; Kumar, A. & Sarma, P.U. (2005). Tryptophan residue is essential for immunoreactivity of a diagnostically relevant peptide epitope of *A. fumigatus*. *Molecular and Cellular Biochemistry*. Vol.275, pp. 223–231, ISSN: 0300-8177

Karinen, R.S. & Krause, A.O.I. (2006). New biocomponents from glycerol. *Applied Catalysis A: General*. Vol.306, pp. 128–133, ISSN: 0926-860X

King, A.J.; He, W.; Cuevas, J.A.; Freudenberger, M.; Ramiaramanana, D. & Graham, I.A. (2009). Potential of Jatropha curcas as a source of renewable oil and animal feed. *Journal of Experimental Botany*. Vol.60, No.10, pp. 2897–2905, ISSN: 1460-2431

Krzyzaniak, A.; Burova, T.; Haertlé, T. & Barciszewski, J. (1998). The structure and properties of napin-seed storage protein from rape (*Brassica napus* L.). *Food/Nahrung*. Vol.42, pp. 201-204, ISSN:0027-769X

Kumar, A. & Sharma, S. (2008). An evaluation of multipurpose oil seed crop for industrial uses (Jatropha curcas L.): A review. *Industrial crops and products*. Vol. 28. pp. 1–10, ISSNA: 0926-6690

Lestari, D.; Mulder, W. & Sanders, J. (2010). Improving *Jatropha curcas* seed protein recovery by using counter current multistage extraction. *Biochemical Engineering Journal*. Vol.50, pp. 16–23, ISSN: 1369-703X

Leung, D.Y.C.; Wu, X.; Leung, M.K.H. (2010). A review on biodiesel production using catalyzed transesterification. *Applied Energy*, 87, 1083–1095, ISSN: 0306-2619

Li, C.-Y.; Devappa, R. K.; Liu, J.-X.; Lv, J.-M.; Makkar, H.P.S. & Becker, K. (2010). Toxicity of *Jatropha curcas* phorbol esters in mice. *Food and Chemical Toxicology*. Vol.48, pp. 620–625, ISSN: 0278-6915

Li, S.S.L.; Lin, T.T.S. & Ford, M.D. (1977). Isolation and characterization of a low molecular weight seed protein from *Ricinus communis*. *Biochimica et Biophysica Acta*. Vol.492, pp. 364-369, ISSN: 0304-4165

Lin, J.; Yan, F.; Tang, L. & Chen, F. (2003). Antitumor effects of curcin from seeds of Jatropha curcas. *Acta Pharmacologica Sinica*. Vol. 24, No.3, pp. 241 -246, ISSN:1671-4083.

Lin, J.; Zhou, X.; Wang, J.; Jiang, P.; Tang K. (2010). Purification and characterization of curcin, a toxic lectin from the seed of *Jatropha curcas*. *Preparative Biochemistry & Biotechnology*, Vol.40, pp. 107–118, ISSN: 1082-6068.

Machado, O.L.T. & Silva, J.G. (1992). An allergenic 2S storage protein from *Ricinus communis* seeds which is part of the albumin precursor predict by c-DNA data. *Brazilian Journal of Medical and Biological Research*. Vol.25, pp. 567-582, ISSN: 1678-4510

Machado, O.L.T.; Marcondes, J.A.; de Souza-Silva, F.; Hansen, E.; Ribeiro, P.D.; Vericimo, M., et al. (2003). Characterization of allergenic 2S albumin isoforms from Ricinus communis seeds. *Allergologie*. Vol.26, pp. 45–51, ISSN:0344-5062

Maciel, F.M.; Laberty, M.A.; Oliveira, N.D.; Felix, S.P.; Soares, A.M.S.; Verícimo, M.A. & Machado, O.L.T. (2009). A new 2S albumin from Jatropha curcas L. seeds and assessment of its allergenic properties. *Peptides*. Vol.30, No.12, pp. 2103-2107, ISSN: 0196-9781

Makkar, H.; Maes, J.; De Greyt, W. & Becker, K. (2009). Removal and Degradation of Phorbol Esters during Pre-treatment and Transesterification of Jatropha curcas Oil. *Journal of the American Oil Chemists´Society*. Vol.86, pp. 73–181, ISSN: 0003-021X

Martín, C.; Moure, A.; Martín, G.; Carrillo, E.; Domínguez, H. & Parajó, J. C. (2010). Fractional characterisation of jatropha, neem, moringa, trisperma, castor and candlenut seeds as potential feedstocks for biodiesel production in Cuba. *Biomass and Bioenergy*. Vol.34, pp. 34533 - 34538, ISSN: 0961-9534

Martinez-Herrera, J.; Siddhuraju, P.; Francis, G.; Davila-Ortiz, G. & Becker, K. (2006) Chemical composition, toxic/antimetabolic constituents, and effects of different treatments on their levels, in four provenances of Jatropha curcas L. from Mexico. *Food Chemistry*. Vol.96, pp. 80–89, ISSN: 0308-8146

Monsalve, R. I. & Rodriguez, R. (1990). Purification and characterization of proteins from the 2S fraction seeds of the Brassicaceae family. *Journal of Experimental Botany*. Vol.41, pp. 89-94, ISSN: 1460-2431

Monsalve, R. I.; Lopez-Otin, C.; Villalba, M. & Rodríguez, R. (1991). A new distinct group of 2S albumins from rapeseed. Amino acid sequence of two low molecular weight napins. *FEBS Letters*. Vol.295, pp. 207-210, ISSN: 0014-5793

Moreno, F.J. (2007) Gastrointestinal digestion of food allergens: Effect on their allergenicity. *Biomedicine & Pharmacotherapy*. Vol.61, pp. 50 - 60, ISSN:0753-3322

Morais & Silva, R. D. (2008). Produção de biodiesel em pequena escala com aproveitamento de co-produtos. III Congresso Brasileiro de Mamona - Mini-curso. Agosto 04-07, Salvador-BA, CD-ROM.

Morra, M.J. & Borek, V. (2010). Glucosinolate preservation in stored Brassicaceae seed meals. *Journal of Stored Products Research*. Vol.46, pp. 98 -102, ISSN:0022-474X

Moshkin, V. A. Castor. New Delhi: Amerind, 1986. 315p.

Murphy, D.J. (1999). Is rapeseed really an allergenic plant? Popular myths versus scientific realities. *Immunology Today*. Vol.20, No.11, pp. 511-514, ISSN:0167-5699

Na, D. H.; Cho, C. K.; Youn, Y. S.; Choi, Y.; Lee, K. R.; Yoo, S. D. & Lee, K. C. (2004). Capillary electrophoresis to characterize ricin and its subunits with matrix-assisted laser desorption/ionization time-of-flight mass spectrometry. *Toxicon*. Vol.43, pp. 329–335, ISSN: 0041-0101

Nascimento, V.V; Castro H.C.; Abreu, P.A.; Oliveira, A.E.A.; Fernandez J.H.; Araujo J.S.; Machado, O.L.T. (2011) *In Silico* Structural Characteristics and α-Amylase Inhibitory Properties of Ric c 1 and Ric c 3, Allergenic 2S Albumins from *Ricinus communis* Seeds. *J. Agric. Food Chem.*, Vol.59, pp. 4814-4821, ISSN: 1520-5118.

Nass, L. L.; Pereira, P. A. A.; Ellis, D. (2007). Biofuels in Brazil: An Overview. *Crop Science*. Vol.47, pp. 2228- 2237, ISSN:0011-183X

Ngah, W.S.W.; Hanafiah, M.A.K.M. (2008). Removal of heavy metal ions from wastewater by chemically modified plant wastes as adsorbents: A review. *Bioresource Technology*. Vol.99, pp. 3935–3948, ISSN: 0960-8524

Ozcimen, D. Karaosmanoglu, F. (2004). Production and characterization of bio-oil and biochar from rapeseed cake. *Renewable Energy*. Vol.29, pp. 779-787, ISSN: 0960-1481

Ogunniyi, D.S. (2006). Castor oil: a vital industrial raw material. *Bioresource Technology*. Vol.97,pp. 1086–1091, ISSN: 0960-8524

Olsnes, S.; Wesche, J.; Falsnes, P.O. (1999). Binding, uptake, routing and translocation of toxins with intracellular sites of action. In: Alouf J.E.; Freer, J.H. *The Comprehensive Sourcebook of Bacterial Toxins.* Academic Press: London, p. 73-93,

Olsnes, S. & Kozlov, J. (2001). Ricin. *Toxicon.* Vol.39, No.11, pp. 1723-1728, ISSN: 0041-0101

Olsnes, S. (2004). The history of ricin, abrin and related toxins. *Toxicon.* Vol.44, pp. 361 - 370, ISSN: 0041-0101

Olsnes, S.; Wesche, J.; Falsnes, P.O. (1999). Binding, uptake, routing and translocation of toxins with intracellular sites of action. In: Alouf J.E.; Freer, J.H. *The Comprehensive Sourcebook of Bacterial Toxins.* Academic Press: London, pp. 73-93, ISBN: 0120884453

Openshaw, K. (2000). A review of *Jatropha curcas*: an oil plant of unfulfilled promise. *Biomass & Bioenergy.* Vol.19, pp. 1–15, ISSN: 0961-9534

Palosuo, T.; Panzani, R.C.; Singh, A.B.; Ariano, R.; Alenius, H. & Turjanmaa, K. (2002). Allergen cross-reactivity between proteins of the latex from *Hevea brasiliensis*, seeds and pollen of *Ricinus communis*, and pollen of *Mercurialis annua*, members of the Euphorbiaceae family. *Allergy and Asthma Proceedings.* Vol.23, pp. 141-147, ISSN: 1539-6304

Pandey, A.; Soccol, C. R. & Mitchell, D. (2000). New developments in solid state fermentation: I-bioprocesses and products. *Process Biochemistry.* Vol.35, pp. 1153 - 1169, ISSN: 1359-5113

Panigrahi, S.; Francis, B. J.; Cano, L. A., Burbage, M. B. (1984). Toxicity of Jatropha curcas seeds from Mexico to rats and mice. *Nutrition Reports International.* Los Altos, US. Vol.29, pp. 1089- 1099, ISSN: 0029-6635

Pantoja-Uceda, D.; Bruix, M.; Giménez-Gallego, G.; Rico, M. & Santoro, J. (2003). Solution Structure of RicC3, a 2S Albumin Storage Protein from *Ricinus communis. Biochemistry.* Vol.42, pp. 13839-13847, ISSN:0006-2960

Pantoja-Uceda, D.; Palomares, O.; Bruix, M.; Villalba, M.; Rodríguez, R.; Rico, M. & Santoro, J. (2004). Solution Structure and Stability against Digestion of rproBnIb, a Recombinant 2S Albumin from Rapeseed: Relationship to Its Allergenic Properties. *Biochemistry.* Vol.43, pp. 16036-16045, ISSN:0006-2960

Pastuszewska, B.; Jablecki, G.; Buraczewska, L.; Dakowski, P.; Taciak, M.; Matyjek , R. & Ochtabinska, A. (2003). The protein value of differently processed rapeseed solvent meal and cake assessed by in vitro methods and in tests with rats. *Animal Feed Science and Technology.* Vol.106, pp. 175–188, ISSN: 0377-8401

Petisco, C.; García-Criado, B.; Vázquez-de-Aldana, B.R.; Haro, A. & García-Ciudad, A., (2010). Measurement of quality parameters in intact seeds of Brassica species using visible and near-infrared spectroscopy. *Industrial Crops and Products.* Vol.32, pp. 139–146, ISSN: 0926-6690

Prueksakorn, K. & Gheewala, S.H. (2008). Full chain energy analysis of biodiesel from *Jatropha curcas* L. in Thailand. *Environmental Science & Technology.* Vol.42, pp. 3388–3393, ISSN: 0013-936X

Puttaraj, S.; Bhagya S.; Murthy, K. N. & Singh, N. (1994). Effect of detoxification of castor seed (Ricinus communis) protein isolate on its nutritional quality. *Plant Foods for Human Nutrition.* Vol.46, pp. 63-70, ISSN:1573-9104

Raab, B.; Leman, H. & Schwenke, K.D. (1992). Comparative study of the protein patterns of some rapeseed (Brassica napus L.) varieties by means of polyacrylamide gel electrophoresis and high-performance liquid chromatography. *Food/Nahrung*. Vol.36, Issue 3, pp. 239–247, ISSN:0027-769X

Rahmanpour, S.; Backhouse, D. & Nonhebel, H.M. (2010). Reaction of glucosinolate-myrosinase defence system in *Brassica* plants to pathogenicity factor of *Sclerotinia sclerotiorum*. *European Journal of Plant Pathology*. Vol.128, pp. 429–433, ISSN:0929-1873

Rakshit, K.D.; Darukeshwara, J.; Rathina Raj, K.; Narasimhamurthy, K.; Saibaba, P. & Bhagya, S. (2008). Toxicity studies of detoxified Jatropha meal (*Jatropha curcas*) in rats. *Food and Chemical Toxicology*. Vol.46, pp. 3621–3625, ISSN: 0278-6915

Ramachandran, S.; Singh, S.K.; Larroche, C.; Soccol, C.R. & Pandey, A. (2007). Oil cakes and their biotechnological applications - a review. *Bioresource Technology*. Vol.98, pp. 2000–2009, ISSN: 0960-8524

Rao, P.V.L.; Jayaraj, R.; Bhaskar, A.S.B.; Kumar, O.; Bhattacharya, R. & Saxena, P. (2005). Mechanism of ricin-induced apoptosis in human cervical cancer cells. *Biochemical Pharmacology*. Vol.69, pp. 855–865, ISSN: 0006-2952

Regente, M. & La Canal, L. (2001). Are storage 2S albumins also defensive proteins? *Physiological and Molecular Plant Pathology*. Vol.59, pp. 275-276, ISSN: 0885-5765

Rico, M.; Bruix, M.; González, C.; Monsalve, R.I. & Rodríguez, R. (1996). 1H NMR assignment and global fold of napin BnIb, a representative 2S albumin seed protein. *Biochemistry*. Vol.35, pp. 15672- 15682, ISSN:0006-2960

Robotham, J.M.; Teuber, S.S.; Sathe, S.K. & Roux, K.H. (2002). Linear IgE epitope mapping of the English walnut (Juglans regia) major food allergen Jug r 1. *Journal of Allergy and Clinical Immunology*. Vol.109, pp. 143–149, ISSN: 0091-6749

Rutenber, E. & Robertus, J.D. (1991). Structure of ricin B-chain at 2.5 Å resolution. *Proteins: Structure, Function and Genetics*. Vol.10, pp. 260–269, ISSN:1097-0134.

Schmidt, I.; Renard, D.; Rondeau, D.; Richomme, P.; Popineau, Y. & Augustin-Viguier Axelos, M. (2004). Detailed Physicochemical Characterization of the 2S Storage Protein from Rape (*Brassica napus* L.). *Journal of Agricultural Food Chemistry*. Vol.52, pp. 5995-6001, ISSN: 0021-8561

Seo, J-H.; Kim, J-H.; Lee, J-W.; Yoo, Y-C.; Kim, M.R.; Park, K-S. & Byun, M-W. (2007). Ovalbumin modified by gamma irradiation alters its immunological functions and allergic responses. *International Immunopharmacology*. Vol.7, pp. 464–472, ISSN:1567-5769

Sehgal, P.; Khan, M.; Kumar, O. & Vijayaraghavan, R. (2010). Purification, characterization and toxicity profile of ricin isoforms from castor beans. *Food and Chemical Toxicology* vol.48, pp 3171–3176, ISSN: 0278-6915

Sharief, F.S. & Li, S.S. (1982). Aminoacid sequence of small and large subunits of seed storage protein from Ricinus communis. *The Journal of Biological Chemistry*. Vol.257, pp. 14753-14759, ISSN: 0021-9258

Sharma, Y.C. & Singh, B. (2009). Development of biodiesel: Current scenario. *Renewable and Sustainable Energy Reviews*. Vol.13, pp. 1646–1651, ISSN: 1364-0321

Shewry, P.R.; Napier, J.A. & Tatham, A.S. (1995). Seed storage proteins: structures and biosynthesis. *The Plant Cell*. Vol.7, No.7, pp. 945-956, ISSN: 1040-4651

Sicherer, S. H. & Leung, D. Y. M. (2009). Advances in allergic skin disease, anaphylaxis, and hypersensitivity reactions to foods, drugs, and insects in 2008. *Journal of Allergy and Clinical Immunology*. Vol.127, Issue 2, pp. 326-335, ISSN: 0091-6749

Silva Jr., J.G.; Machado, O.L.T.; Izumi, C.; Pandovan, J.C.; Chait, B.T.; Mirza, U.A. & Greene, L.J. (1996). Aminoacid sequence of New 2S Albumin from Ricinus communis Whith is Part of 29-Kda Precursor Protein. *Archives of Biochemistry and Biophysics.* Vol.336, pp. 10-18, ISSN: 0003-9861

Singh, A.; Panzani, R. C. & Singh, A.B. (1997). Specific IgE to castor bean (*Ricinus communis*) pollen in the sera of clinically sensitive patients to seeds. *Journal of investigational allergology & clinical immunology.* Vol.7, pp. 169-174, ISSN: 1018-9068

Singh, A.B. & Kumar, P. (2003). Aeroallergens in clinical practice of allergy in India. An overview. *Annals of Agricultural and Environmental Medicine.* Vol.10, pp. 131-136, ISSN:1232-1966

Singh, R.N.; Vyas, D.K.; Srivastava, N.S.L. & Narra, M. (2008). SPRERI experience on holistic approach to utilize all parts of Jatropha curcas fruit for energy. *Renewable Energy.* Vol.33, Issue 8, pp. 1868–1873, ISSN: 0960-1481

Singh, S.P. & Singh, D. (2010). Biodiesel production through the use of different sources and characterization of oils and their esters as the substitute of diesel: A review.Renewable and Sustainable Energy Reviews. Vol. 14, pp. 200–216, ISSN: 1364-0321

Spies, J.R. & Coulson, E.J. (1943). The chemistry of allergens VIII. Isolation and properties of an active protein-polysaccharidic fration, CB-1A, from castor bean. *Journal of the American Chemical Society.* Vol.65, pp. 1720- 1725, ISSN: 0002-7863

Stirpe, F. & Batelli, M. G. (2006). Ribosome-inactivating proteins: progress and problems. *Cellular and Molecular Life Sciences.* Vol.63, pp. 1850-1866, ISSN: 1420-682X

Stirpe, F.; Pession-Brizzi, A.; Lorenzoni, E.; Strocchi, P.; Montanaro, L. & Sperti, S. (1976). Studies on the proteins from the seeds of *Croton tigliumi* and of *Jatropha curcas.* Toxic properties and inhibition of protein synthesis in vitro. *Biochemical Journal.* Vol.156, pp. 1-6, ISSN: 0264-6021

Swiatkiewicz, S.; Koreleski, J.& Arczewska-Wlosek, A. (2010). Egg performance, egg quality, and nutrient utilization in laying hens fed diets with different levels of rapeseed expeller cake. *Agricultural and Food Science.* Vol.19, pp. 233–239, ISSN: 1459-6067

Thelen, J.J. (2009). Proteomics tools and resources for investigating protein allergens in oilseeds. *Regulatory Toxicology and Pharmacology.* Vol.54, pp. 41–45, ISSN: 0273-2300

Thorpe S.C.; Kemeny D.M.; Panzani R.C.; McGurl B. & Lord J.M. (1988). Identification of the major allergens in castor bean seeds. *Journal of Allergy and Clinical Immunology.* Vol.82, pp. 67–72, ISSN: 0091-6749

Tokarnia, C. H. & Döbereiner, J. (1997). Imunidade cruzada pelas sementes de *Abrus precatorius* e *Ricinus communis* em bovinos. *Pesquisa Veterinária Brasileira.* Vol.17, No.1, pp. 25-35, ISSN 0100-736X

Trinidade, A.; Kumar, S.; Haji, M.; Shakeel, M. & Leong, P. (2010). The prevalence of oilseed rape hypersensitivity in a mixed cereal farming population. *Clinical Otolaryngology.* Vol.35, pp. 13–17, ISSN:1749-4478

Tripathi, M.K. & Mishra, A.S. (2007). Glucosinolates in animal nutrition: A review. *Animal Feed Science and Technology.* Vol.132, pp. 1–27, ISSN: 0377-8401

Vallverdú, A.; Garciaortega, P.; Martinez, J.; Martinez, A.; Esteban, M.I.; Demolina, M.; Fernandeztavora, L.; Fernandez, J.; Bartolome, B.; Palacios, R. (1997). *Mercurialis annua*: Characterization of main allergens and crossreactivity with other species.

International Archives of Allergy and Immunology. Vol.112, pp. 356-364, ISSN: 1018-2438

Vaz, A. F.M.; Costa, R.M.P.B.; Coelho, L.C.B.B.; Oliva, M.L.V.; Santana, L.A.; Melo, A. M.M.A. & Correia, M.T.S. (2011) Gamma irradiation as an alternative treatment to abolish allergenicity of lectins in food. *Food Chemistry*. Vol.124, pp. 1289–1295, ISSN: 0308-8146

Weinstein, I. B.; Lee, L. S; Fisher, P. B.; Mufson, A. & Yamasaki, H. (1979). Action of phorbol esters in cell culture: mimicry of transformation, altered differentiation, and effects on cell membranes. *Journal of Supramolecular Structure*. Vol.12, Issue 2, pp. 195–208, ISSN: 0161-3294.

Welch, J.; Jones, M.G.; Cullinan, P.; Coates, O.A. & Newman Taylor, A.J. (2000). Sensitization to oilseed rape is not due to cross-reactivity with grass pollen. *Clinical and Experimental Allergy*. Vol.30, Issue 3, pp. 370-375, ISSN: 0954-7894

Youle, R.J. & Huang, A.H. (1978). Evidence that the castor bean allergens are the albumin storage proteins in the proteins bodies of castor bean. *Plant Physiology*. Vol.61, No.6, pp. 1040-1042, ISSN: 1677-0420

Zanetti, F.; Vamerali, T. & Mosca, G. (2009). Yield and oil variability in modern varieties of high-erucic winter oilseed rape (Brassica napus L. var. oleifera) and Ethiopian mustard (Brassica carinata A. Braun) under reduced agricultural inputs. *Industrial Crops and Products*. Vol.30, pp. 265–270, ISSN: 0926-6690

Zhang, W. J.; Xu, Z. R.; Zhao, S. H.; Sun, J. Y. & Yang, X. (2007). Development of a microbial fermentation process for detoxification of gossypol in cottonseed meal. *Animal Feed Science and Technology*. Vol.135, pp. 176-186, ISSN: 0377-8401

4

Emissions of Diesel - Vegetable Oils Mixtures

Charalampos Arapatsakos
Department of Production and Management Engineering,
Democritus University of Thrace, Xanthi
Greece

1. Introduction

The industrialization of society, the introduction of motorized vehicles and the explosion of the population are factors contributing toward the growing air pollution problem. Moreover, the exhaust from burning fuels in automobiles, homes and industries is a major source of pollution in the air. Apart from the anthropogenic sources of air pollution there are natural sources as well. Natural sources related to dust from natural source, usually large areas of land with little or no vegetation, the smoke and carbon monoxide from wildfires, volcanic activity etc. Air pollution not only affects the air we breathe, but it also impacts the land and the water. The human health effects of poor air quality are far reaching, but principally affect the body's respiratory system and the cardiovascular system. The human health effects caused by air pollution may range from subtle biochemical and physiological changes to difficulty breathing. It can also cause deaths, aggravated asthma, bronchitis, emphysema, lung and heart diseases to human beings. There are several many types of air pollutant [1,2]. These include smog, acid rain, the greenhouse effect and holes in the ozone layer. The atmospheric conditions such as the wind, rain, stability affect the transportation of the air pollutant [3,4]. Furthermore, depending on the geographical location temperature, wind and weather factors, pollution is dispersed differently [5,6]. For instance, the wind and rain may effectively dilute pollution to relatively safe concentrations despite a fairly high rate of emissions. In contrast when atmospheric conditions are stable relatively low emissions can cause buildup of pollution to hazardous levels.

The quality of fuel affects diesel engine emissions (HC, CO, NOx and particulate emissions) very strongly. The fuel that is used in diesel engines is a mixture of hydrocarbons and its boiling temperature is approximately 170°C to 360°C [4]. Diesel fuel emissions composition and characteristics depend on mixture formation and combustion. In order to compare the quality of fuels the following criteria are tested: ketene rating, density, viscosity, boiling characteristics, aromatics content and sylph content. For environmental compatibility, the fuel must have low density, low content of aromatic compounds, low sylph content and high ketene rating [6,7,8].

One of the most important and renewable sources of energy is biomass. Biomass as a renewable source of energy refers to living and recently dead biological material that can be used as fuel or for industrial production. Some examples of biomass fuels are wood, crops, manure and some garbage. Biomass is a renewable energy source due to photosynthesis. Concretely, with the photosynthesis is committed the solar energy and is changed in chemical (energy). At the combustion of biomass the committed solar energy is changed in thermo while the dioxide of coal (CO_2) returns in the atmosphere, while the inorganic elements that are contained in the ash, enrich the soil with nutritious elements. Nowadays, the use of biomass, covers approximately 4% of the total energy which is consumed in USA and 45% of the renewable sources of energy [9,10,11]. The most common source of biomass is the wood. For thousands of years people have burned wood for heating and cooking. Another source of biomass is our garbage that comes from plant or animal products. Moreover, various materials of plant origin, as agricultural remains (e.g. straw), material of animal origin, remains from veterinary surgeon units as well as remains of fishery and their sub products, urban waste etc. Wood waste or garbage can be burned to produce steam for making electricity or to provide heat to industries and homes. Biomass can be used for the production of liquid fuel (called biofuel) which is used for the transportation to many countries of Europe, USA etc. [12,13,14]. Bio-diesel is also produced from oily plants (soya, sunflower) animal greases, products of carcasses, and used oils. Some of biomass advantages which make it an attractive source of energy are the following:

1. Reduction of air pollutants. The combustion of biomass has null balance of dioxide of coal ($CO_{2,}$) does not contribute in the phenomenon of green house, because the quantities of dioxide of coal ($CO_{2,}$) that are released at the combustion of biomass are committed again by the plants for the creation of biomass.
2. Zero existence of sulphur in biomass contributes considerably in the restriction of emissions of dioxide of sulphur ($SO_{2,}$) that is in charge of the acid rain.
3. Reduction of dependence from imported fuels, improvement of commercial balance, in the guaranty of energy supply and in the saving of exchange.
4. Sources are commonly available.
5. Sources are locally produced, consequently it increases the occupation to the agriculture places with the use of alternatives cultures (several kinds of cane, sorghum), as well as the creation of alternative markets for the traditional cultures (sunflower etc.) and withholding of population in their hearths.
6. Increase of Biomass production can often mean the restoration of waste land.

Biofuels are liquid or gas fuels which are produced from the biomass. Biomass can replace the conventional mineral fuels, totally or partial in the engines [15].

The major issue is how a four-stroke diesel engine behaves on the side of pollutants and operation, when it uses mixed fuel of diesel - vegetable oils.

2. Instrumentation and experimental results

In the experiment stage has been used directly used vegetable oil (used sunflower oil that emanated from cooking) in the mixture of diesel in to a four - stroke diesel engine. Specifically it has been used diesel, mixture diesel-5% used vegetable oil (u5), diesel-10 used vegetable oil (u10), diesel-20% used vegetable oil (u20), diesel-30% used vegetable oil (u30), diesel-40% used vegetable oil (u40), diesel-50% used vegetable oil (u50) in a four-stroke diesel air-cooled engine named Ruggerini type RD-80, volume 377cc, and power 8.2hp/3000rpm, who was connected with a pump of water centrifugal. Measurements were made when the engine was function on 1000, 1500, 2000 and 2500rpm.

During the experiments, it has been counted:

- The percent of CO
- The ppm of HC

- The ppm of NO
- The percent of smoke

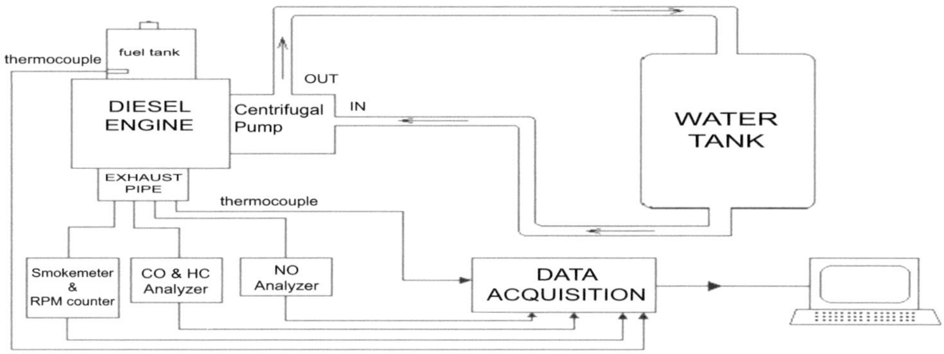

Fig. 1. Experimental Layout

The measurement of rounds/min of the engine was made by a portable tachometer (Digital photo/contact tachometer) named LTLutron DT-2236. Smoke was measured by a specifically measurement device named SMOKE MODULE EXHAUST GAS ANALYSER MOD 9010/M, which has been connected to a PC unit. The CO and HC emissions have been measured by HORIBA Analyzer MEXA-324 GE. The NO emissions were measured by a Single GAS Analyser SGA92-NO.

2.1 Used vegetable oil

The experimental results are shown at the following tables and figures [16]:

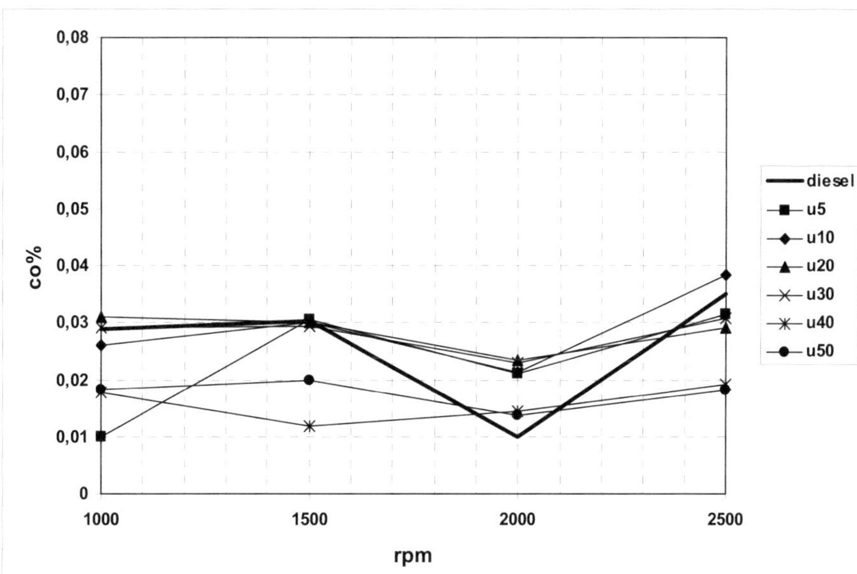

Fig. 2. The CO variation on different rpm regarding to the mixture

rpm	CO %						
	diesel	u5	u10	u20	u30	u40	u50
1000	0,02898	0,01000	0,026081	0,030985	0,029143	0,017823	0,018223
1500	0,03039	0,03059	0,030043	0,029979	0,029310	0,011818	0,019767
2000	0,01000	0,02108	0,021379	0,023500	0,023059	0,014483	0,013624
2500	0,03508	0,03145	0,038315	0,029120	0,030713	0,019111	0,018298

Table 1. The CO average value variation on different rpm regarding to the mixture

rpm	HC (ppm)						
	diesel	u5	u10	u20	u30	u40	u50
1000	2,535343	8,844156	5,653105	5,246253	5,124364	2,147903	2,974304
1500	13,31714	24,99127	12,87527	13,15385	9,358621	2,934461	6,714588
2000	7,131223	8,326797	12,67026	9,195652	13,79747	5,267241	4,936681
2500	10,96128	16,63420	17,30454	16,94635	6,706013	6,598698	6,759574

Table 2. The HC average value variation on different rpm regarding to the mixture

rpm	NO (ppm)						
	diesel	u5	u10	u20	u30	u40	u50
1000	518,210	771,001	696,827	495,603	380,361	349,140	207,760
1500	739,366	754,126	913,037	771,607	723,381	872,06	582,908
2000	762,155	834,334	520,485	760,936	839,268	928,337	720,505
2500	795,461	946,349	518,287	710,402	864,585	674,432	847,835

Table 3. The NO average value variation on different rpm regarding to the mixture

rpm	% smoke						
	diesel	u5	u10	u20	u30	u40	u50
1000	3,262370	4,870779	5,966167	16,43362	12,26745	15,7298	11,32741
1500	7,100651	8,174236	5,768602	7,652778	5,56423	9,206977	13,05011
2000	5,688865	7,619826	4,704957	6,151304	4,948101	4,351724	9,59869
2500	29,00617	23,21970	25,67279	16,86674	14,59399	17,48286	15,87915

Table 4. The % smoke average value variation on different rpm regarding to the mixture

From figure 2 it is clear that the more constant behaviour appears in the mixture u40, while the best behaviour is appears in the case diesel/1500rpm. From figure 3 it can be noticed the biggest reduction of HC regarding to diesel in case of mixture u40. From figure 4 it can be noticed the biggest reduction of NO regarding to diesel in the case of mixture u40. From figure 5 it can be seen the biggest reduction for u40 until the case u40/1000rpm. From the above figures it is clear that the use of different mixtures can constitute changes to CO, HC, NO and smoke too. It is also important the fact that there was no changes in the rounds of the engine, as well as in the supply of water at the use of mixtures. Finally as far as the consumption is concerned, did not observed changes with the use of different mixtures.

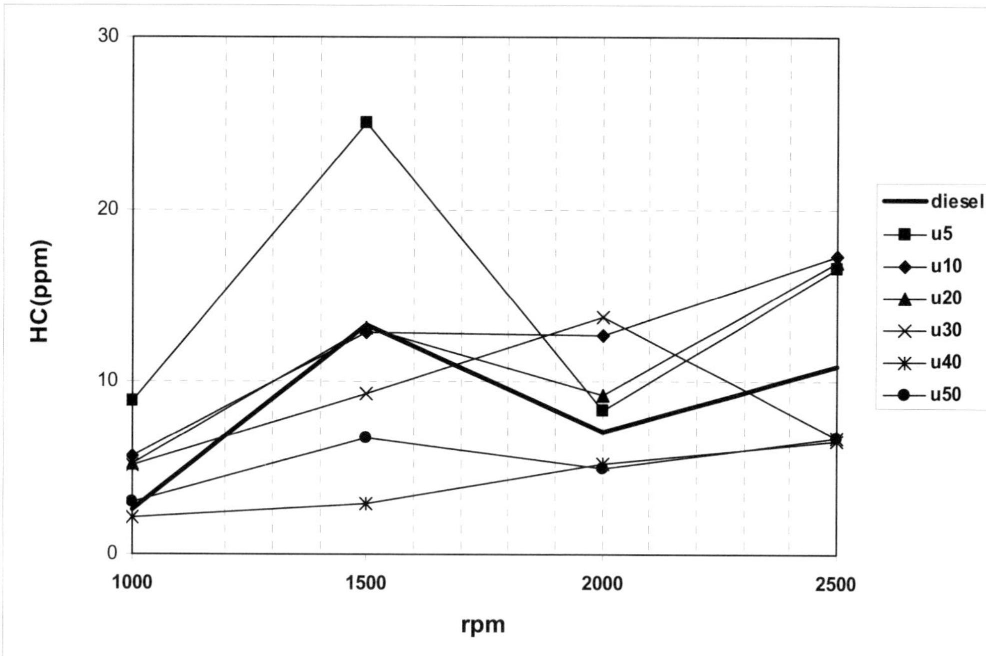

Fig. 3. The HC variation on different rpm regarding to the mixture

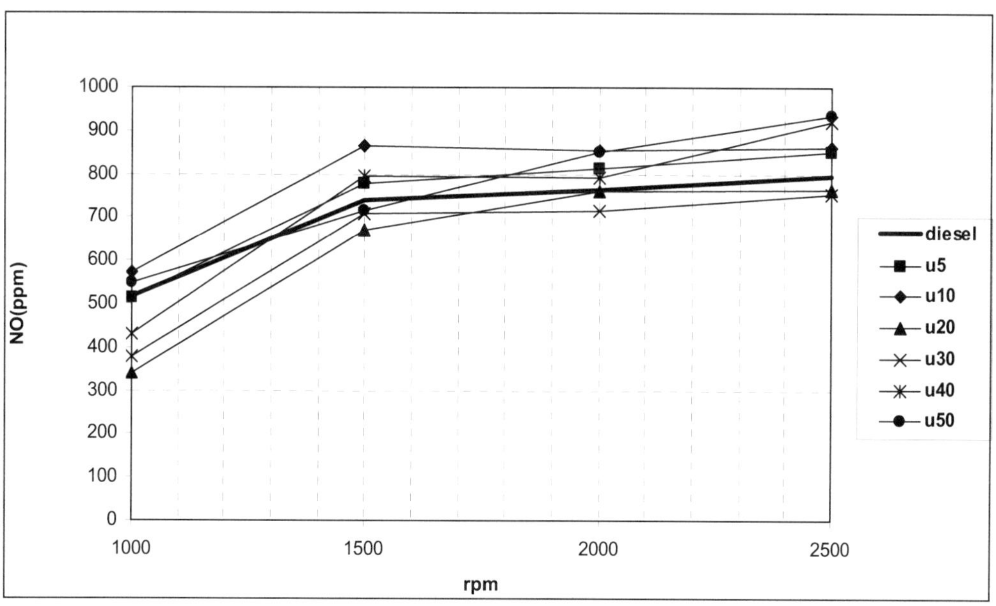

Fig. 4. The NO variation on different rpm regarding to the mixture

The use of mixtures of diesel-vegetable oil has as result change of gas emissions with better behaviour in the mixture u40. It is important, that is not presented reduction of power of engine from the combustion of the mixtures.

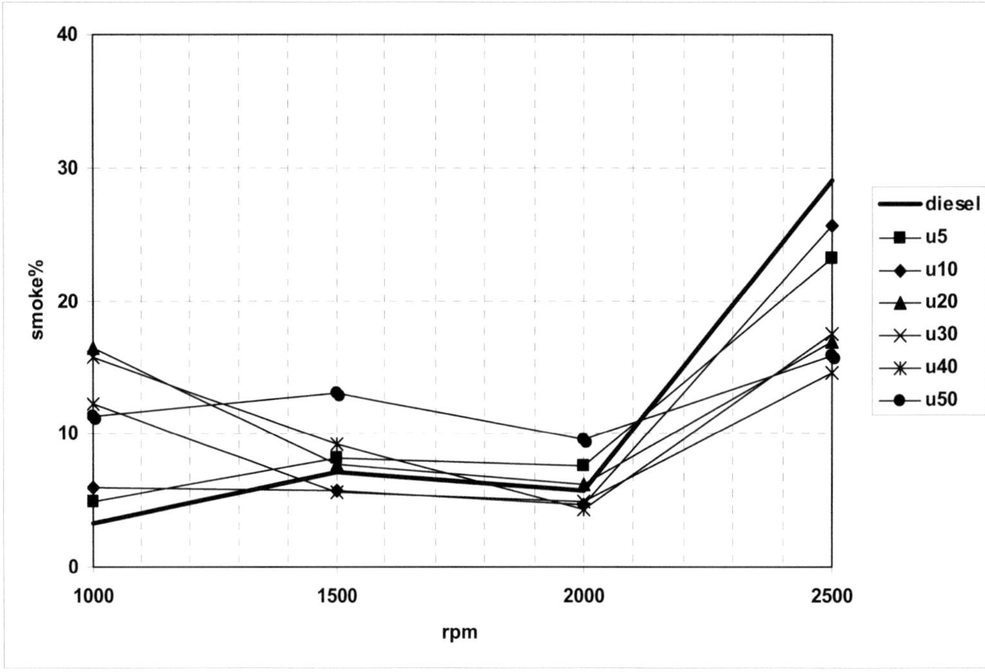

Fig. 5. The smoke variation on different rpm regarding to the mixture

2.2 Maize oil

In the experiment stage has been used directly maize oil in the mixture of diesel in to a four – stroke diesel engine. Specifically it has been used diesel, mixture diesel-5% maize oil (k5), diesel-10% maize oil (k10), diesel-20% maize oil (k20), diesel-30% maize oil (k30), diesel-40% maize oil (k40), diesel-50% maize oil (k50) in a four-stroke diesel engine [17]:

rpm	CO %						
	diesel	k5	k10	k20	k30	k40	k50
1000	0,0289	0,0310	0,0309	0,0309	0,0319	0,0397	0,0345
1500	0,0303	0,0302	0,0304	0,0311	0,0345	0,0211	0,0288
2000	0,01	0,0280	0,0232	0,0284	0,0274	0,0281	0,0219
2500	0,0350	0,0244	0,0317	0,0296	0,0324	0,0305	0,0292

Table 5. The CO average value variation on different rpm regarding to the mixture

rpm	HC (ppm)						
	diesel	k5	k10	k20	k30	k40	k50
1000	2,535	14,937	6,244	10,326	3,406	5,358	9,167
1500	13,31	21,485	9,236	17,997	14,718	0,449	17,197
2000	7,131	3,184	13,970	15,965	8,402	8,502	12,913
2500	10,961	16,347	18,884	23,556	30,551	7,451	17,712

Table 6. The HC average value variation on different rpm regarding to the mixture

rpm	NO (ppm)						
	diesel	k5	k10	k20	k30	k40	k50
1000	518,210	771,001	696,827	495,603	380,361	349,140	207,760
1500	739,366	754,126	913,037	771,607	723,381	872,06	582,908
2000	762,155	834,334	520,485	760,936	839,268	928,337	720,505
2500	795,461	946,349	518,287	710,402	864,585	674,432	847,835

Table 7. The NO average value variation on different rpm regarding to the mixture

rpm	% smoke						
	diesel	k5	k10	k20	k30	k40	k50
1000	3,262	12,722	7,301	7,488	16,623	7,200	26,232
1500	7,100	10,924	5,487	6,547	14,850	12,141	24,035
2000	5,688	18,679	4,001	6,588	9,936	14,071	18,884
2500	29,006	28,282	21,848	15,730	17,579	13,438	14,265

Table 8. The % smoke average value variation on different rpm regarding to the mixture

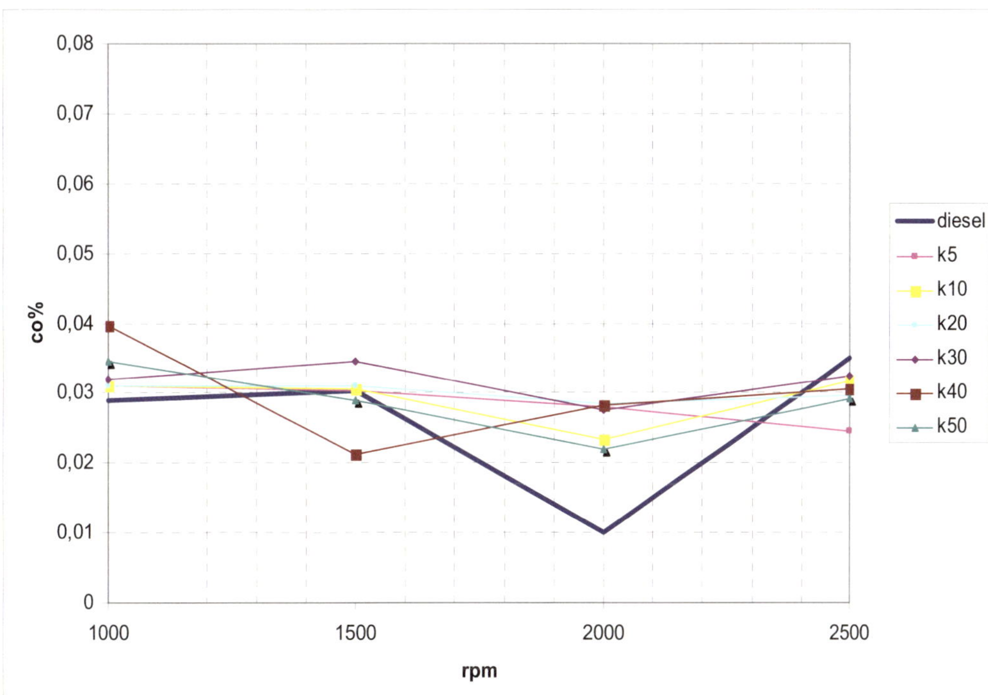

Fig. 6. The CO variation on different rpm regarding to the mixture

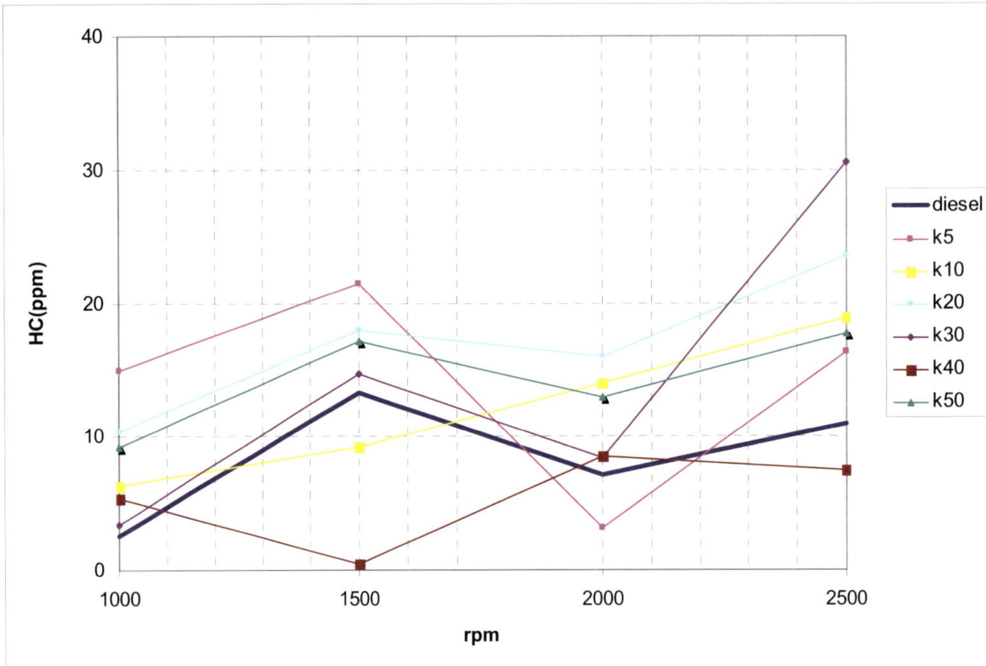

Fig. 7. The HC variation on different rpm regarding to the mixture

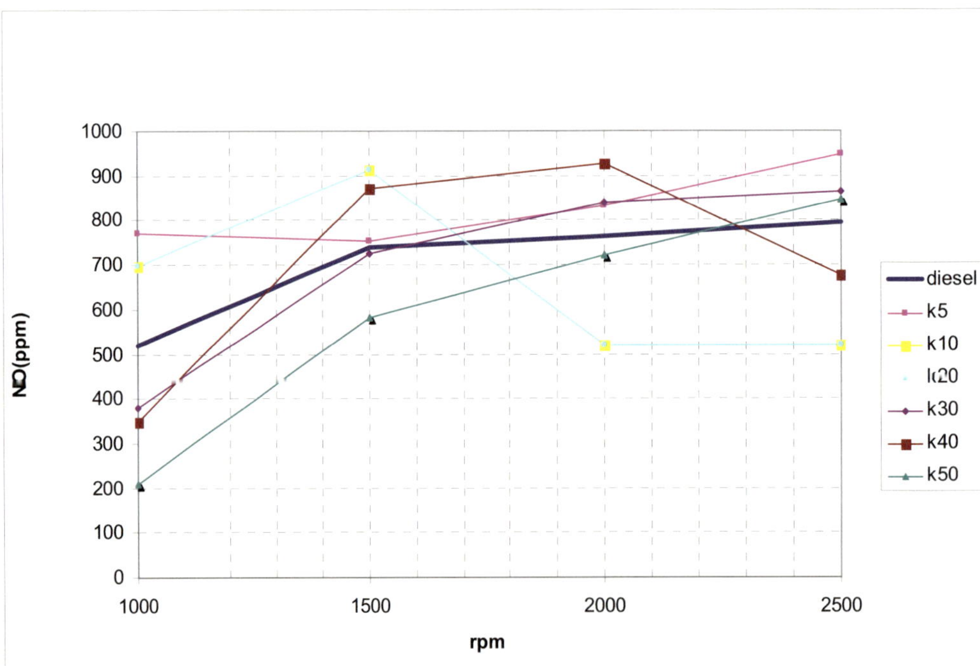

Fig. 8. The NO variation on different rpm regarding to the mixture

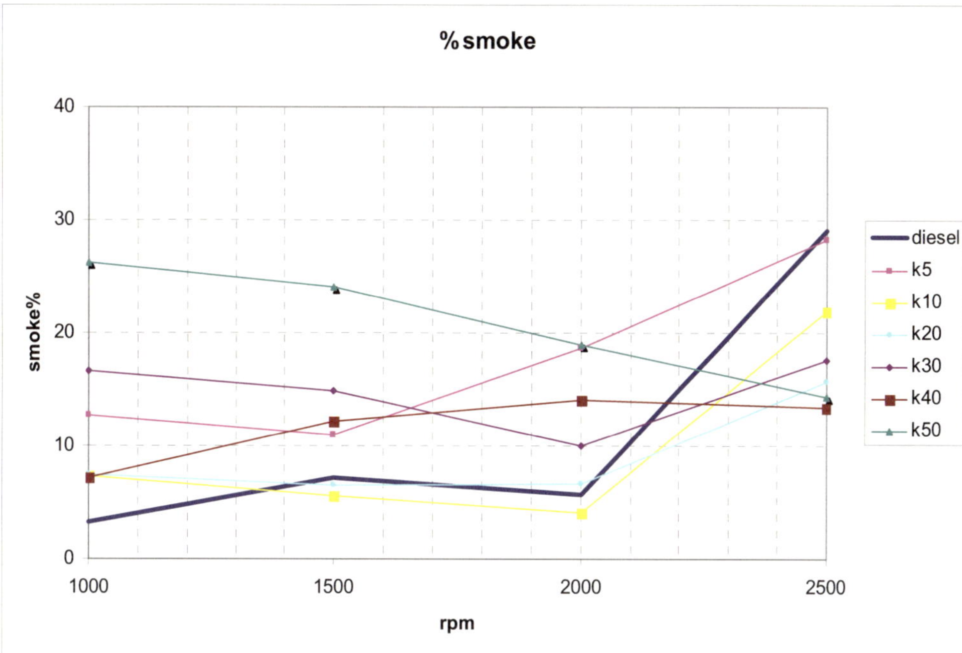

Fig. 9. The smoke variation on different rpm regarding to the mixture

From figure 6 it is clear that when the maize oil is increased on the fuel regarding to diesel, it appears an increase of CO, except in the case k40/1500rpm. From figure 7 it can be noticed the biggest reduction of HC regarding to diesel in case of k40/1500rpm. From figure 8 it can be noticed the biggest reduction of NO regarding to diesel in the case of k20/2000-2500rpm. From figure 9 it can be noticed the biggest reduction for k10/1500-2000rpm. From the above figures it is clear that the use of different mixtures can constitute changes to CO, HC, NO and smoke too. It is also important the fact that there was no changes in the rounds of the engine, as well as in the supply of water at the use of mixtures. Finally as far as the consumption is concerned, did not observed changes with the use of different mixtures. The use of mixture of diesel and maize oil has the following impacts:

- About CO it can be noticed that when the maize oil is increased on the fuel regarding to diesel, it appears a decrease of CO, except in the case k40/1500rpm.
- About HC it can be noticed the biggest reduction of HC regarding to diesel in case of k40/1500rpm
- The biggest reduction of NO regarding to Diesel is noticed in the case of k20/2000-2500rpm.
- The smoke it can be noticed the biggest reduction for k10/1500-2000rpm

2.3 Cotton oil

In the experiment stage has been used directly cotton oil in the mixture of diesel in to a four – stroke Diesel engine and not elaborated in the figure of bio-diesel. Specifically it has been used diesel, mixture diesel- 10% cotton oil(B10), diesel- 20% cotton oil(B20), diesel- 30% cotton oil (B30), diesel- 40% cotton oil (B40), diesel- 50% cotton oil (B50) in a four-stroke diesel engine [18]:

The experimental results are shown at the following tables and figures:

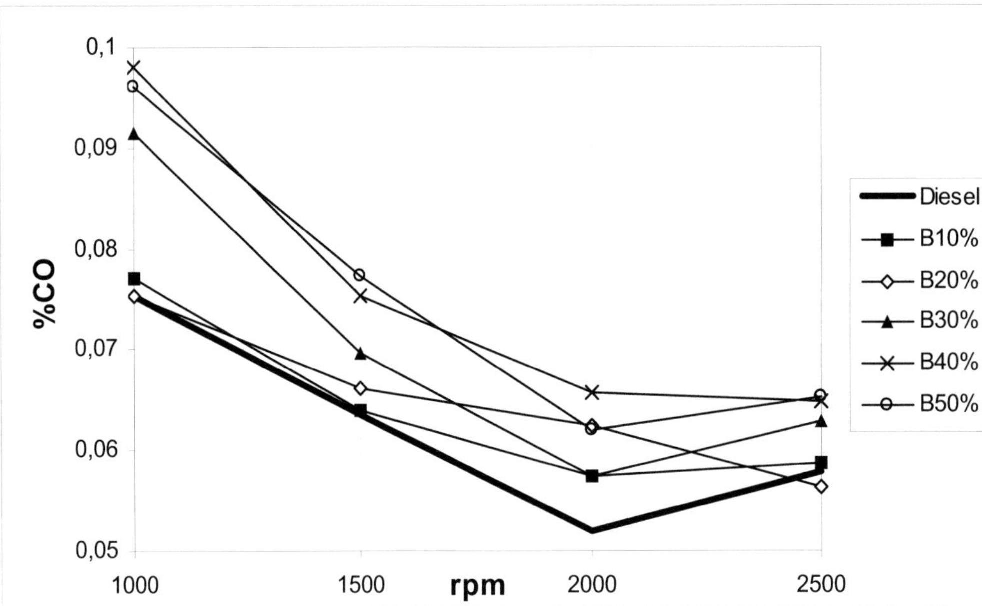

Fig. 10. The CO variation on different rpm regarding to the mixture

From figure 10 it is clear that when the cotton oil is increased on the fuel regarding to Diesel, it appears an increasement of CO.

rpm	% CO					
	Diesel	B10	B20	B30	B40	B50
1000	0,075	0,076	0,075	0,091	0,098	0,095
1500	0,063	0,064	0,066	0,069	0,075	0,077
2000	0,052	0,057	0,062	0,057	0,065	0,061
2500	0,057	0,058	0,056	0,062	0,064	0,065

Table 9. The CO average value variation on different rpm regarding to the mixture

rpm	HC (ppm)					
	Diesel	B10	B20	B30	B40	B50
1000	30,78	35,86	39,04	39,05	14,86	46,64
1500	62,86	41,18	35,59	48,74	53,84	51,34
2000	125,52	83,84	101,38	109,07	76,42	142,94
2500	78,26	84,93	169,34	103,64	167,82	105,80

Table 10. The HC average value variation on different rpm regarding to the mixture

rpm	NO (ppm)					
	Diesel	B10	B20	B30	B40	B50
1000	439,67	471,17	464,34	361,59	318,85	320,47
1500	649,65	660,83	626,78	611,71	565,26	522,16
2000	710,41	688,75	679,64	687,06	710,18	798,96
2500	868,88	930,50	919,53	919,08	987,35	947,80

Table 11. The no average value variation on different rpm regarding to the mixture

rpm	%smoke					
	Diesel	B10	B20	B30	B40	B50
1000	7,72	5,76	6,36	13,89	12,88	13,35
1500	5,81	3,16	5,41	10,72	12,17	13,62
2000	5,24	3,62	4,45	7,59	7,28	7,70
2500	10,98	7,94	9,93	7,92	9,62	9,01

Table 12. The %smoke average value variation on different rpm regarding to the mixture

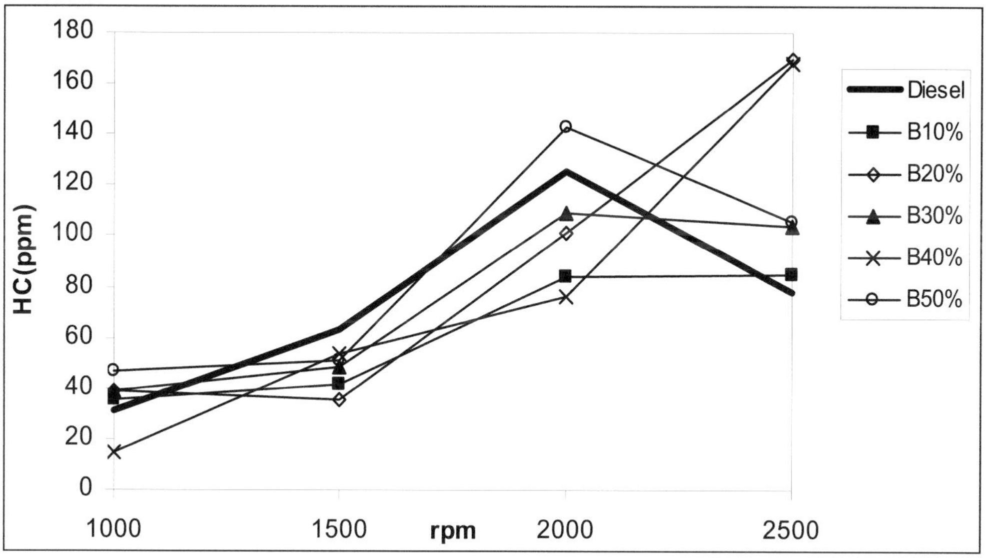

Fig. 11. The HC variation on different rpm regarding to the mixture

From figure 11 it can be noticed the biggest reduction of HC regarding to Diesel in case of the mixture B20/1500 rpm and in the case of the mixture B40/2000 rpm.
From figure 12 it can be noticed the biggest reduction of NO regarding to Diesel in the cases of the mixture B40/1000 rpm, B50/1000 rpm and B50/1500 rpm too.
From figure 13 it can be seen the reduction of smoke regarding to Diesel in case of the mixture B10 and B20 at all rounds per minute. It can also be noticed the reduction of smoke in the case of B30, B40, B50/2500 rpm. Finally it can be seen an increasement of the mixture B30, B40, B50 at all rounds regarding to Diesel. From the above figures it is clear that the use of different mixtures can constitute changes to CO, HC, NO and smoke too.

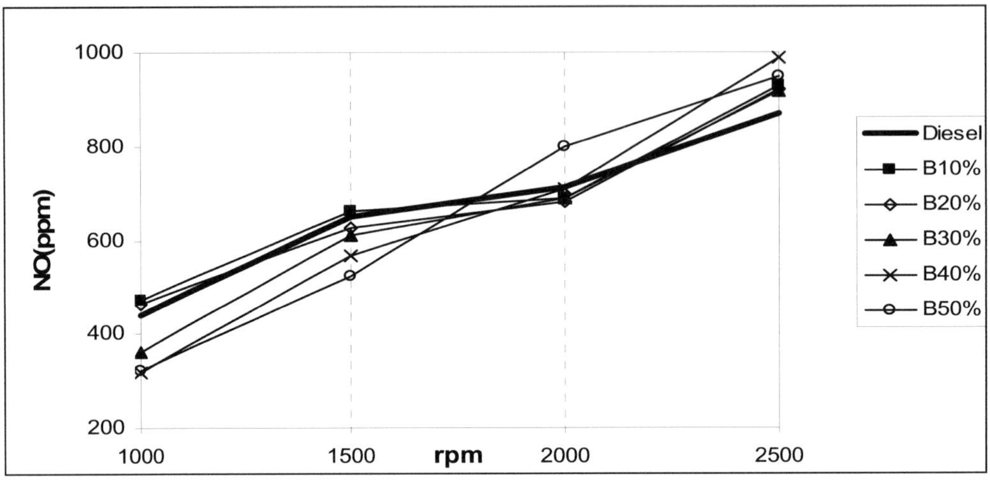

Fig. 12. The NO variation on different rpm regarding to the mixture

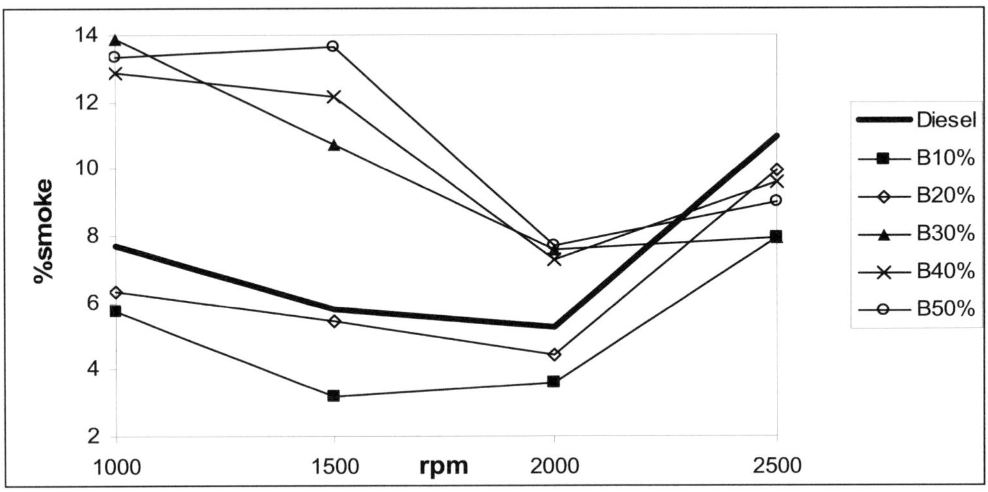

Fig. 13. The smoke variation on different rpm regarding to the mixture

It is also important the fact that there was no changes in the turns of engine, as well as in the supply of water at the use of mixtures. Finally as far as the consumption is concerned, did not exist changes with the use of different mixtures. The use of mixture of Diesel and Cotton Oil has the following impacts:

- About CO it can be noticed an increasement when the cotton oil is used as a fuel.
- About HC it can be noticed a reduction at 1500 rpm and particularly bigger reduction in the use of B20. It also appears reduction of the HC for all the mixture at 2000 rpm with the exception of B50. Finally about the HC, for all the mixture at 2500 rpm is observed increase of HC regarding to Diesel.
- About NO has been noticed a reduction at 1000 rpm and 1500 rpm for all the mixtures. A small reduction appeared for all the mixtures at 2500 rpm with the exception of B50, regarding to Diesel. Finally about the NO for all the mixtures appeared increase at 2500 rpm regarding to Diesel.

- About the smoke it can be noticed a reduction of the mixture of B20 and B10, but it appears an increasement for all other mixture in any round regarding to Diesel, with the exception of 2500 rpm, in where all the mixture appear a reduction.

2.4 Olive seed oil

In the experiment stage has been used directly cotton oil in the mixture of diesel in to a four - stroke Diesel engine. Specifically it has been used diesel, mixture diesel-5% olive seed oil (Pyrin5%), diesel-10% olive seed oil (Pyrin10%), diesel-20% olive seed oil (Pyrin20%), diesel-30% olive seed oil (Pyrin30%), diesel-40% olive seed oil (Pyrin40%), diesel-50% olive seed oil (Pyrin50%) in a four-stroke diesel engine [19]:
The experimental results are shown at the following tables and figures:

rpm	CO %						
	diesel	Pyrin 5%	Pyrin 10%	Pyrin 20%	Pyrin 30%	Pyrin 40%	Pyrin 50%
1000	0,056	0,056	0,054	0,060	0,053	0,053	0,048
1500	0,055	0,044	0,038	0,055	0,040	0,041	0,036
2000	0,043	0,038	0,031	0,050	0,031	0,036	0,030

Table 13. The CO average value variation on different rpm regarding to the mixture

rpm	HC (ppm)						
	diesel	Pyrin 5%	Pyrin 10%	Pyrin 20%	Pyrin 30%	Pyrin 40%	Pyrin 50%
1000	31,783	35,237	77,922	152,830	13,023	16,799	12,508
1500	38,001	48,434	79,198	165,479	22,954	24,870	22,860
2000	38,338	71,585	97,513	208,166	60,209	37,725	47

Table 14. The HC average value variation on different rpm regarding to the mixture

	NO (ppm)						
rpm	diesel	Pyrin 5%	Pyrin 10%	Pyrin 20%	Pyrin 30%	Pyrin 40%	Pyrin 50%
1000	518,210	415,212	375,075	392,478	372,681	473,620	362,663
1500	739,366	730,361	677,793	703,549	673,198	729,462	758,413
2000	762,155	790,676	738,929	805,702	825,376	938,210	880,990

Table 15. The NO average value variation on different rpm regarding to the mixture

rpm	%smoke						
	diesel	Pyrin 5%	Pyrin 10%	Pyrin 20%	Pyrin 30%	Pyrin 40%	Pyrin 50%
1000	9,990	12,605	14,787	12,717	11,018	9,932	16,278
1500	7,363	11,967	10,594	13,715	12,575	13,285	19,673
2000	6,634	14,212	12,201	14,131	14,098	17,528	23,359

Table 16. The %smoke average value variation on different rpm regarding to the mixture

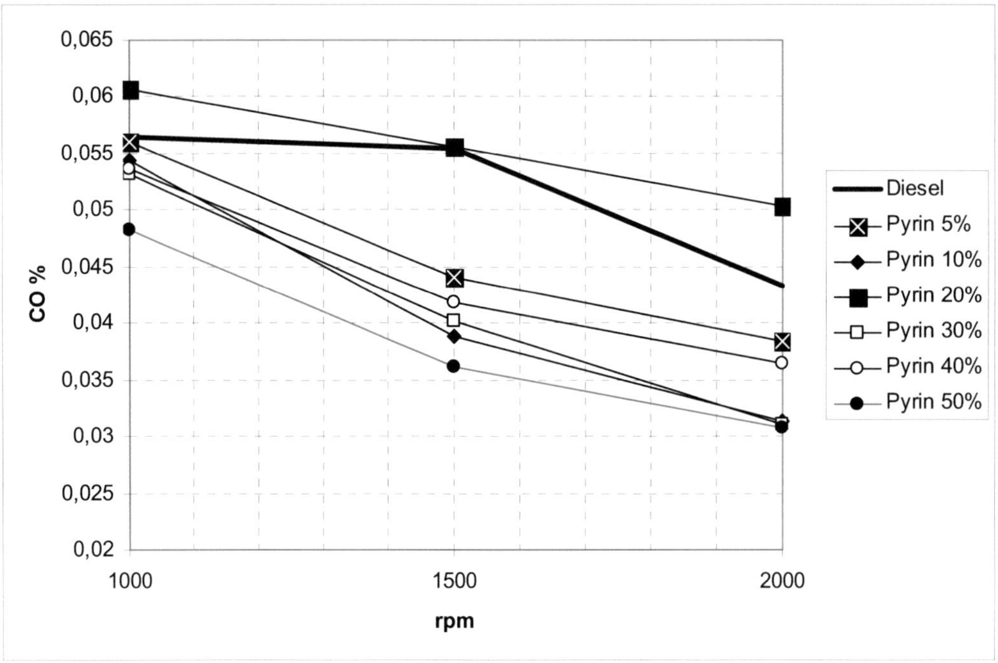

Fig. 14. The CO variation on different rpm regarding to the mixture

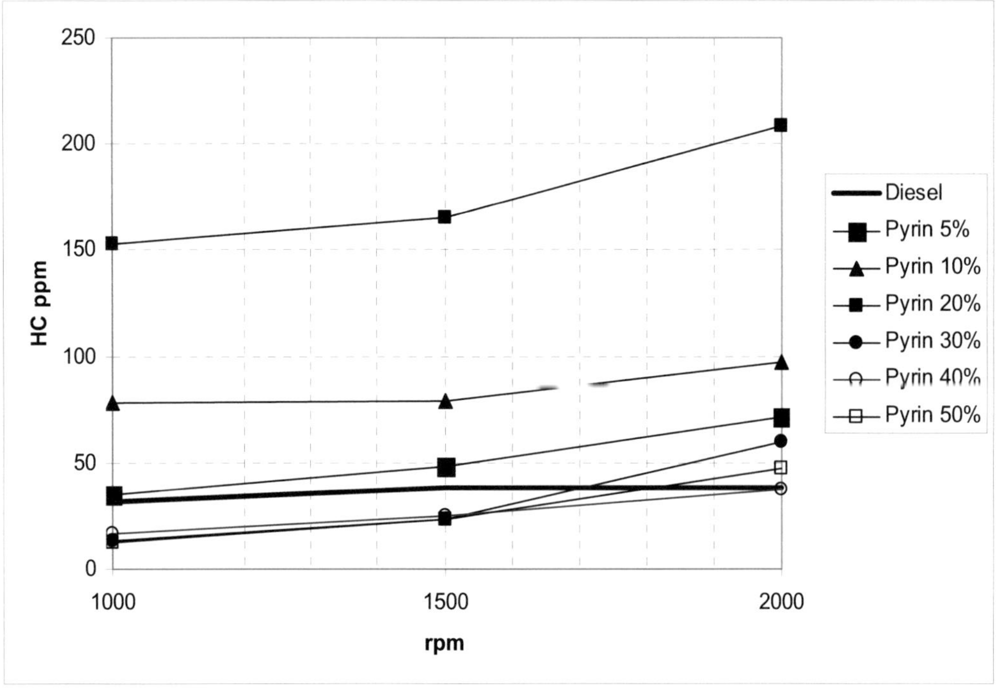

Fig. 15. The HC variation on different rpm regarding to the mixture

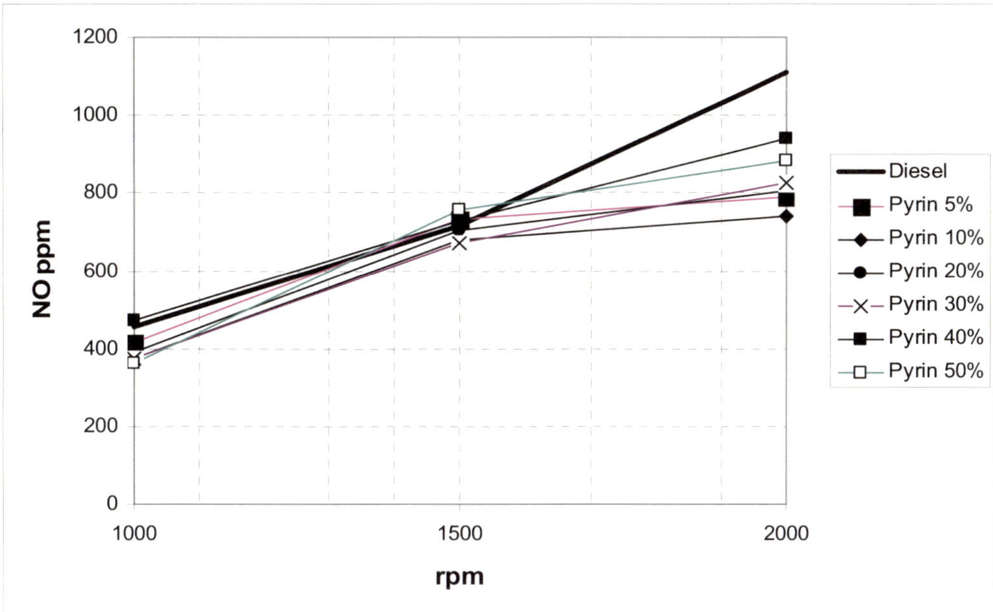

Fig. 16. The NO variation on different rpm regarding to the mixture

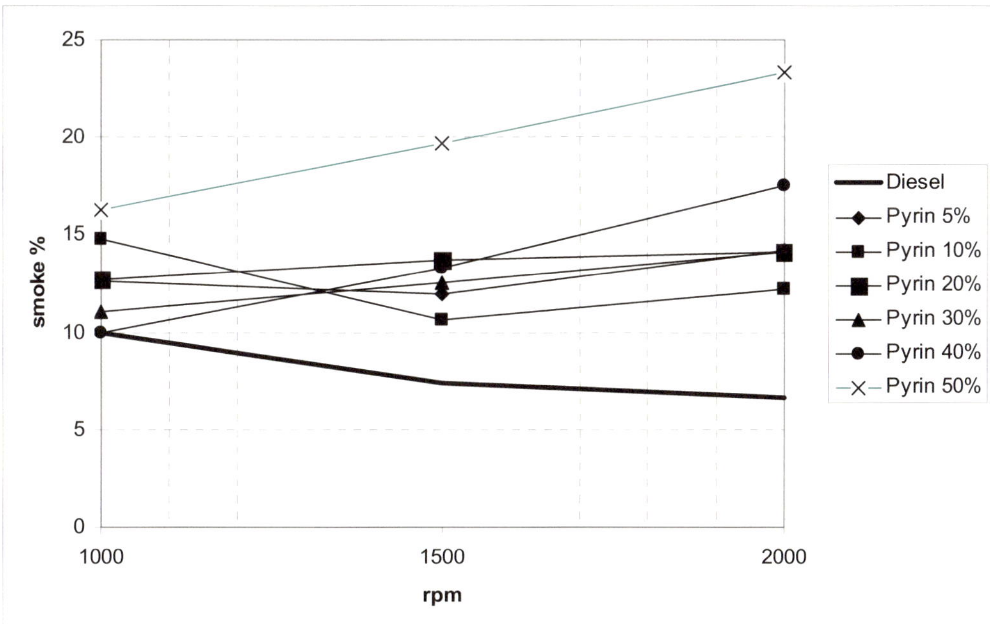

Fig. 17. The smoke variation on different rpm regarding to the mixture

From figure 14 it is clear that when the olive seed oil is increased on the fuel regarding to diesel, it appears a decrease of CO. From figure 15 it can be noticed the biggest reduction of HC regarding to diesel in case of pyrin50%. From figure 16 it can be noticed the biggest reduction of NO regarding to diesel in the case of pyrin10%/2000rpm. From figure 17 it can be noticed that the best behaviour appears on diesel. From the above figures it is clear that the use of different mixtures can constitute changes to CO, HC, NO and smoke too. It is also important the fact that there was no changes in the rounds of the engine, as well as in the supply of water at the use of mixtures. Finally as far as the consumption is concerned, did not observed changes with the use of different mixtures. The use of mixture of diesel and olive seed oil has the following impacts:

- About CO it can be noticed when the olive seed oil is increased on the fuel regarding to diesel, it appears a decrease of CO
- About HC it can be noticed the biggest reduction of HC regarding to diesel in case of pyrin50%
- The biggest reduction of NO regarding to diesel in the case of pyrin10%/2000rpm.
- The smoke it can be noticed that the best behaviour appears on diesel.

2.5 Soy oil

In the experiment stage has been used directly soy oil in the mixture of diesel in to a four - stroke Diesel engine. Specifically it has been used Diesel, mixture Diesel-5% soy oil (S5), Diesel-10% soy oil (S10), Diesel-20% soy oil (S20), Diesel-30% soy oil (S30), Diesel-40% soy oil (S40), Diesel-50% soy oil (S50) in a four-stroke diesel engine [20]:
The experimental results are shown at the following tables and figures:

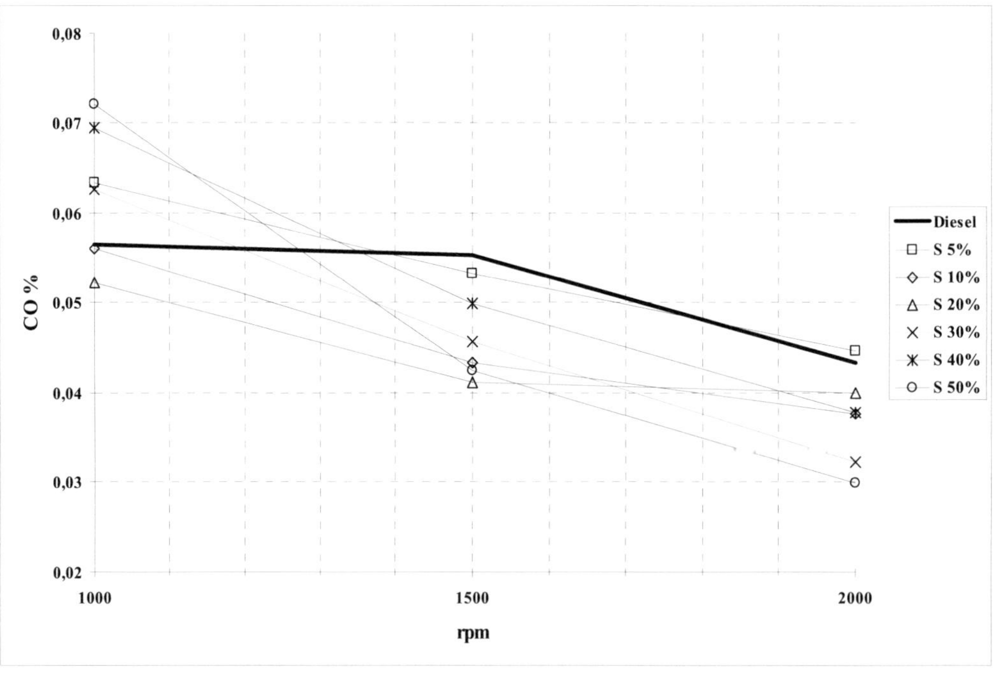

Fig. 18. The CO variation on different rpm regarding to the mixture

From figure 18 it is clear that when the soy oil is increased on the fuel regarding to diesel, it appears a decrease of CO, except in the cases S5,30,40,50/1000rpm.

rpm	HC (ppm)						
	Diesel	S5	S10	S20	S30	S40	S50
1000	31,78	21,15	21,88	8,28	5,76	54,61	28,01
1500	38,00	24,30	51,65	9,16	5,80	55,53	30,04
2000	38,33	23,70	89,90	28,68	22,34	84,88	67,47

Table 17. The CO average value variation on different rpm regarding to the mixture

rpm	NO (ppm)						
	Diesel	S5	S10	S20	S30	S40	S50
1000	454,2	387,6	397,5	416,1	414,8	341,0	277,9
1500	715,3	739,8	743,6	720,9	758,8	718,8	651,1
2000	1109,6	621,7	829,6	808,2	915,6	919,8	920,2

Table 18. The HC average value variation on different rpm regarding to the mixture

rpm	% smoke						
	Diesel	S5	S10	S20	S30	S40	S50
1000	9,99	8,72	9,41	11,61	14,26	18,32	24
1500	7,36	8,23	8,43	9,87	13,02	18,21	17,84
2000	6,63	6,25	7,70	8,08	11,27	17,21	20,5

Table 19. The NO average value variation on different rpm regarding to the mixture

rpm	CO %						
	Diesel	S5	S10	S20	S30	S40	S50
1000	0,056	0,063	0,056	0,052	0,062	0,069	0,072
1500	0,055	0,053	0,043	0,041	0,045	0,049	0,042
2000	0,043	0,044	0,037	0,04	0,032	0,037	0,029

Table 20. The %smoke average value variation on different rpm regarding to the mixture

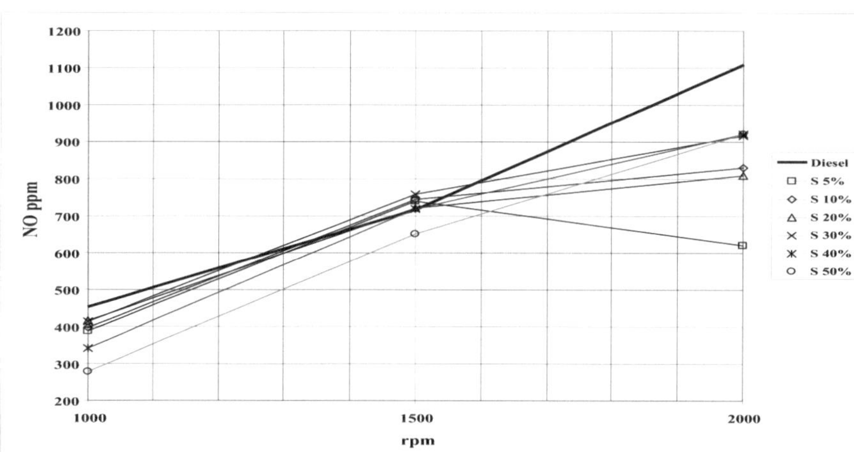

Fig. 19. The HC variation on different rpm regarding to the mixture

From figure 19 it can be noticed the biggest reduction of HC regarding to diesel in case of the mixtures S5, S20 and the mixture S40.

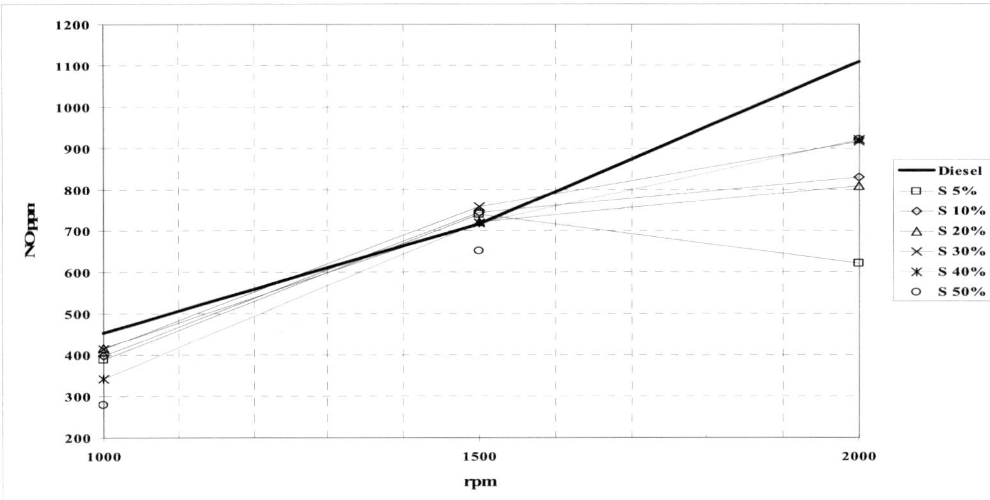

Fig. 20. The NO variation on different rpm regarding to the mixture

From figure 20 it can be noticed the biggest reduction of NO regarding to Diesel in the case of the mixture S50.

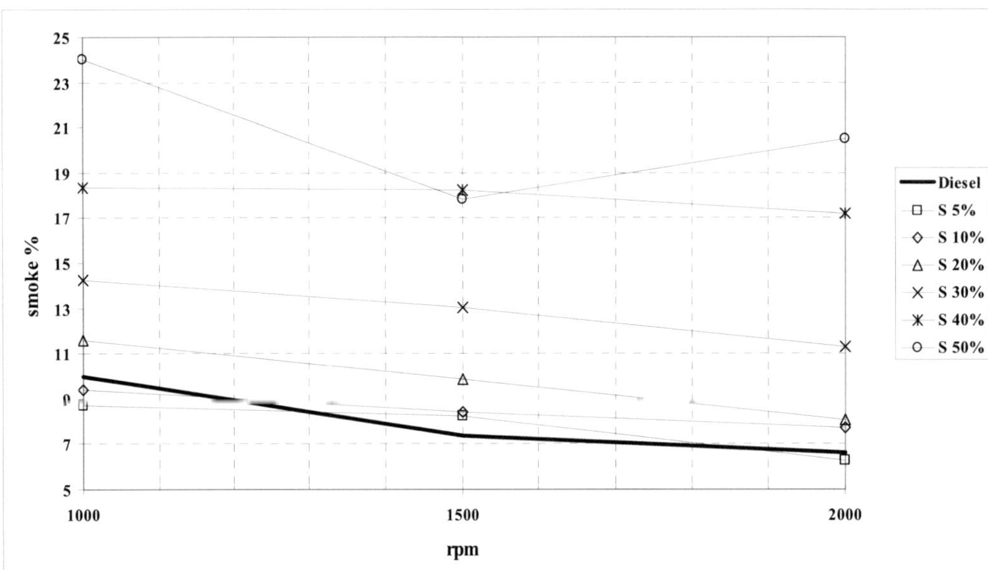

Fig. 21. The smoke variation on different rpm regarding to the mixture

From figure 21 it can be seen the increase of smoke regarding to diesel for all the mixtures. From the above figures it is clear that the use of different mixtures can constitute changes to CO, HC, NO and smoke too. It is also important the fact that there was no changes in the rounds of the engine, as well as in the supply of water at the use of mixtures. Finally as far as the consumption is concerned, did not observed changes with the use of different mixtures. The use of mixture of diesel and soy oil has the following impacts:

- About CO it can be noticed that when the soy oil is increased on the fuel regarding to diesel, it appears a decrease of CO, except in the cases S5,30,40,50/1000rpm.
- About HC it can be noticed the biggest reduction of HC regarding to diesel in case of the mixtures S5, S20 and the mixture S40.In the case of S30 appears the maximum increase of HC in relation to diesel.
- The biggest reduction of NO regarding to Diesel is noticed in the case of the mixture S50.
- The smoke is increased regarding to diesel for all the mixtures. Except the cases S5,50/1000rpm.

3. References

[1] Aldritton L, Monastersky R., Eddy A, Hall M, Shea E. (1992). *Our Ozone Shield Reports to the Nation on Our Changing Planet.* Fall 1992, University Cooperation for Atmospheric research office for interdisciplinary studies Boulder, Colorado.

[2] Arapatsakos C, (2000). Air and water influence of two stroke outboard engine using gasoline - ethanol mixtures, *Transaction of SAE*, Book SP-1565.

[3] Arapatsakos C, Karkanis A, Sparis P. Environmental Contribution of Gasoline- Ethanol Mixtures, *WSEAS Transactions on Environment and Development*, Issue 7, Volume 2, July 2006.

[4] Arapatsakos C, Karkanis A, Sparis P. Gas emissions and engine behavior when gasoline-alcohols mixtures are used, *Journal of Environmental Technology*, Vol. 24, pp. 1069-1077.

[5] Arapatsakos C. Testing the tractor engine using diesel – ethanol mixtures under full load conditions, *International Journal of Heat & Technology*, Vol. 19, n.1, 2001.

[6] Arapatsakos C, Christoforidis D, Karkanis A. The gas emissions variation of diesel engine from the combustion of used vegetable oils, *Proceedings of International Conference on Energy and Environment*, Cambridge 2009.

[7] Arapatsakos C., Christoforidis D, Karkanis A, Mitroulas D, Teka C. (2007). Test results from the use of cotton oil mixtures as fuel in a four-stroke engine, *International journal of Energy and Environment*, Issue3 Vol. 1, 2007.

[8] Arapatsakos C, Christoforidis D, Karkanis A, Mitroulas D. Fuel of diesel – olive seed oil mixture' *International journal of energy* Issue 3 vol 2, 2008.

[9] Arapatsakos C, Christoforidis D, Karkanis A, Mitroulas K. Soy oil as fuel in a four stroke engine, *Journal of wseas transactions on environment and development* 2008.

[10] Arapatsakos C, Christoforidis D, Sarantitis G, Giannopoulos D. Fuel mixtures of diesel-maize oil *International journal of energy* Issue 3 vol 2, 2008

[11] Schafer F., Basshuysen R. V., (1995). *Reduced Emissions and Fuel Consumption in Automobile Engines* Published by SAE.

[12] Menrad H, Haselhorst M. Alcohol fuels, *Monograph.* Springer, New York, ISBN 3211816968, 1981.

[13] Harrington A.; Shishu C. A Single-Cylinder Engine Study of the Effects of Fuel Type, Fuel Stoichiometry and Hydrogen-to-Carbon Ratio on CO, NO and HC Exhaust Emissions, *SAE-Paper* 730476

[14] Owen K, Coley T, (1995). Automotive Fuels Reference Book, (Second Edition), Published by SAE.

[15] Mitchell B. (1989) .The greenhouse effect and climate change. *Reviews of Geophysics* 27.

[16] Arapatsakos C, Karkanis A, Sparis P, Gas emissions and engine behaviour when gasoline -alcohol mixtures are used, *Environmental technology* 24 (2003) 1069-1077.

[17] Siddharth S., Green Energy-Anaerobic Digestion. Converting Waste to Electricity, *WSEAS Transactions on Environment and Development*, Issue 7, Volume 2, July 2006.

[18] Swedish Motor Fuel Technology Co., *Alcohols and alcohol blends as motor fuels*, Vol. II B, p.8:39, STU information No 580,1986.

[19] Maxwell T, Jones C, (1995). Alternative fuels: Emissions, *Economics and Performance* Published by SAE.

[20] Schenewerk E, Automatic DRAC LMFBR to Speed Licensing and Mitigate CO2, *WSEAS Transactions on Environment and Development*, Issue 7, Volume 2, July 2006.

5

Biofuels and Ecosystem Carbon Balance Under Global Change

Christina Schädel and Yiqi Luo
Department of Botany and Microbiology, University of Oklahoma
USA

1. Introduction

Terrestrial ecosystems are estimated to contain 3000 Pg of organic carbon (C) of which more than two thirds are stored in soils (Jobbágy and Jackson 2000). Total soil organic carbon (SOC) consists of different C pools with intrinsic turnover rates ranging from less than a year to thousands of years (Amundson 2001). The processes that drive the cycling of soil C are C inputs from net primary productivity (NPP=GPP-ecosystem respiration) and outputs through C decomposition (Fig. 1). New organic matter is the product of NPP which is transferred to soils in the form of litter and presents the largest C input to soils. Carbon output to the atmosphere is mainly driven by soil respiration (consisting of autotrophic and heterotrophic respiration) which is the second largest driver of C in the global C cycle (Fig. 1). Annually, soils release 98 ± 12 Pg C to the atmosphere which has increased yearly by 0.1 Pg C between 1989-2008 (Bond-Lamberty and Thomson 2010) and which yearly exceeds the current rate of fossil fuel combustion by a factor of 10. These large numbers show that even slight changes in the soil C and soil C cycling are highly relevant to the global C cycle mainly because of their potential to sequester or release CO_2 (Trumbore 1997). C sequestration denotes the transfer of C from atmospheric CO_2 into long-lived pools (e.g. woody biomass, recalcitrant soil C pools) without reemitting it immediately.

Although the soil is a dynamic system C input and output need to be balanced in order to keep the SOC pool at equilibrium (Fig. 2a). If C input is smaller than C output a depletion of the SOC pool occurs which can result in large releases of CO_2 to the atmosphere (Fig. 2b). On the other hand, if C input to the soil exceeds C output additional SOC can be sequestered in soils. SOC is not only an important C sink within the terrestrial C budget it also strongly influences soil fertility and soil quality which in return is needed for plant growth (Lal 2004; Cruse et al. 2010).

Global change denotes all human-caused changes to the atmosphere, hydrosphere, pedosphere and biosphere (Körner 2003). The increasing CO_2 concentration in the atmosphere is one of the most drastic global change components that directly affect plants and the ecosystems they live in (IPCC 2007). Secondary effects of higher CO_2 concentrations are climate warming causing tertiary effects such as extended growing seasons, shifts in species composition and alterations in precipitation patterns. Elevated CO_2 directly affects plants through photosynthesis and as photosynthesis at the current level of CO_2 concentration is not yet CO_2-saturated there is leeway for more carbon fixation and with it the possibility of more C storage in terrestrial ecosystems. On the other hand climate warming affects almost all aspects of carbon cycling and enhanced C fluxes potentially feed back to the atmosphere causing the so called positive feedback to climate change (Luo 2007).

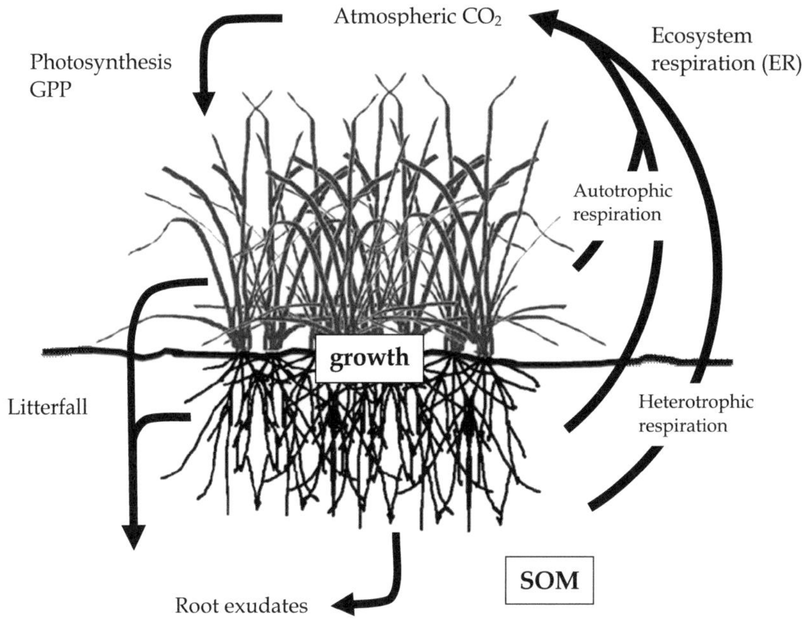

Fig. 1. Schematic diagram of terrestrial ecosystem C processes. GPP, gross primary production, SOM, soil organic matter

Identifying possible strategies for mitigating climate change by reducing increases in atmospheric CO_2 has put a strong focus on growing biofuel feedstock for alternative energy. Biofuels are generated through the combustion of biomass, usually grain or cellulosic-based feedstock. Biofuel feedstock production can help offset C emissions from fossil fuels but continuous biomass harvesting involves the removal of large quantities of C inducing a disequilibrium in the ecosystem's C balance (Fig. 2b; Luo and Weng 2011). There is thus an urgent need to investigate the impacts of biofuel feedstock harvesting on an ecosystem's C balance and its feedback to climate change (Luo et al. 2009)

This chapter focuses on key issues related to biofuel feedstock harvesting and ecosystem C balance under global change.

2. Terrestrial ecosystem C cycle

Carbon enters a terrestrial ecosystem through photosynthetic uptake (GPP=gross primary production) and is either incorporated into biomass, respired through ecosystem respiration (ER=autotrophic + heterotrophic respiration) or secreted into the soil via litter and root exudates (Fig. 1; Chapin et al. 2009). The main terrestrial ecosystem C cycling processes are photosynthesis and C decomposition and hence are the main components to calculate net ecosystem productivity (NEP=GPP-ER). Once C has entered the plant system it is allocated to above- and belowground tissues with the partitioning being strongly dependent on nutrient availability. The partitioning of C allocation within the plant influences how much C goes into aboveground biomass (leaves, stems) and how much enters the soil (coarse and fine roots). Belowground primary productivity can account for up to 33% of the annual net primary productivity (Jackson et al. 1997; Norby et al. 2004) which assigns roots an important role in the net ecosystem C balance. Dead plant material from above- and belowground litter is decomposed by microorganisms releasing CO_2 to the atmosphere and providing energy for microbial biomass growth. Microbial biomass together with organic residuals of dead plant material and dead microbes form the soil organic matter (SOM, Luo and Zhou 2006) of which carbon represents 50% by weight. C decomposition from short- and long-lived C pools is one of the most important factors in regulating terrestrial ecosystem C cycles.

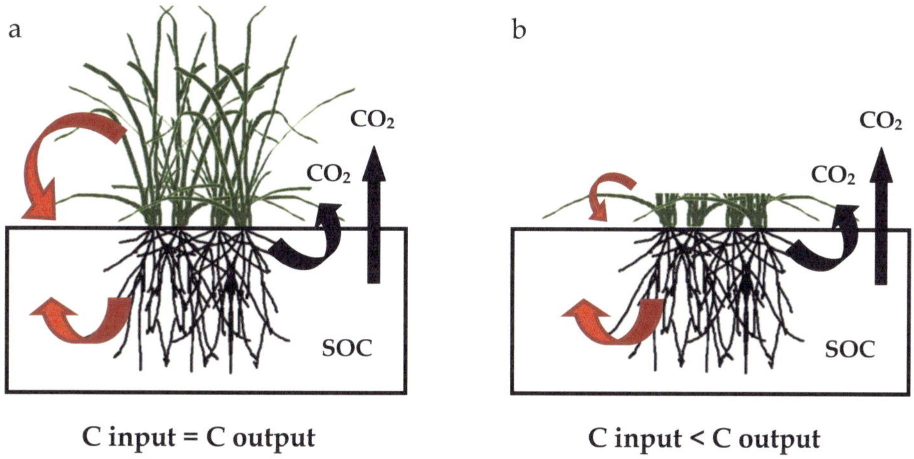

Fig. 2. Soil C dynamics for a) a natural grassland and b) a harvested field for biofuel feedstock. Red arrows represent C input to the soil and black arrows C output

3. Global change and C processes

Global change is one of the most important issues that this century is going to face. Global atmospheric CO_2 concentrations gradually increased from pre-industrial concentrations of 280 ppm to 390 ppm in 2011 (Mauna Loa, Hawaii; Tans 2011) and continue to increase, the main causes being deforestation, land use change and the burning of fossil-fuels (IPCC 2007). As a consequence to the increasing CO_2 and other green house gas concentrations in the atmosphere, global temperatures have increased by 0.74°C over the last century and predictions for the end of this century suggest a further increase in temperatures by another 1.1°C - 6.4°C (IPCC 2007). Temperature affects almost all chemical and biological processes (Shaver et al. 2000) and increasing global temperatures will likely enhance C fluxes between pools. As primary productivity and C decomposition are directly affected by changes in temperature ecosystem C uptake will be highly influenced by global warming. One of the major uncertainties of global warming is a possible positive feedback between the global carbon cycle and climate change but interactions among processes and fluxes are most uncertain making predictions difficult (Shaver et al. 2000; Luo 2007). Global change also includes altered precipitation regimes as a direct result to climate warming with a higher frequency in extreme events (rainfall extreme, drought) being very likely (IPCC 2007). Furthermore, human activities have increased the rate and magnitude of N deposition (Galloway et al. 2004) and since N availability strongly influence C processes future ecosystem C sequestration largely depends on N availability (Luo et al. 2004; Luo et al. 2006; Field et al. 2007). Global change includes various factors and their multiple interactions have smaller or larger impacts on ecosystem C processes and understanding those interactive effects on ecosystem processes becomes crucial in predicting possible carbon-climate feedbacks (Luo et al. 2008).

3.1 Impacts of global change on C processes

C influx through photosynthesis is stimulated by increased CO_2 concentrations (Curtis and Wang 1998; Norby et al. 1999; Ainsworth and Long 2005) often leading to increasing biomass growth but results are highly variable between species and plant age (Norby et al. 1999; Körner et al. 2005). Higher CO_2 concentrations could also induce higher C sequestration in soils (Jastrow et al. 2005; Lichter et al. 2005) but results strongly varied with CO_2 experimental facility, ecosystem type and N treatment (Luo et al. 2006). Highest C sequestration rates in soils were found in studies with N fertilization (Hungate et al. 2009) and most likely reflect effects of elevated CO_2 and N on plant growth as well as interactions between CO_2 and N on soil C decomposition (Reich et al. 2006). Increased soil C efflux under elevated CO_2 was reported from most CO_2-enrichment studies which was attributed to additional substrate from greater plant growth under elevated CO_2 available to soil microorganisms (Zak et al. 2000).

Increasing temperatures affect most biochemical processes such as C influx and C decomposition. Results are highly variable and range from no change, increases and decreases in photosynthesis, plant growth, primary production and soil respiration (Luo 2007). Generally, C decomposition reacts more to higher temperatures (Melillo et al. 2002; Zhou et al. 2007) than does C influx although the response of increasing soil C decomposition to warming attenuates over time (Rustad et al. 2001). The overall response of C decomposition to climate warming could be a net release of CO_2 from soils which might result in a positive feedback between C cycles and climate warming within this century (Friedlingstein et al. 2006). Furthermore, climate warming affects regulatory mechanisms of ecosystem C processes such as increased length of growing season, changes in water dynamics, species composition and nutrient availability (Luo 2007).

Precipitation patterns are changing in its frequency and intensity as a consequence of climate warming. Increased precipitation mostly increases C uptake via photosynthesis as well as increases soil respiration (Zhou et al. 2009). Drier conditions as they occur during heat waves reduce gross primary productivity and result in large net releases of CO_2 (Ciais et al. 2005; Arnone et al. 2008). This net release in CO_2 can offset C sequestration from previous years and turn ecosystems into C sources which will then contribute to a positive carbon-climate feedback. Besides the direct effects of increasing or decreasing water availability there are indirect effects of water availability through other climate change factors. Elevated CO_2 usually leads to less water loss through stomata (Medlyn et al. 2001; Leuzinger and Körner 2007) whereas a warmer climate increases evaporation leading to more negative water balances in the soil. As primary production in many ecosystems is largely limited by water availability any changes in precipitation will have substantial impacts on ecosystem C dynamics (Heimann and Reichstein 2008).

Nitrogen is one of the most limiting nutrients for plant growth (Vitousek and Howarth 1991) and N fertilization is widely used to improve plant productivity. N addition can increase plant growth by 30-50% although there are large differences between plant functional types and biomes (LeBauer and Treseder 2008; Xia and Wan 2008). Interactions between C and N cycles are important as N has the potential to sequester C when plant growth is not restricted by the lack of available N. Aboveground biomass is more stimulated by additional N availability than belowground biomass reducing C input to the soil (Lu et al. 2011). Litter quality usually increases with N addition which enhances decomposition of this high-quality litter (Knorr et al. 2005) causing less accumulation of C in the soil. Results on soil C storage under N addition are controversial and likely vary between natural ecosystems and agricultural ecosystems (Pregitzer et al. 2008; Liu and Greaver 2010; Lu et al. 2011). As climate change is a multi-factor process we need to investigate interactions between those climate change factors and expect dynamics that we cannot investigate by only looking at one factor.

4. Impacts of biofuel feedstock harvesting on C cycles

4.1 Biomass removal and SOC

Crop and plant residues (usually deriving from corn, barley, oat, sorghum, soybean, sunflower and wheat) are considered to be free products for biofuel production as they are left behind after harvest and seem not be of any use (Lal 2004). Plant residues are high-cellulosic feedstock (high concentration of cellulose and hemicelluloses) which are suitable for ethanol production but are not in direct competition to food compared to starch-rich biomass from grain. Complete biomass removal might impact soil quality, SOC content and soil water content more than a partial removal. The amount of corn stover (referring to all aboveground plant material such as stalk, leaves, cobs and husk) residue that can be removed without endangering SOC contents of soils was estimated to be 25% of total removal (Blanco-Canqui and Lal 2007). If more than 25% of corn stover was removed soils showed decreasing SOC contents.

Most biofuel feedstock derives from monocultures such as corn, soybeans, switchgrass or sugarcane and is grown on fertile soils. However, studies have shown that more bioenergy can be produced from switch-grass mixtures or high-diversity grasslands than from monocultures (Tilman et al. 2006; Wang et al. 2010). Studies have shown that even though aboveground biomass of high-diversity grasslands was removed annually soils still sequestered carbon. The full life-cycle of these high-diversity grasslands results in a net sequestration of atmospheric CO_2, which makes these high-diversity grasslands so called carbon negative biofuels.

The partitioning of C allocation within the plant influences how much C goes into harvestable biomass and how much enters the soil. Depending on the vegetation type of the biofuel feedstock C allocation strongly varies and influences the amount of C stored in soils. Root architecture and their vertical distribution contribute to soil C maintenance with less C stored in deeper soil layers. Deep soil layers though store the highest amounts of highly stable organic C compounds and are less affected by freshly added litter input. It was suggested that in the absence of fresh organic carbon input the stability of organic compounds in deep soil layers is maintained as the available energy required to breakdown recalcitrant SOC is not provided (Fontaine et al. 2007). Therefore biomass removal for biofuel production could protect deep soil recalcitrant C pools from decomposition.

Root systems play as much a role in the build-up and maintenance of SOC as does the amount of residue removed. Especially graminoids (to which most biofuel crops belong to) have larger ratios of belowground biomass to aboveground biomass and therefore allocate more C through belowground biomass to the soil than through aboveground biomass which gives them a higher C sequestration potential (De Deyn et al. 2008). It was even suggested that belowground biomass is more important for C sequestration than aboveground biomass as studies showed that changes in SOC pools positively correlated with the quantity of belowground biomass input but not with input of aboveground biomass (Russell et al. 2009; Lu et al. 2011). Balesdent and Balabane (1996) measured root-derived C in maize cultivated soils and found that although the shoot to root ratio was only 0.5 root-derived C was 1.5 times higher than aboveground-derived C (from stalks and leaves). Furthermore, root litter of grasses is of lower quality and therefore less easily decomposable compared to aboveground litter due to lower N but higher lignin concentration (Vivanco and Austin 2006). This higher recalcitrance of plant litter slows down the litter decay process and increases the amount of C stored in the soil (Sartori et al. 2006; Johnson et al. 2007).

4.2 Biofuel feedstock harvest and global change

The sustainability of biofuel feedstock harvest under global change needs to be evaluated in order to quantify changes in the net ecosystem C balance as well as assess a possible positive feedback to climate change. Biofuel feedstock harvest and the coherent changes in the C balance can be evaluated from experimental studies that use clipping or biomass harvesting to remove aboveground biomass (Luo et al. 2009). One study that combined the effects of climate warming and biomass feedstock harvesting on ecosystem C dynamics was conducted in the Southern Great Plains, USA, which is considered to be a major region for biofuel feedstock production (Luo et al. 2009). Temperatures were increased by 2°C and biomass was clipped annually. On average, data of nine years showed increased net primary productivity (NPP) under warming and even higher values in the combination treatment of warming and clipping. Although warming increased soil respiration rates clipping showed a decreasing trend in soil respiration. Yearly biomass removal reduced the C input to soils which was clearly demonstrated by higher losses of soil C in the clipped plots compared to the unclipped plots (Fig. 3). In both clipped treatments losses in soil C after nine years were more than twice as high as they were for the unclipped plots. Additionally, warming enhanced soil C loss resulting in the highest loss of soil C under clipping and warming treatment (Fig. 3). These results clearly show that biofuel feedstock harvest in combination with warmer temperatures results in the highest loss in soil C.

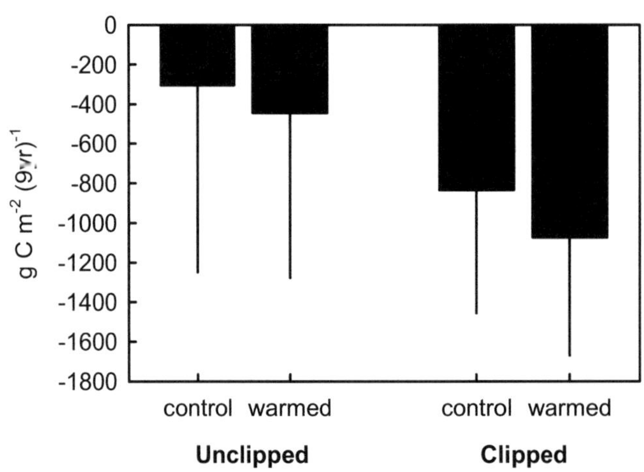

Fig. 3. Change in soil C content between 1999-2008. Values are means of 5 plots ± 1 se

4.3 Clipping-induced erosion under global change

Changes in land use through alteration of land coverage and disturbance of soil structure result in changes in soil moisture which can induce higher soil erosion rates (Lal 2004). Generally, plant coverage protects the soil from soil erosion by intercepting rainfall and runoff. Plant cover, plant height, rooting characteristics and other plant related parameters are important factors in reducing soil erosion rates (Wilhelm et al. 2007; Johnson et al. 2010). If aboveground biomass is removed for biofuel feedstock harvest more bare ground will increase temperatures as well as surface runoff and thus accelerate soil erosion (Schlesinger et al. 1990; Zuazo and Pleguezuelo 2008). Cover and type of vegetation can therefore affect soil erosion and potentially lead to a net source of C by soil erosion induced loss of SOC.

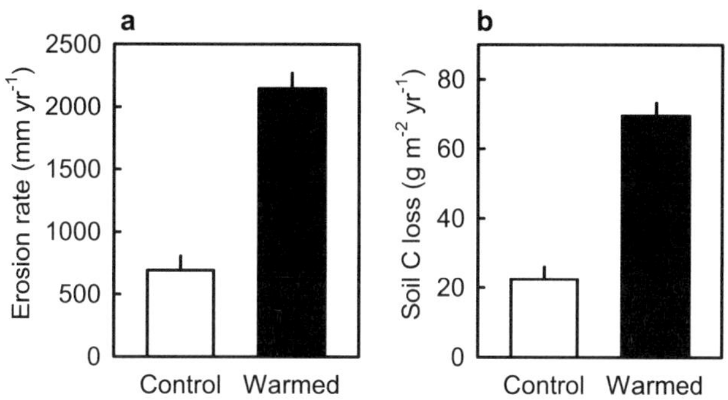

Fig. 4. a) Yearly erosion rate in the clipped subplots, b) yearly soil C loss in the clipped subplots. Values are means of 16 measurements per treatment ± SD. Redrawn with permission from *Global Change Biology Bioenergy*, Xue et al. 2011

It is well known that biomass removal on a continuous basis results in increased soil erosion but it is not well known how a warmer climate might amplify C loss from soils through erosion. The only study, we are aware of, that combines the effects of biomass removal and climate warming on soil erosion rates was conducted in a tallgrass prairie in the Southern Great Plains, USA (Xue et al. 2011). In a multiyear experiment (since 1999) grassland was warmed (+2°C) on a whole ecosystem-level and half the plots were clipped in order to mimic biofuel feedstock harvest. One side effect of warming was a reduction in soil moisture which was even greater in the clipped plots. Clipping-induced relative soil erosion rate was threefold increased under the warming scenario (Fig. 4a). These high erosion rates resulted in high losses of SOC (Fig. 4b). The stronger response to the warming treatment in the clipped plots was ascribed to lower soil moisture in the clipped plots as evaporation from the soil surface was increased when biomass was removed. Some of the consequences of higher erosion rates are reduced soil fertility, degraded soil structure and reduced SOC, all being enhanced by biomass removal. The soil that is most affected by erosional processes is the SOC-rich upper soil level making erosion a net source of C to the atmosphere (e.g. Lal 2003).

5. Interactive effects of biofuel feedstock harvesting and global change

5.1 Biofuel feedstock harvesting and NECB

Soils and their C stocks will be affected by land use change and by manipulations in the substrate supply but more importantly changes in the soil C budget will potentially affect the net ecosystem C balance (Fargione et al. 2008; Sanderson 2008; Luo et al. 2009) and consequentially contribute to the overall terrestrial C-cycle feedback.

Ecosystems can function as C sources or C sinks and their role in the global C cycle becomes even more important with global change as ecosystems either release or absorb atmospheric CO_2 and with it enhance or mitigate climate warming (Chapin et al. 2006). Net ecosystem production (NEP) is a measure of gross primary productivity (GPP) minus ecosystem respiration and mostly coincides with the net ecosystem C balance (NECB) unless C in other forms than CO_2 or dissolved organic C moves in or out of the system (Chapin et al. 2006; Lovett et al. 2006). Therefore, NECB is the net estimate of C accumulation (positive NECB) or C loss (negative NECB) in any system. If an ecosystem's net C balance is positive the ecosystem functions as a C sink by sequestering C. In contrast, a negative NECB implies C release to the atmosphere and any ecosystem showing a negative balance functions as a C source. NECB can be applied on short-term or long-term scales and to any spatial scale which makes it a very useful parameter for cross-scale comparisons (Chapin et al. 2006). To fully estimate the impact of biofuel feedstock removal on ecosystems under global change the net ecosystem C balance needs to be calculated to estimate a feedback of biomass removal to climate change. So far there are not many studies that measure the impacts of biofuel feedstock harvest on the net ecosystem C balance under global change. Nevertheless this is important as biofuels are supposed to help mitigate climate change by reducing CO_2 release from fossil fuels. But if biofuel feedstock harvest has large negative impacts on the net ecosystem C balance this mitigation strategy might not help reduce CO_2 release to the atmosphere.

5.2 NECB under elevated CO_2

Elevated atmospheric CO_2 generally increases above- as well as belowground biomass and also enhances soil C storage although the extent to which C is stored in soils is largely dependent on N availability (Luo et al. 2006). Belowground biomass often shows a higher response to elevated CO_2 therefore increasing C input to soils (Luo et al. 2006). C accumulation in plant and soil pools reflects increased C input into ecosystems that usually decreases litter quality and with it decomposability. Decreasing decomposability also derives from increased mycorrhizal growth under elevated CO_2 that enhances physical protection through formation of intra-aggregate or organomineral complexes to protect organic matter from microbial decomposition (Rillig 2004). Large fractions of the C accrued in soils under elevated CO_2 derive from increased belowground biomass growth which is not affected by biomass removal. Nevertheless there are some factors that need to be considered when making predictions about net ecosystem C balances for biofuel feedstock harvest under elevated CO_2. It is not yet clear whether there will be a down-regulation of CO_2 stimulation of photosynthesis and with it in plant growth and other C processes under persistent CO_2 stimulation (Long et al. 2004;). Photosynthetic acclimation was alleviated in grassland when plants were harvested but only under high N availability (Ainsworth et al. 2003). Low N conditions resulted in some acclimation of photosynthetic capacity. It seems that all responses of C processes under elevated CO_2 are strongly dependent on N availability. However, when only considering the global change factor elevated CO_2, biofuel feedstock harvest might still allow for C sequestration in soils resulting in a positive net ecosystem C balance.

5.3 NECB under climate warming

Unlike elevated CO_2 that primarily influences C uptake through photosynthesis warming affects almost all chemical and biological processes. Furthermore, warming involves some secondary effects on ecosystems such as extended growing seasons, change in species composition and drier conditions. Hence, it is not surprising that ecosystem warming experiments have produced inconsistent results regarding plant growth, soil respiration and net ecosystem production. Nevertheless the most important biomass fraction for C sequestration under biofuel feedstock harvest is the belowground biomass which was positively stimulated under warming and harvesting scenarios (Luo et al. 2009). This positive interaction was ascribed to over-compensatory mechanisms of plant physiological processes to biomass removal (Owensby et al. 2006). As belowground biomass growth is enhanced under warmer conditions the C loss through biomass removal might be less important for the net ecosystem C balance than the gain in C through increased belowground biomass. On the other hand continuous biomass removal increases soil erosion rates (Xue et al. 2011) which is accompanied by high losses of soil C. Even higher erosion rates occur when biomass removal takes place under warmer conditions as the soil dries out more easily leaving unstable soil structures favoring soil erosion. Therefore, biomass harvesting of natural grassland (Luo et al. 2009) in combination with warming resulted in a more negative net ecosystem C balance than for the warming treatment alone (Fig. 5). The more negative C balance is mainly due to high soil C losses (Fig. 4) as C input to soils was smaller than the C lost through CO_2 release and soil erosion. Thus, over-compensatory belowground biomass growth was not enough to offset soil C loss under warming and clipping. This long-term experiment shows that growing biofuel feedstock for harvesting under climate warming puts an additionally strain on the ecosystem C balance and does not help to sequester more C in order to reduce CO_2 release to the atmosphere.

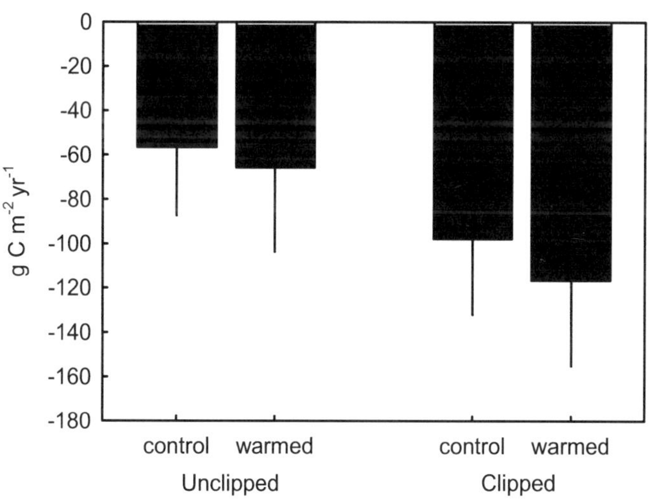

Fig. 5. Net ecosystem C balance calculated per year for the period of 2000-2008. Values are means of 6 plots ± 1 se

5.4 NECB and change in precipitation

Changes in precipitation as a consequence of global change include more frequent extreme precipitation and drought events which likely have large effects on ecosystem processes (Weltzin et al. 2003). Precipitation is an important factor in shaping ecosystem C dynamics as aboveground biomass and soil respiration linearly increase with mean annual precipitation but belowground biomass and soil C content remain rather constant (Zhou et al. 2009). As was shown for the Southern Great Plains in the USA no change in belowground C allocation is more important to the net ecosystem C balance than higher aboveground plant growth since this higher aboveground litter input was compensated by higher litter decomposition. A more positive net ecosystem C balance therefore seems plausible under wetter conditions. On the other hand warming induced drought suppresses net primary productivity and turns ecosystems into net sources of carbon dioxide (Ciais et al. 2005; Arnone et al. 2008). If additionally biomass is removed the net ecosystem C balance could become even more negative contributing more to a positive carbon-climate feedback.

5.5 NECB and N addition

N addition strongly influences ecosystem C processes through photosynthesis and biomass production and therefore has large impacts on the net ecosystem C balance. Generally N addition increases C input to soil through increased aboveground litter input (Liu and Greaver 2010). With higher N availability plants invest less C into belowground biomass as roots can more easily acquire N. Furthermore, higher N availability strongly influences the shoot to root ratio and root litter flux to soil decreases (Liu and Greaver 2010). If additionally C from aboveground biomass is not returned to soil due to biofuel feedstock harvest total C input to the soil will decrease and a negative net ecosystem C balance is very likely.

6. Conclusion

Growing biofuels for alternative energy can help mitigate increasing atmospheric CO_2 concentration; however continuous biofuel feedstock harvest will influence the whole ecosystem C balance possibly resulting in a positive feedback to climate change. Ecosystem C processes are strongly influenced by global change factors and their interactive effects are very complex and not yet well understood. An overall response of biomass feedstock removal on the net ecosystem C balance under global change is therefore still speculative but we know that global change factors that enhance root biomass have a more positive effect on the net ecosystem C balance when biomass is continuously removed than factors that enhance aboveground biomass. Increased CO_2 concentration in the atmosphere has the potential to increase belowground C storage especially when N and other nutrients are not limiting. Climate warming on the other hand seems to reduce soil C storage as C decomposition and C losses through soil erosion under biofuel feedstock harvest are higher. Responses to changes in precipitation are very variable but drier conditions result in a more negative ecosystem C balance if biomass is continuously removed. This effect could be neutralized again under elevated CO_2 as stomatal conductance and evapotranspiration decline thus decreasing the plant water use. N availability is a crucial factor for optimized plant growth and C storage but high N addition can also reduce belowground biomass and thus C input to soils. If additionally all biomass is removed there will be an even smaller C input into soil. One way to alleviate strong impacts of biomass harvest on C-cycling might be to harvest at a later time as harvesting after plant senescence showed to reduce C and N losses although biomass yield might be slightly lower (Heaton et al. 2009; Niu et al. 2010).

In conclusion, this chapter showed that biofuel harvesting has large impacts on the net ecosystem C balance which are likely enhanced under global change. More information on interactive effects of multiple global change factors is still needed to fully estimate the impacts of biofuel feedstock harvest on net ecosystem C balance and any possible feedback to climate change.

7. References

Ainsworth, E. A.; Davey, P. A.; Hymus, G. J.; Osborne, C. P.; Rogers, A.; Blum, H.; Nösberger, J. & Long, S. P. (2003). Is stimulation of leaf photosynthesis by elevated carbon dioxide concentration maintained in the long term? A test with *Lolium perenne* grown for 10 years at two nitrogen fertilization levels under Free Air CO_2 Enrichment (FACE). *Plant Cell and Environment*, 26, 5, 705-714

Ainsworth, E. A. & Long, S. P. (2005). What have we learned from 15 years of free-air CO_2 enrichment (FACE)? A meta-analytic review of the responses of photosynthesis, canopy properties and plant production to rising CO_2. *New Phytologist*, 165, 2, 351-371

Amundson, R. (2001). The carbon budget in soils. *Annual Review of Earth and Planetary Sciences*, 29, 535-562

Arnone, J. A. et al. (2008). Prolonged suppression of ecosystem carbon dioxide uptake after an anomalously warm year. *Nature*, 455, 7211, 383-386

Balesdent, J. & Balabane, M. (1996). Major contribution of roots to soil carbon storage inferred from maize cultivated soils. *Soil Biology & Biochemistry*, 28, 9, 1261-1263

Blanco-Canqui, H. & Lal, R. (2007). Soil and crop response to harvesting corn residues for biofuel production. *Geoderma*, 141, 3-4, 355-362

Bond-Lamberty, B. & Thomson, A. (2010). Temperature-associated increases in the global soil respiration record. *Nature*, 464, 7288, 579-582

Chapin, F. S.; McFarland, J.; McGuire, A. D.; Euskirchen, E. S.; Ruess, R. W. & Kielland, K. (2009). The changing global carbon cycle: linking plant-soil carbon dynamics to global consequences. *Journal of Ecology*, 97, 5, 840-850

Chapin, F. S. et al. (2006). Reconciling carbon-cycle concepts, terminology, and methods. *Ecosystems*, 9, 7, 1041-1050

Ciais, P. et al. (2005). Europe-wide reduction in primary productivity caused by the heat and drought in 2003. *Nature*, 437, 7058, 529-533

Cruse, R. M.; Cruse, M. J. & Reicosky, D. C. (2010). Soil quality impacts of residue removal for biofuel feedstock. In: *Soil quality and biofuel production*. Lal, R. and Stewart, B. A., (Ed.) CRC Press, Taylor & Francis Group, ISBN 978-1-4398-0073-7, Boca Raton, FL

Curtis, P. S. & Wang, X. Z. (1998). A meta-analysis of elevated CO_2 effects on woody plant mass, form, and physiology. *Oecologia*, 113, 3, 299-313

De Deyn, G. B.; Cornelissen, J. H. C. & Bardgett, R. D. (2008). Plant functional traits and soil carbon sequestration in contrasting biomes. *Ecology Letters*, 11, 5, 516-531

Fargione, J.; Hill, J.; Tilman, D.; Polasky, S. & Hawthorne, P. (2008). Land clearing and the biofuel carbon debt. *Science*, 319, 5867, 1235-1238

Field, C. B.; Lobell, D. B.; Peters, H. A. & Chiariello, N. R. (2007). Feedbacks of terrestrial ecosystems to climate change. *Annual Review of Environment and Resources*, 32, 1-29

Fontaine, S.; Barot, S.; Barré, P.; Bdioui, N.; Mary, B. & Rumpel, C. (2007). Stability of organic carbon in deep soil layers controlled by fresh carbon supply. *Nature*, 450, 7167, 277-U210

Friedlingstein, P. et al. (2006). Climate-carbon cycle feedback analysis: Results from the (CMIP)-M-4 model intercomparison. *Journal of Climate*, 19, 14, 3337-3353

Galloway, J. N. et al. (2004). Nitrogen cycles: past, present, and future. *Biogeochemistry*, 70, 2, 153-226

Heaton, E. A.; Dohleman, F. G. & Long, S. P. (2009). Seasonal nitrogen dynamics of Miscanthus x giganteus and Panicum virgatum. *Global Change Biology Bioenergy*, 1, 4, 297-307

Heimann, M. & Reichstein, M. (2008). Terrestrial ecosystem carbon dynamics and climate feedbacks. *Nature*, 451, 7176, 289-292

Hungate, B. A.; van Groenigen, K. J.; Six, J.; Jastrow, J. D.; Lue, Y. Q.; de Graaff, M. A.; van Kessel, C. & Osenberg, C. W. (2009). Assessing the effect of elevated carbon dioxide on soil carbon: a comparison of four meta-analyses. *Global Change Biology*, 15, 8, 2020-2034

IPCC (2007). *Climate Change 2007: The physical science basis. Contribution of working group I to the fourth assessment report of the Intergovernmental Panel on Climate Change,* Cambridge University Press, Cambridge, United Kingdom and New York, NY, USA

Jackson, R. B.; Mooney, H. A. & Schulze, E. D. (1997). A global budget for fine root biomass, surface area, and nutrient contents. *Proceedings of the National Academy of Sciences of the United States of America*, 94, 14, 7362-7366

Jastrow, J. D.; Miller, R. M.; Matamala, R.; Norby, R. J.; Boutton, T. W.; Rice, C. W. & Owensby, C. E. (2005). Elevated atmospheric carbon dioxide increases soil carbon. *Global Change Biology*, 11, 12, 2057-2064

Jobbágy, E. G. & Jackson, R. B. (2000). The vertical distribution of soil organic carbon and its relation to climate and vegetation. *Ecological Applications*, 10, 2, 423-436

Johnson, J. M. E.; Papiernik, S. K.; Mikha, M. M.; Spokas, K. A.; Tomer, M. D. & Weyers, S. L. (2010). Soil processes and residue harvest management. In: *Soil Quality and Biofuel Production*. Lal, R. and Stewart, B. A., (Ed.): 1-44, CRC Press, Taylor & Francis Group, ISBN 978-1-4398-0073-7, Boca Raton, FL

Johnson, J. M. F.; Barbour, N. W. & Weyers, S. L. (2007). Chemical composition of crop biomass impacts its decomposition. *Soil Science Society of America Journal*, 71, 1, 155-162

Knorr, M.; Frey, S. D. & Curtis, P. S. (2005). Nitrogen additions and litter decomposition: A meta-analysis. *Ecology*, 86, 12, 3252-3257

Körner, C. (2003). Ecological impacts of atmospheric CO_2 enrichment on terrestrial ecosystems. *Philosophical Transactions of the Royal Society of London Series a-Mathematical Physical and Engineering Sciences*, 361, 1810, 2023-2041

Körner, C. et al. (2005). Carbon flux and growth in mature deciduous forest trees exposed to elevated CO2. *Science*, 309, 5739, 1360-1362

Lal, R. (2003). Soil erosion and the global carbon budget. *Environment International*, 29, 4, 437-450

Lal, R. (2004). Soil carbon sequestration impacts on global climate change and food security. *Science*, 304, 5677, 1623-1627

LeBauer, D. S. & Treseder, K. K. (2008). Nitrogen limitation of net primary productivity in terrestrial ecosystems is globally distributed. *Ecology*, 89, 2, 371-379

Leuzinger, S. & Körner, C. (2007). Water savings in mature deciduous forest trees under elevated CO_2. *Global Change Biology*, 13, 12, 2498-2508

Lichter, J.; Barron, S. H.; Bevacqua, C. E.; Finzi, A. C.; Irving, K. E.; Stemmler, E. A. & Schlesinger, W. H. (2005). Soil carbon sequestration and turnover in a pine forest after six years of atmospheric CO_2 enrichment. *Ecology*, 86, 7, 1835-1847

Liu, L. L. & Greaver, T. L. (2010). A global perspective on belowground carbon dynamics under nitrogen enrichment. *Ecology Letters*, 13, 7, 819-828

Long, S. P.; Ainsworth, E. A.; Rogers, A. & Ort, D. R. (2004). Rising atmospheric carbon dioxide: Plants face the future. *Annual Review of Plant Biology*, 55, 591-628

Lovett, G.; Cole, J. & Pace, M. (2006). Is net ecosystem production equal to ecosystem carbon accumulation? *Ecosystems*, 9, 1, 152-155

Lu, M.; Zhou, X.; Luo, Y.; Yang, Y.; Fang, C.; Chen, J. & Li, B. (2011). Minor stimulation of soil carbon storage by nitrogen addition: A meta-analysis. *Agriculture, Ecosystems & Environment*, 140, 1-2, 234-244

Luo, Y. et al. (2004). Progressive nitrogen limitation of ecosystem responses to rising atmospheric carbon dioxide. *Bioscience*, 54, 8, 731-739

Luo, Y. & Weng, E. (2011). Dynamic disequilibrium of the terrestrial carbon cycle under global change. *Trends in Ecology & Evolution*, 26, 2, 96-104

Luo, Y. & Zhou, X. (2006). *Soil respiration and the environment*, Academic Press, ISBN 978-0-12-088782-8, San Diego, CA, USA

Luo, Y. Q. (2007). Terrestrial carbon-cycle feedback to climate warming. *Annual Review of Ecology Evolution and Systematics*, 38, 683-712

Luo, Y. Q. et al. (2008). Modeled interactive effects of precipitation, temperature, and CO_2 on ecosystem carbon and water dynamics in different climatic zones. *Global Change Biology*, 14, 9, 1986-1999

Luo, Y. Q.; Hui, D. F. & Zhang, D. Q. (2006). Elevated CO_2 stimulates net accumulations of carbon and nitrogen in land ecosystems: A meta-analysis. *Ecology*, 87, 1, 53-63

Luo, Y. Q.; Sherry, R.; Zhou, X. H. & Wan, S. Q. (2009). Terrestrial carbon-cycle feedback to climate warming: experimental evidence on plant regulation and impacts of biofuel feedstock harvest. *Global Change Biology Bioenergy*, 1, 1, 62-74

Medlyn, B. E. et al. (2001). Stomatal conductance of forest species after long-term exposure to elevated CO_2 concentration: a synthesis. *New Phytologist*, 149, 2, 247-264

Melillo, J. M. et al. (2002). Soil warming and carbon-cycle feedbacks to the climate system. *Science*, 298, 5601, 2173-2176

Niu, S. L.; Sherry, R. A.; Zhou, X. H.; Wan, S. Q. & Luo, Y. Q. (2010). Nitrogen regulation of the climate-carbon feedback: evidence from a long-term global change experiment. *Ecology*, 91, 11, 3261-3273

Norby, R. J.; Ledford, J.; Reilly, C. D.; Miller, N. E. & O'Neill, E. G. (2004). Fine-root production dominates response of a deciduous forest to atmospheric CO_2 enrichment. *Proceedings of the National Academy of Sciences of the United States of America*, 101, 26, 9689-9693

Norby, R. J.; Wullschleger, S. D.; Gunderson, C. A.; Johnson, D. W. & Ceulemans, R. (1999). Tree responses to rising CO_2 in field experiments: implications for the future forest. *Plant Cell and Environment*, 22, 6, 683-714

Nowak, R. S.; Ellsworth, D. S. & Smith, S. D. (2004). Functional responses of plants to elevated atmospheric CO_2 - do photosynthetic and productivity data from FACE experiments support early predictions? *New Phytologist*, 162, 2, 253-280

Owensby, C. E.; Ham, J. M. & Auen, L. M. (2006). Fluxes of CO_2 from grazed and ungrazed tallgrass prairie. *Rangeland Ecology & Management*, 59, 2, 111-127

Pregitzer, K. S.; Burton, A. J.; Zak, D. R. & Talhelm, A. F. (2008). Simulated chronic nitrogen deposition increases carbon storage in Northern Temperate forests. *Global Change Biology*, 14, 1, 142-153

Reich, P. B.; Hungate, B. A. & Luo, Y. Q. (2006). Carbon-nitrogen interactions in terrestrial ecosystems in response to rising atmospheric carbon dioxide. *Annual Review of Ecology Evolution and Systematics*, 37, 611-636

Rillig, M. C. (2004). Arbuscular mycorrhizae and terrestrial ecosystem processes. *Ecology Letters*, 7, 8, 740-754

Russell, A. E.; Cambardella, C. A.; Laird, D. A.; Jaynes, D. B. & Meek, D. W. (2009). Nitrogen fertilizer effects on soil carbon balances in Midwestern US agricultural systems. *Ecological Applications*, 19, 5, 1102-1113

Rustad, L. E. et al. (2001). A meta-analysis of the response of soil respiration, net nitrogen mineralization, and aboveground plant growth to experimental ecosystem warming. *Oecologia*, 126, 4, 543-562

Sanderson, M. A. (2008). Upland switchgrass yield, nutritive value, and soil carbon changes under grazing and clipping. *Agronomy Journal*, 100, 3, 510-516

Sartori, F.; Lal, R.; Ebinger, M. H. & Parrish, D. J. (2006). Potential soil carbon sequestration and CO_2 offset by dedicated energy crops in the USA. *Critical Reviews in Plant Sciences*, 25, 5, 441-472

Schlesinger, W. H.; Reynolds, J. F.; Cunningham, G. L.; Huenneke, L. F.; Jarrell, W. M.; Virginia, R. A. & Whitford, W. G. (1990). Biological feedbacks in global desertification. *Science*, 247, 4946, 1043-1048

Shaver, G. R. et al. (2000). Global warming and terrestrial ecosystems: A conceptual framework for analysis. *Bioscience*, 50, 10, 871-882

Tans, T. (2011). "Trends in atmospheric carbon dioxide." from http://www.esrl.noaa.gov/gmd/ccgg/trends/global.html.

Tilman, D.; Hill, J. & Lehman, C. (2006). Carbon-negative biofuels from low-input high-diversity grassland biomass. *Science*, 314, 5805, 1598-1600

Trumbore, S. E. (1997). Potential responses of soil organic carbon to global environmental change. *Proceedings of the National Academy of Sciences of the United States of America*, 94, 16, 8284-8291

Vitousek, P. M. & Howarth, R. W. (1991). Nitrogen limitation on land and in the sea-how can it occur. *Biogeochemistry*, 13, 2, 87-115

Vivanco, L. & Austin, A. T. (2006). Intrinsic effects of species on leaf litter and root decomposition: a comparison of temperate grasses from North and South America. *Oecologia*, 150, 1, 97-107

Wang, D. A. N.; Lebauer, D. S. & Dietze, M. C. (2010). A quantitative review comparing the yield of switchgrass in monocultures and mixtures in relation to climate and management factors. *GCB Bioenergy*, 2, 1, 16-25

Weltzin, J. F. et al. (2003). Assessing the response of terrestrial ecosystems to potential changes in precipitation. *BioScience*, 53, 10, 941-952

Wilhelm, W. W.; Johnson, J. M. E.; Karlen, D. L. & Lightle, D. T. (2007). Corn stover to sustain soil organic carbon further constrains biomass supply. *Agronomy Journal*, 99, 6, 1665-1667

Xia, J. Y. & Wan, S. Q. (2008). Global response patterns of terrestrial plant species to nitrogen addition. *New Phytologist*, 179, 2, 428-439

Xue, X.; Luo, Y.; Zhou, X.; Sherry, R. & Jia, X. (2011). Climate warming increases soil erosion, carbon and nitrogen loss with biofuel feedstock harvest in tallgrass prairie. *GCB Bioenergy*, DOI: 10.1111/j.1757-1707.2010.01071.x

Zak, D. R.; Pregitzer, K. S.; King, J. S. & Holmes, W. E. (2000). Elevated atmospheric CO_2, fine roots and the response of soil microorganisms: a review and hypothesis. *New Phytologist*, 147, 1, 201-222

Zhou, X.; Talley, M. & Luo, Y. (2009). Biomass, litter, and soil respiration along a precipitation gradient in southern great plains, USA. *Ecosystems*, 12, 8, 1369-1380

Zhou, X.; Wan, S. Q. & Luo, Y. Q. (2007). Source components and interannual variability of soil CO_2 efflux under experimental warming and clipping in a grassland ecosystem. *Global Change Biology*, 13, 4, 761-775

Zuazo, V. H. D. & Pleguezuelo, C. R. R. (2008). Soil-erosion and runoff prevention by plant covers. A review. *Agronomy for Sustainable Development*, 28, 1, 65-86

6

Biofuel Combustion Emissions - Chemical and Physical Smoke Properties

Chuen-Yu Chan[1], Guenter Engling[2], Xuefang Sang[1] and Ting Zhang[1]
[1]*School of Environmental Science and Engineering, Sun Yat-Sen University, Guangzhou,*
[2]*Department of Biomedical Engineering and Environmental Sciences, National Tsing Hua University, Hsinchu*
[1]*China*
[2]*Taiwan*

1. Introduction

Biofuels have recently gained much attention, mainly as alternative fuels for applications in energy generation and transportation. The utilization of biofuels in such controlled combustion processes has the great benefit of not further depleting the limited resources of fossil fuels, yet it is associated with emissions of greenhouse gases and smoke particles similar to traditional combustion processes, i.e., those of fossil fuels. On the other hand, a vast amount of biofuels is subject to combustion in small-scale processes, such as for heating and cooking in residential dwellings, as well as in agricultural operations, such as for crop residue removal and land clearing. In addition, large amounts of biomass are consumed annually during forest and savanna fires in many parts of the world. These types of burning processes are typically uncontrolled and unregulated. Consequently, the emissions from such processes may be substantially larger compared to industrial-type operations. Aside from direct effects on human health, especially due to a sizeable fraction of the smoke emissions remaining inside residential homes, the smoke particles and gases released from uncontrolled biofuel combustion impose significant effects on regional and global climate. Estimates have shown the majority of carbonaceous airborne particulate matter to be derived from the combustion of biofuels and biomass. The resulting "clouds" of carbonaceous aerosol particles nowadays span vast areas across the Globe. Aside from the negative health impacts and influence on global climate, these smoke particles affect biogeochemical cycles and regional air quality, which is also associated with severe economic impacts.

Whereas emissions from industrial operations and traffic have been fairly well characterized, smoke released during combustion of biofuels is poorly understood in terms of its chemical composition and physical properties. Biofuel combustion generates smoke particles which are predominantly of carbonaceous nature, consisting of an organic carbon (OC) and an elemental carbon (EC) fraction, the latter of which is at times mistakenly referred to as black carbon (BC) or soot. While the OC and EC fractions can be quantified by various methods, there is a large gap in our knowledge regarding the specific composition of OC in biofuel smoke particles. In fact, OC is composed of thousands of individual organic compounds with a wide range of chemical and physical properties. Recent advances in the speciation of the OC fraction in smoke aerosol generated from biofuel combustion provide some new insights into the chemical and physical characteristics of such particles. For instance, it is now understood that biomass smoke particles contain a sizeable portion of higher molecular weight substances as well as polar organic compounds. However, much effort is still needed to better characterize emissions from biofuel combustion, which has to include source and near-source emissions measurements as well as better characterization of ambient aerosol influenced by biofuel/biomass smoke.

This chapter will give an overview of the current state of knowledge regarding the chemical and physical properties of smoke aerosol released from biofuel combustion, by providing selected key references, and point out future research needs and directions.

2. Biofuel usage in Asia and China

In Asia, biofuel emissions are very substantial and have significant influence on regional air quality. Streets et al. (2003) estimated that the major biofuel emissions in Asia arose from the combustion of woods, animal waste (dung) and agricultural waste, and the high biofuel emission regions were mainly located in central and East China, Southeast Asia, and South Asia by spatial and rural population allocation. The average annual biofuel consumption in Asia was estimated to be 730 Tg from both anthropogenic and natural sources, with 45, 34 and 20% accounted for by forest burning, crop residue open burning and grassland/savanna burning, respectively. When allocated to countries, it was found that China contributed 25%, India 18%, Indonesia 13%, and Myanmar 8% of the total consumption. Regionally, forest fires in Southeast Asia dominated.

Tropical Southeast Asia is an active biofuel emission region as a result of increasing deforestation and agricultural activities (Stott, 1988; Christopher and Kimberly, 1996; Dwyer et al., 1998), including East-Central India and the region comprised by Thailand, Burma, Laos, Cambodia and Vietnam (Christopher and Kimberly, 1996). March and April constitute the intensive burning season in this region (Stott, 1988). The intensive fire activity resulting from burning of agricultural waste and shifting cultivation is clearly reflected by the fire hot spots derived from the Along Track Scanning Radiometer (ATSR) on board of a European Space Agency (ESA) satellite (Figure 1); these fires usually reach their full strength in March or April (Figure 2). The amount of biofuels burned in all tropical Asia is very large, which was estimated at about one-half of the amount burned in tropical America, and about one-third of the amount burned in tropical Africa (Liu et al., 1999).

China has a large rural population whose major energy source has been biofuels (crop residues, fuel woods, etc.) for the last several decades. It is not uncommon to see burning of wood and crop residues in kitchens and stoves in the countryside, and even in the surrounding regions of wealthy areas, such as Guangdong Province and Beijing. In addition, biofuel burning is often used as a convenient way of clearing vegetated areas in China (Figure 3). Based on the crop output data from 2001 to 2005, Yang et al. (2008) estimated that the generated annual average amount of crop residue was 3.04×10^{6} t, and about 43% of this was burned in the field. According to the stastics of Guangdong Province, the annual consumption of fuel wood in Guangdong Province is about 5.13-6.00 Tg, and 30%-40% of the produced straws is used as biofuel. $PM_{2.5}$ mass concentrations derived from rice straw combustion can reach as much as 3557 Tg. There have been several literature reports of biofuel/biomass burning contributions to ambient air in China (Zhang et al., 2008; Zhang et al., 2010; Sang et al., 2011).

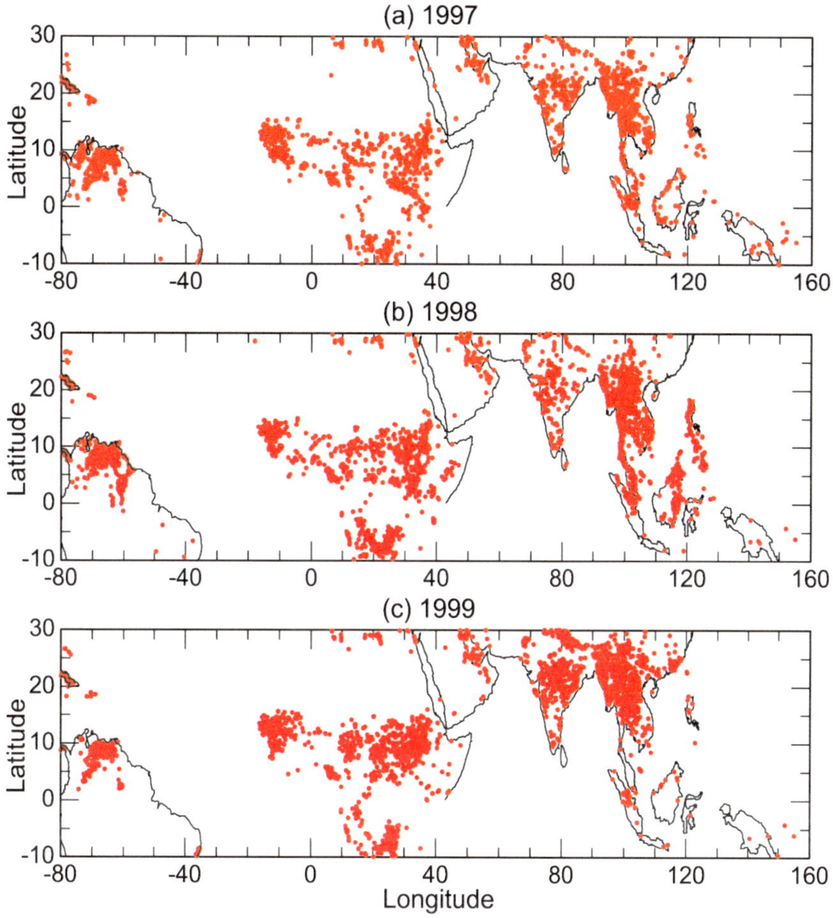

Fig. 1. The geographical distribution of fire hot spots in the tropics derived from ATSR data

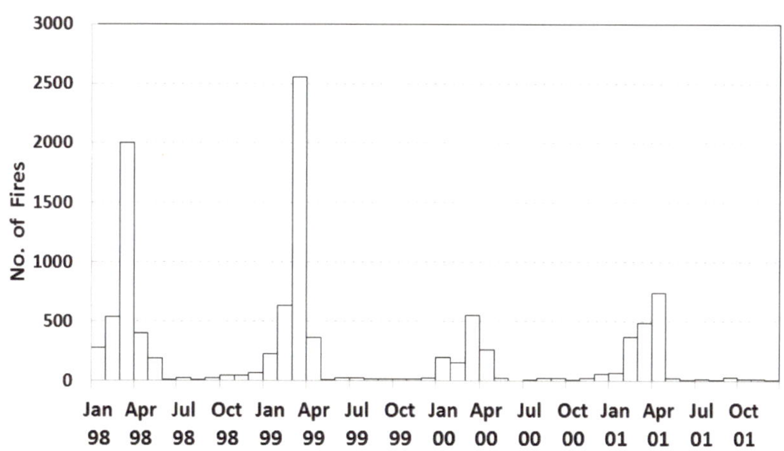

Fig. 2. Monthly variability of fire hot spots in the southeast Asian subcontinent (30 °N, 90 °E - 5 °N, 115 °E)

Fig. 3. Photos showing storage of fuel wood in a typical household in a village of south China (left) and burning of crop residues (right) in a sugarcane field of western Guangdong Province

3. Combustion process

Biomass/biofuel burning can be divided into four types: forest fires, savanna or grassland fires, burning of crop residues in the field and domestic biofuel combustion. Here we refer to biofuels mainly in respect to biomass burned in domestic stoves and ovens for cooking and heat generation, in contrast to biomass that is openly burned on a larger scale, such as in wildland fires. The main structural components of biomass/biofuels are the biopolymers cellulose (40-50%), hemicelluloses (20-30%) and lignins (15-35%) (Sergejewa, 1959; Petterson, 1984). Cellulose, a linear polymer composed of 7000-12000 D-glucose monomers, is the elementary fibrils and could form larger fiber structures (Sergejewa, 1959). Hemicelluloses, consisting of only about 100-200 sugar monomers, are mixtures of polysaccharides derived from glucose, mannose, galactose, xylose, arabinose, 4-O-methylglucuronic acid (4-OMGA), and galacturonic acid (Sergejewa, 1959; Pharham and Gray, 1984) and are less structured than cellulose molecules. The biofuel combustion processes could be summarized as the heating, flaming and smoldering phases. At the heating stage, biofuels are being hydrolyzed, oxidized, dehydrated and pyrolyzed to form tarry substances, volatiles and highly reactive carbonaceous char (Roberts, 1970; Shafizaden, 1984). When reaching the required temperature of the volatiles and tarry substance, the flaming combustion phase commences, which could provide enough energy for the gasification of the biofuel substrate, propagation of the fire and char formation until the combustible volatile flux drops below the minimum level required for the propagation of flaming combustion. Then the smoldering process starts and is best described as the gradual oxidation of the reactive char (solid phase combustion). Table 1 shows the characteristics of various combustion processes during the different combustion phases. The gas and particle-phase chemical species contained in the smoke released during biomass/biofuel include a large number of compounds with a wide range of chemical and physical properties, depending on biofuel type and combustion conditions. As it is beyond the scope of this chapter to give a comprehensive overview of the chemical smoke constituents, the reader is referred to some key literature (Andreae and Merlet, 2001; Hays et al., 2002; Christian et al., 2003; Akagi et al., 2010), while we will focus the discussion here on source-specific compounds, i.e., molecular tracers for biomass/biofuel combustion.

Combustion Stage		Process	Process Characteristics
Flamming Stage	Solid Phase	Drying/Distilling Process	Water and volatile contents are removed or diffused into the inner layers of the bulk material
		Pyrolysis Process	Starts at about 400 K Below 450 K the process is endothermic Above 450 K the process is exothermic Dehydrocellulose decompostion takes place
		Glowing Combustion	Starts at about 800 K if oxygen is present Resulting in char being oxidized
	Gas Phase	The Flame	The emitted volatiles are converted to combustion products of low-molecular weight
Smoldering Stage		Smoldering Process	A low-temperature process Takes place at concentrations of oxygen as low as 5% Can proceed over days under conditions of high moisture

Table 1. Different combustion stages and the characteristics of different combustion processes

4. Molecular tracers for biomass burning processes

During the combustion, the cellulose molecules decompose by two pathways. When the temperature is <300 degrees C, biofuels are depolymerized, fragmented and oxidized to char. During the second pathway, i.e. > 300 degrees C, bond cleavage by transglycosylation, fission and disproportionation reactions give rise to the formation of levoglucosan, accompanied by its stereoisomers, mannosan (Man) and galactosan (Gal). (Simoneit et al., 1999; Schmidl et al., 2008b; Engling et al., 2009; Fabbri et al., 2009). Due to reasonable atmospheric stability with no decay over 10 days in acidic conditions, levoglucosan has been widely used as a molecular marker for biomass burning processes (Fraser and Lakshmanan, 2000), although it could be oxidized when exposed to gas phase hydroxyl radicals (OH) (Hennigan et al., 2010), nitrate (NO_3) or sulfate (SO_4) radicals (Hoffmann et al., 2010). Combustion of other materials (e.g., fossil fuels) or biodegradation and hydrolysis of cellulose does not produce any levoglucosan.

The typical bulk chemical composition of smoke particles derived from agricultural residues/fuels is shown in Figure 4 by the example of rice straw smoke particles. While OC being the predominant species in the carbonaceous fraction, chloride and potassium are the key components in the ionic fraction of smoke aerosol found in source emissions studies. As such, biomass burning plumes are generally characterized by high water-soluble potassium content, specifically enriched in the fine mode. Thus, potassium has also been used as source tracer to estimate the contributions of biomass burning smoke to the ambient aerosol burden (e.g., Duan et al., 2004). However, other sources, such as sea salt, mineral dust and meat cooking, contribute additional potassium to atmospheric PM (Lawson and Winchester, 1979; Morales et al., 1996; Schauer et al., 2002). This may cause a certain bias in the quantitative estimation of contributions from biomass burning emissions when using potassium as source tracer, although a correction for sea-salt contributions is possible. Potassium/levoglucosan ratios which could be utilized for the identification of open/stove fires are below 0.2 for wood combustion in fire places and ovens, while they approach 0.5 for open fires (Fine et al., 2001; Fine et al., 2002; Fine et al., 2004a; Puxbaum et al., 2007).

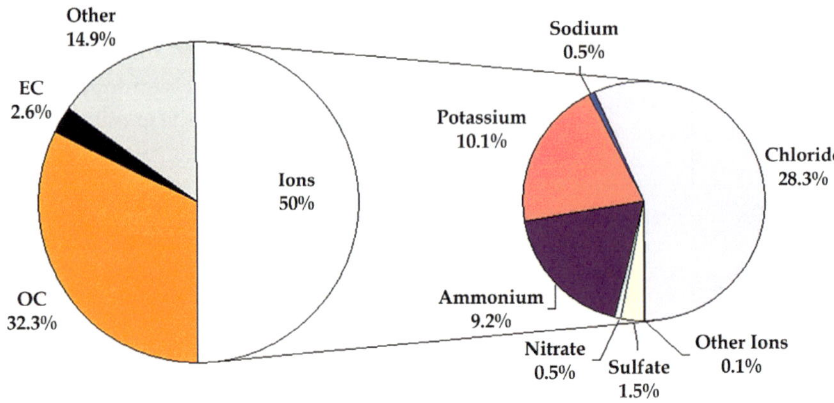

Fig. 4. Typical chemical composition of smoke particles derived from rice straw burning

Table 2 gives a summary of ambient concentrations of levoglucosan, mannosan and galactosan reported for rural, suburban and urban regions around the world. Anhydrosugar concentrations at rural sites have been observed with the highest levels, reaching thousands of ng/m^3, while they were in the hundreds ng/m^3 levels in suburban and urban locations. The variability in these data is mainly influenced by the biofuel usage patterns and potential smoke transport.

Location	**Season**	**Particle Size**	**Levoglucosan ($ng\ m^{-3}$) (Range (ave)/Ave±S.D.)**	**Mannosan ($ng\ m^{-3}$) (Range (ave)/Ave±S.D.)**	**Galactosan ($ng\ m^{-3}$) (Range (ave)/Ave±S.D.)**	**Reference**
Rural	Dry	$PM_{2.5}$	1182 - 6900 (2460)	6.0 - 371 (126)	2.3 - 148 (55.4)	Graham et al., 2002
Rural	Dry	$PM_{2.5}$	40 - 2660 (1180)	1.7 - 127 (49.5)	1.6 – 44.6 (22.7)	Graham et al., 2002
Rural	Dry	$PM_{2.5}$	446 - 4106 (2006)	21 - 259 (116)	7.6 – 61.5 (31)	Zdrahal et al., 2002
Rural	Dry	$PM_{2.5}$	1182 - 6900 (2460)	6 - 371 (126)	2 – 148 (55)	Simoneit et al., 2004
Rural	Dry	$PM_{2.5}$	284 - 7485 (2222)	23.7 - 543 (152)	7.7 - 261 (58.7)	Decesari et al., 2006
Rural	Dry	$PM_{2.5}$	763 - 7903 (3698)	34.0 - 345 (151)	16.4 - 193 (80.3)	Decessari et al., 2006
Suburban	Winter	PM_{10}	134 - 971 (407)	34 - 286 (116)	1 - 7 (2)	Yttri et al., 2007
Suburban	Winter	PM_{10}	232 - 971 (605)	56 - 286 (167)	1.1 - 6.8 (4.0)	Yttri et al., 2007
Suburban	Summer	PM_{10}	n.d. - 151 (47)	n.d. - 42 (10)	n.d. – 7.5 (3)	Yttri et al., 2007

Location	Season	Particle Size	Levoglucosan (ng m^{-3}) (Range (ave)/Ave±S.D.)	Mannosan (ng m^{-3}) (Range (ave)/Ave±S.D.)	Galactosan (ng m^{-3}) (Range (ave)/Ave±S.D.)	Reference
Urban	Winter	PM_{10}	121 - 1133 (477)	17.3 - 153 (66)	4.4 – 44.2 (19.6)	Zdrahal et al., 2002
Urban	Winter	PM_{10}	420	61	25	Pashynska et al., 2002
Urban	Summer	PM_{10}	19.1	3	1	Pashynska et al., 2002
Urban	Winter	PM_{10}	121 - 1133 (477)	17 - 153 (66)	4 - 44 (20)	Simoneit et al., 2004
Urban	Winter	TSP	6 - 56	0.2 - 15	0.6 - 2.4	Simoneit et al., 2004
Urban	Winter	TSP	1162 - 33400 (14460)	154 - 4430 (1422)	84 - 2410 (1014)	Simoneit et al., 2004
Urban	Winter	TSP	1350	108	106	Simoneit et al., 2004
Urban	Winter	PM_{10}	n.d. - 475 (166)	n.d. - 155 (41)	n.d. - 17 (3)	Yttri et al., 2007
Urban	Summer	$PM_{2.5}$	860 - 6090	330 - 1090	130 - 490	Ward et al., 2006
Urban	Fall	PM_{10}	n.d. - 475 (193)	n.d. - 155 (52)	n.d. – 6.9 (1.7)	Yttri et al., 2007
Urban	Yearly	PM_{10}	120 - 160(140)	18 - 44 (31)	5 - 12 (8.5)	Caseiro et al., 2009
Urban	Yearly	PM_{10}	250 - 480 (380)	37 - 114 (84)	14 - 37 (28)	Caseiro et al., 2009
Urban	Yearly	PM_{10}	150 - 220 (193)	27 - 40 (35)	7 - 12 (10)	Caseiro et al., 2009
Urban	Winter	PM_{10}	430 - 1894 (901)	22 - 134 (54)	30 - 186 (96)	Xie et al., 2011
Urban	Spring	PM_{10}	87 - 644 (261)	3.8 - 37 (15)	7.2 - 85 (30)	Xie et al., 2010
Urban	Winter	PM_{1}	422 ± 165	71.2 ± 25.8	19.5 ± 7.67	Krumal et al., 2010
Urban	Winter	$PM_{2.5}$	572 ± 71.3	105 ± 14.1	48.7 ± 2.92	Krumal et al., 2010
Urban	Summer	PM_{10}	15.6 - 472.9			Zhang et al., 2010
Urban	Spring	$PM_{2.5}$	26.2 – 133.7 (36.0)			Sang et al., 2011
Suburban	Spring	$PM_{2.5}$	21.1 – 91.5 (30.0)			Sang et al., 2011

Table 2. Ambient concentrations of anhydrosugars reported in the literature

Biomass type	Combustion type	Location	Particle size	Lev/Man	Lev/Gal	Lev/ (Gal+Man)	Reference
Cereal straw	Chamber burn	China	$PM_{2.5}$	55.7		52.4	Zhang et al., 2007
Rice straw	Chamber burn	Taiwan	$PM_{2.5}$	40	14.0	10.3	Engling et al., 2009
Rice straw	Chamber burn	Bangladesh	$PM_{2.5}$	41.6	25.1	15.6	Sheesley et al., 2003
Sugarcane	Chamber burn	Malaysia	TSP	12.7	12.7	6.4	Oros et al., 2006
Peat	Chamber burn	Sumatra, Indonesia	PM_{10}	11.4	28.1	8.1	Iinuma et al., 2007
Leaves	Open air burning	Lower-Austria	PM10	5.5	1.3	1.0	Schmidl et al., 2008
Pine	Chamber burn	Germany	PM_{10}	3.8	5.0	2.1	Iinuma et al., 2007
Pine	Wildfire	Canada		2.5	10.0	2.0	Otto et al., 2006
Pine	Chamber burn	US	$PM_{2.5}$	3.0	12.6	2.4	Engling et al., 2006a
Spruce	Residential stove	Austria	PM_{10}	3.6	12.6	2.8	Schmidl et al., 2008
White spruce	Residential fireplace	Western US	$PM_{2.5}$	3.9	14.2	3.1	Fine et al., 2004
Douglas fir	Residential fireplace	Western US	$PM_{2.5}$	4.4	22.6	3.7	Fine et al., 2004
Hemlock	Residential fireplace	North-Eastern US	$PM_{2.5}$	3.7	38.7	3.4	Fine et al., 2001
Cottonwood	Chamber burn	US	$PM_{2.5}$	14	23.4	8.7	Engling et al., 2006a
Beech	Residential stove	Austria	PM_{10}	14.6	20.5	8.5	Schmidl et al., 2008
Musasa	Chamber burn	Africa	PM_{10}	22.7	25.0	11.9	Iinuma et al., 2007
White oak	Residential fireplace	Western US	$PM_{2.5}$	12.9	20.4	7.9	Fine et al., 2004
Sugar maple	Residential fireplace	Western US	$PM_{2.5}$	19.8	84.0	16.0	Fine et al.,2004
Red maple	Residential fireplace	North-Eastern US	$PM_{2.5}$	33.2		33.2	Fine et al., 2001
Red oak	Residential fireplace	North-Eastern US	$PM_{2.5}$	35.4	47.7	20.3	Fine et al., 2001

Table 3. The ratios of Lev/Man, Lev/Gal and Lev/(Gal+Man) for various types of biomass

The ratios of levoglucosan to other anhydrosugars in biomass burning smoke particles can be used to identify the specific biomass burning types. For example, levoglucosan to mannosan (Lev/Man) could be used to distinguish the biomass/biofuel types, such as softwood versus hardwood or coniferous versus deciduous wood (Ward et al., 2006; Oliveira et al., 2007; Pio et al., 2008; Schmidl et al., 2008a; 2008b; Engling et al., 2009). Galactosan is usually 10-50 times less abundant in smoke PM than levoglucosan and 1-3 times lower than mannosan levels (Schmidl et al., 2008a). The levoglucosan/galactosan (Lev/Gal) ratio, for example, has been used to distinguish smoke aerosol from leaf and wood burning (Schmidl et al., 2008a). Moreover, levoglucosan to mannosan (Lev/Man) and levoglucosan to mannosan plus galactosan (Lev/(Man+Gal)) ratios were proposed as discriminators of smoke aerosol from lignite and extant biomass due to the lower galactosan content in lignite (Fabbri et al., 2009).

Table 3 summarizes the typical Lev/Man ratios for various biomass/biofuels reported in the literature. Sheesley et al. (2003) reported a Lev/Man ratio in $PM_{2.5}$ for rice straw burning in Bangladesh of 41.6, similar to the ratio of 40 found for rice straw in Taiwan (Engling et al., 2009), while that of mixed cereal straw (wheat, rice and corn) in China was 55.7 (Zhang et al., 2007). Compared with rice straw burning, the ratios of sugarcane and bamboo smoke in the same grass family are relatively low with a range of 5-13, while that of peanuts in the bean family was in the range of sugarcane and bamboo (Oros et al., 2006; Iinuma et al., 2007) (Table 3). The Lev/Man ratio for soft wood (spruce, fir and pine) ranged from 3-6 in the US (Fine et al., 2001; Fine et al., 2004b) and 2.5-4 in Germany, Austria and Canada (Otto et al., 2006; Iinuma et al., 2007; Schmidl et al., 2008b) (Table 3). Fine et al. (2001; 2004b) and Engling et al. (2006a) showed that the Lev/Man ratios for hard wood (oak, maple, beech, cherry and aspen) varied from 13-35, while it was 12.5-22.7 for beech and musasa (Iinuma et al., 2007; Schmidl et al., 2008b) in Austria and Africa (Table 3). Thus, we could conclude that the Lev/Man ratios could be used to at least differentiate soft wood (2-6), hard wood (13-35), and crop residue (40-55.7).

5. Size-resolved composition of biomass burning smoke

The investigation of the size-resolved composition of biofuel burning smoke has recently gained attention in source/near source and ambient studies. Typically, biofuel smoke emissions are characterized by predominantly fine (<2.5 μm aerodynamic diameter) particles (Engling et al., 2006b), which has also been observed in ambient aerosol particles influenced by biomass/biofuel smoke (Wang et al., 2009). For instance, carbonaceous aerosol and biomass smoke markers in particular were found predominantly in submicron particles during a long-range transport episode of wood smoke effecting Yosemite National Park in California, US (Herckes et al., 2006). A temporal variation in PM size distributions suggested a certain dependence on the burning process or atmospheric processing of the smoke particles. In contrast, a substantial mass fraction of the anhydrosugar tracers, including levoglucosan, was recently found in aerosol particles with diameters larger than 10 μm in ambient aerosols (Lee et al., 2008), indicating possible influence by the ambient atmospheric conditions, such as high relative humidity, in addition to unique properties of the biofuel and the specific burning practices.

Likewise, a distinct bimodal distribution was observed with a large fraction of levoglucosan present in a super-coarse mode (>10 μm aerodynamic particle diameter) as well as a fine mode (<0.49 μm aerodynamic particle diameter) in a rice straw field burning study conducted by Engling et al. (2009) (Figure 5). In a more precise size distribution study, Wang et al. (2009) reported that concentrations of particulate matter (PM) mass, n-alkanes, and low molecular weight (LMW) PAHs and levoglucosan showed a unimodal size distribution, peaking at 0.7-1.1 μm during the hazy days impacted by wheat straw burning, and a bimodal distribution, peaking at 0.7-1.1 μm and 4.7-5.8 μm in normal days.

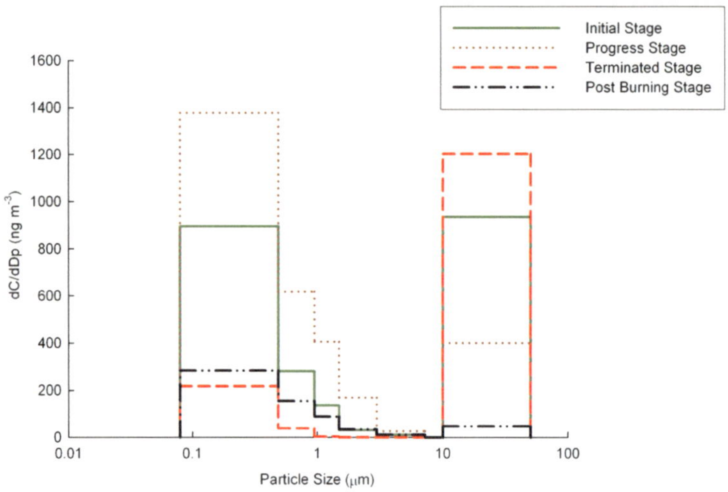

Fig. 5. Levoglucosan size distributions based on 7 particle size ranges (<0.49 μm; 0.49–0.95 μm; 0.95–1.5 μm; 1.5–3 μm; 3–7.2 μm; 7.2–10 μm; and 10–50 μm) in smoke particles generated during field burning of rice straw

6. Chemical analysis methods

Much effort has been put into developing methods for the quantification of biomass burning products and particularly the smoke tracers, such as the anhydrosugars. Both gas chromatographic (GC) and aqueous-phase methods have been reported (Schkolnik and Rudich, 2006). The former methods are the most common ones with good separation and high sensitivity by utilizing mass spectrometric (MS) detectors (Zdrahal et al., 2002), but require complex sample preparation, large amounts of solvents, and expensive equipment. The latter ones, including Electrospray Ionization–Mass Spectrometry (ESI-MS) (Wan and Yu, 2006), Microchip Capillary Electrophoresis (microchip-CE) with Pulsed Amperometric Detection (PAD) (Garcia et al., 2005), Ion-exclusion Chromatography (IEC) (Schkolnik et al., 2005), High Performance Liquid Chromatography (HPLC) (Dye and Yttri, 2005; Dixon and Baltzell, 2006), and High Performance Anion Exchange Chromatography (HPAEC) coupled with PAD or MS (Engling et al., 2006a), have been developed more recently and are, therefore, at present applied less frequently for the quantification of levoglucosan and other biomass/biofuel combustion products. However, these methods are rapidly gaining attention due to their speed and no need for chemical derivatizations (Ma et al., 2010). The IEC-HPLC-PDA method, for instance, is suitable for measuring levoglucosan, inorganic ions and carboxylic acids in a large set of water-extracted aerosols or aqueous samples. HPLC-ESI-MS has been shown to completely separate levoglucosan from its isomers in concentrations ranging from background to polluted levels with short sample preparation, good separation and high sensitivity. However, for detailed organic speciation of smaller sets of samples, GC-MS analysis remains the method of choice to date.

Recently, stable carbon isotope analysis is emerging as a powerful tool to provide additional constraints on the atmospheric budgets, and to increase our understanding of source emissions and ambient aerosols influenced by biomass burning (Goldstein and Shaw, 2003; Huang et al., 2006) and secondary formation processes (Fisseha et al., 2009a). Stable carbon isotopic composition can be determined for both bulk material (e.g., total carbon) and for individual compounds (Hoefs, 1987; Flanagan et al., 2005). However, until recently few studies have applied stable isotope measurements to atmospheric chemistry and particularly for biomass burning aerosols (Rudolph, 2007). The measurement of isotopic ratios for the biomass burning tracer levoglucosan is still not explored because of the high polarity of the sugars and the resulting difficult separation. Martinelli et al. (2002) determined the bulk stable carbon isotopic composition of organic matter in aerosols in order to assess sugar cane sources. Rudolph et al. (1997) and Iannone et al. (2007) presented a new method named gas chromatography coupled to isotope ratio mass spectrometry (GC-C-IRMS) to determine the isotopic ratio of volatile organic carbons (VOCs). Fisseha et al. (2009a) determined the $\delta^{13}C$ values of formic, acetic and oxalic acid in ambient gas and aerosol phases using a wet oxidation method followed by isotope ratio mass spectrometry. The first chamber study of investigating the stable carbon isotopic composition of secondary organic aerosol (SOA) formed from ozonolysis of β-pinene was conducted by Fisseha et al. (2009b). As for biomass burning aerosols, O'Malley et al. (1997) and Czapiewski et al. (2002) determined the isotopic composition of the non-methane hydrocarbons in emissions from biomass burning by using a GC-MS/C/IRMS system.

7. Impact of biomass burning smoke

The influence of smoke emissions from biomass/biofuel burning on the immediate surroundings and on areas downwind of the fire activity can be manifold. In this section, findings from several case studies are used to demonstrate the significant impacts that can be exerted by biomass smoke particles. The importance of the impact of biomass burning in the tropics on atmospheric chemistry and biogeochemical cycles was pointed out in the early 1990s by Curtzen and Andreae (1990). South and Southeast Asia are the two major biomass burning source regions in the world with natural forest fires and human initiated burning activities (Haberle et al., 2001; Pochanart et al., 2003; Radojevic, 2003; Sheesley et al., 2003; Venkataraman et al., 2005; Hasan et al., 2009; Chang and Song, 2010; Ram and Sarin, 2010). Chan et al. (2000) first showed with in-situ sounding measurements, satellite data and trajectory analyses that the frequently observed springtime ozone enhancements in the lower troposphere over Hong Kong were due to photochemical reactions during the transport of ozone precursors originating from the upwind Southeast Asian subcontinent, where intensive biomass burning activities occur during each spring. The enhanced ozone accompanied with a layer of increased biomass burning tracers, such as methyl chloride and carbonaceous aerosol, was shown to further extend to other parts of subtropical south China, the east Asian coast and western Pacific (Chan et al., 2003a,b).

In addition, aircraft and mountain-top measurements have shown that smoke aerosol derived from biomass burning activities in Southeast and East Asia can be transported eastward towards (and across) the Pacific Ocean (Bey et al., 2001; Jacob et al., 2003; Ma et al., 2003b). Ma et al. (2003a) observed biomass burning plumes with enhanced fine particle potassium and CO concentrations originating from Southeast Asia during the experimental period of the Transport and Chemical Evolution over the Pacific (TRACE-P) campaign in

March, 2001. Lin et al. (2010) observed elevated carbon monoxide (CO) mixing ratios in central Taiwan due to biomass burning activities in the Asian continent, including India, the Indochina Peninsula and south Coastal China from January to April 2008. Stohl et al. (2007) predicted that an air pollution plume in the upper troposphere over Europe on 24-25 March 2006 originated from Southern and Eastern Asia with the FLEXPART particle dispersion model. Most recently, it was shown that biomass (rice straw) smoke generated in the Philippines could be transported to southeast coastal China and can contribute to 16-28% of the ambient OC burden in the background atmosphere during spring (Zhang et al., 2011).

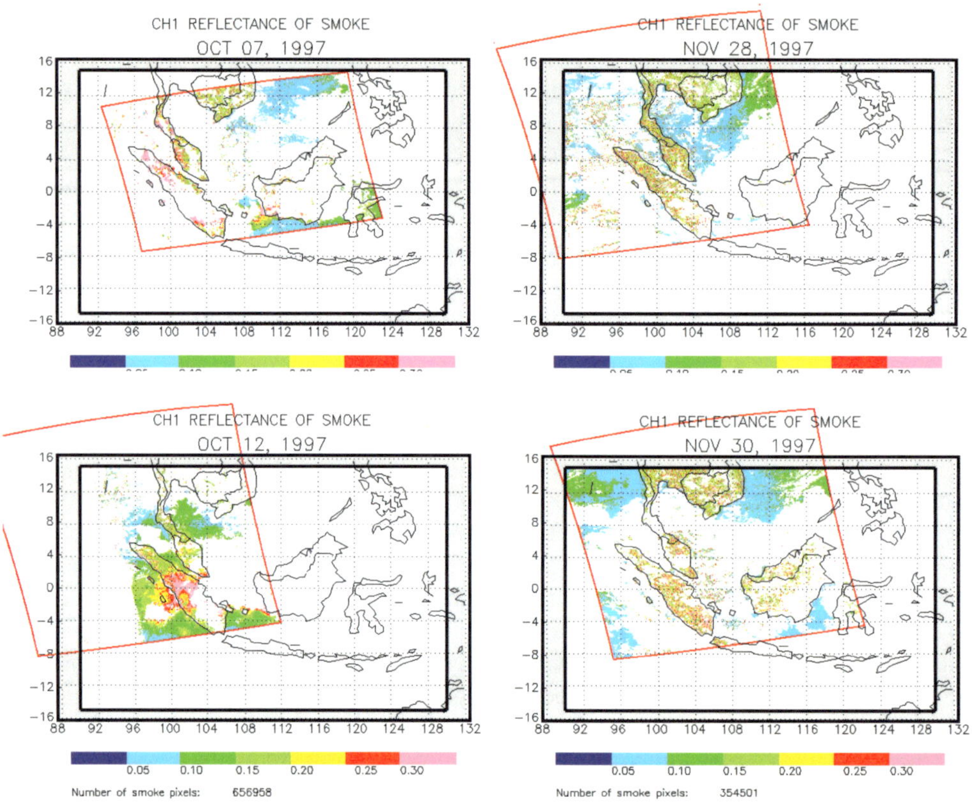

Fig. 6. Smoke pixels estimated from AVHRR on (left) October 7 and 12, and (right) November 28 and 30, 1997 during the Indonesian forest fire period in 1997. The borders indicate the coverage area of the satellite images

During the extreme El Nino period in 1997, when agricultural burning went out of control and resulted in widespread forest fires in Indonesia, Chan et al. (2003b) showed that the smoke aerosol can span over large gographical regions to high latitudes of south China (Figure 6), while Thompson et al. (2001) reported that it can reach longitudially as far as to the Indian Ocean. Chan et al. (2003b) further showed with evidence form in-situ ozonesonde measurements and empirical formulation results that such large-scale biomass burning can result in significant changes in atmospheric composition and radiative forcing in tropical and subtropical Asia and the western Pacific. Furthermore, Wang et al. (2007b) reported that plumes of biomass burning aerosols in South Asia had been extended to the Indian Ocean and the western Pacific Ocean.

The Tibetan Plateau is the largest plateau in the world, which exerts profound effects on the regional and global radiative budget and climate (Lau et al., 2006; Wang et al., 2006). However, scarce data of trace gases and aerosols were observed in this region, let alone biomass burning smoke aerosol. Chan et al. (2006) showed that pollution from active fire regions of Southeast Asia and South Asia had relatively strong impact on the abundance of O_3, trace gases and aerosols in the background atmosphere of the Tibetan Plateau. According to the characteristic levoglucosan/mannosan (Lev/Man) ratios, Sang et al. (2011) identified for the first time that a mountain site in the Tibetan Plateau was affected by long-range transported biomass burning smoke derived from soft wood and crop residue burning in South/Southeast Asia, while a suburban site was mainly affected by local (residential) soft-wood burning. At a remote mountain site in the southeastern part of the Tibetan Plateau during spring, Engling et al. (2011) showed a substantial regional build-up of BC and other aerosol components during the dry period, accompanied by fire activities and transport of pollution from the nearby regions of Southeast Asia and the northern part of the Indian Peninsula (Figure 7). Moreover, BC and aerosol mass concentrations during episodic events were found to be comparable to those reported for certain large Asian cities, mainly due to influence from biomass/biofuel smoke.

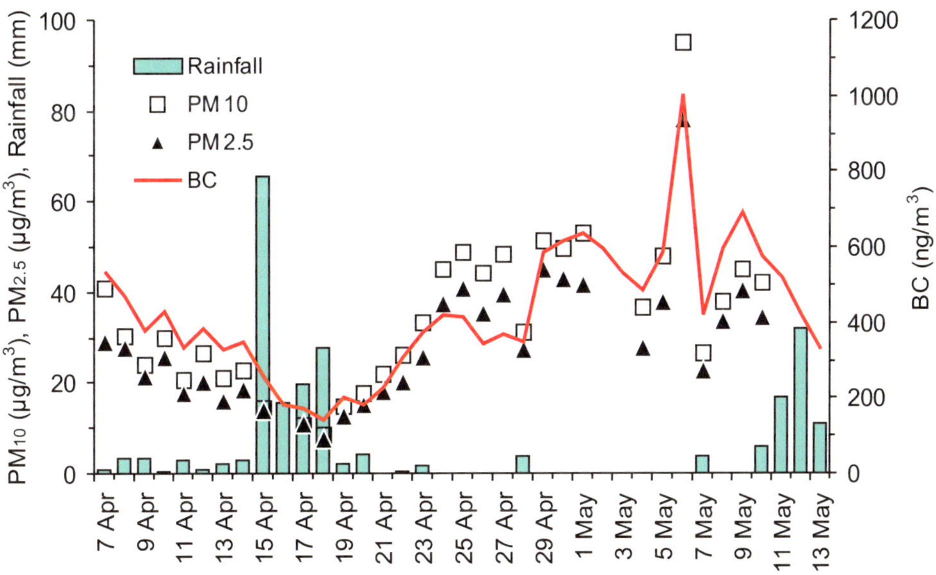

Fig. 7. Daily average concentrations of $PM_{2.5}$, PM_{10}, black carbon and rainfall at a remote mountain site in the southeastern Tibetan Plateau at Tengchong during April-May 2004

In the highly developed Pearl River Delta, biomass smoke contributes a sizeable portion of the ambient aerosol mass as well, as shown by high concentrations of the biomass burning gas-phase tracer CH_3Cl (Chan et al., 2003a). The biomass burning smoke contributions to fine particles were 3-19% (Wang et al., 2007a) and to organic carbon in PM_{10} were 7.0-14% (Zhang et al., 2010) in Guangzhou. Aerosols in Beijing were heavily influenced by different kinds of biofuel burning all year long. The wheat harvest season in summer is the most intensive period, while biomass smoke influence could be detected in spring (due to field preparation burning) and autumn as well (burning of maize residue and fallen dead leaves) (Duan et al., 2004). The contributions from biofuel burning were 18–38% and 14–32% to the $PM_{2.5}$ and PM_{10} organic carbon in Beijing, respectively (Zhang et al., 2008).

8. Conclusions

The combustion of biomass/biofuels for agricultural residue removal and domestic use (for cooking and heating) is a major source of smoke emissions, in addition to large-scale savanna and forest fires, on a global scale. The Asian continent in particular is a major source region of smoke aerosol. As most of these burning processes occur with little/no control and at low combustion efficiency, the amount of smoke emitted and the resulting effects on air quality and global climate are substantial. While importnat advances have been made lately, by conducting detailed source emissions studies and using novel chemical analysis methods for smoke particle characterization, the uncertainty in the estimates of biofuel smoke emissions and their environmental effects remains rather large. It is, therefore, critical to assess the particle-size dependent chemical composition and physical as well as optical properties of biomass/biofuel smoke particles in future source and ambient studies.

9. Acknowledgement

This study was supported through a key project of the Natural Science Foundation of Guangdong Province, China (No. 825102501000002) and the National Natural Science Foundation of China (No. 40875075) and a joint fund of the National Natural Science Foundation of China and Natural Science Foundation of Guangdong Province, China (No. U0833001).

10. References

Akagi, S. K., R. J. Yokelson, C. Wiedinmyer, M. J. Alvarado, J. S. Reid, T. Karl, J. D. Crounse and Wennberg, P. O. (2010). Emission factors for open and domestic biomass burning for use in atomospheric models. *Atmospheric Chemistry and Physics Discussions* 10, pp. 27523-27602, doi: 10.5194/acpd-10-27523-2010.

Andreae, M. O. and Merlet, P. (2001). Emission of trace gases and aerosols from biomass burning. *Global Biogeochemical Cycles* 15, pp. 955-966.

Bey, I., Jacob, D. J., Logan, J. A. and Yantosca, R. M. (2001). Asian chemical outflow to the Pacific in spring: Origins, pathways, and budgets. *J. Geophys. Res.* 106, pp. 23097-23113, doi: 10.1029/2001jd000806.

Caseiro, A., Bauer, H., Schmidl, C., Pio, C. A. and Puxbaum, H. (2009). Wood burning impact on PM10 in three Austrian regions. *Atmospheric Environment* 43, pp. 2186-2195i.

Chan, C. Y. and Chan, L. Y. (2000). Effect of meteorology and air pollutant transport on ozone episodes at a subtropical coastal Asian city, Hong Kong. *Journal of Geophysical Research* 105, pp. 20707-20724, doi: 10.1029/2000jd900140.

Chan, C. Y., Chan, L. Y., Chang, W. L., Zheng, Y. G., Cui, H., Zheng, X. D., Qin, Y. and Li, Y. S. (2003a). Characteristics of a tropospheric ozone profile and implications for the origin of ozone over subtropical China in the spring of 2001. *Journal of Geophysical Research* 108, Doi: 10.1029/2003jd003427.

Chan, C. Y., Chan, L. Y., Harris, J. M., Oltmans, S. J., Blake, D. R., Qin, Y., Zheng, Y. G. and Zheng, X. D. (2003b). Characteristics of biomass burning emission sources, transport, and chemical speciation in enhanced springtime tropospheric ozone profile over Hong Kong. *Journal of Geophysical Research* 108, Doi: 10.1029/2001JD001555.

Chan, C. Y., Wong, K. H., Li, Y. S., Chan, L. Y. and Zheng, X. D. (2006). The effects of Southeast Asia fire activities on tropospheric ozone, trace gases and aerosols at a remote site over the Tibetan Plateau of Southwest China. *Tellus B* 58, pp. 310-318, doi: DOI 10.1111/j.1600-0889.2006.00187.x.

Chang, D. and Song, Y. (2010). Estimates of biomass burning emissions in tropical Asia based on satellite-derived data. *Atmospheric Chemistry and Physics* 10, pp. 2335-2351.

Christian, T. J., Kleiss, B., Yokelson, R. J., Holzinger, R., Crutzen, P. J., Hao, W. M., Saharjo, B. H. and Ward, D. E. (2003). Comprehensive laboratory measurements of biomass-burning emissions: 1. Emissions from Indonesian, African, and other fuels. *Journal of Geophysical Research* 108.

Christopher, D. and Kimberly, E. (1996). Survey of fires in Southeast Asia and India during 1987. *Global Biomass Burning* 2, pp. 663-670.

Crutzen, P. J. and Andreae, M. O. (1990). Biomass Burning in the Tropics - Impact on Atmospheric Chemistry and Biogeochemical Cycles. *Science* 250, pp. 1669-1678.

Czapiewski, K. V., Czuba, E., Huang, L., Ernst, D., Norman, A. L., Koppmann, R. and Rudolph, J. (2002). Isotopic composition of non-methane hydrocarbons in emissions from biomass burning. *Journal of Atmospheric Chemistry* 43, pp. 45-60.

Decesari, S., Fuzzi, S., Facchini, M. C., Mircea, M., Emblico, L., Cavalli, F., Maenhaut, W., Chi, X., Schkolnik, G., Falkovich, A., Rudich, Y., Claeys, M., Pashynska, V., Vas, G., Kourtchev, I., Vermeylen, R., Hoffer, A., Andreae, M. O., Tagliavini, E., Moretti, F. and Artaxo, P. (2006). Characterization of the organic composition of aerosols from Rondonia, Brazil, during the LBA-SMOCC 2002 experiment and its representation through model compounds. *Atmospheric Chemistry and Physics* 6, pp. 375-402.

Dixon, R. W. and Baltzell, G. (2006). Determination of levoglucosan in atmospheric aerosols using high performance liquid chromatography with aerosol charge detection. *Journal of Chromatography A* 1109, pp. 214-221, doi.

Duan, F., Liu, X., Yu, T. and Cachier, H. (2004). Identification and estimate of biomass burning contribution to the urban aerosol organic carbon concentrations in Beijing. *Atmospheric Environment* 38, pp. 1275-1282, doi.

Dwyer, E., Grégoire, J. M. and Malingreau, J. P. (1998). A global analysis of vegetation fires using satellite images: Spatial and temporal dynamics. *Ambio* 27, pp. 175-181.

Dye, C. and Yttri, K. E. (2005). Determination of Monosaccharide Anhydrides in Atmospheric Aerosols by Use of High-Performance Liquid Chromatography

Combined with High-Resolution Mass Spectrometry. *Analytical Chemistry* 77, pp. 1853-1858, doi: 10.1021/ac049461j.

Engling, G., Carrico, C. M., Kreldenweis, S. M., Collett, J. L., Day, D. E., Malm, W. C., Lincoln, E., Hao, W. M., Iinuma, Y. and Herrmann, H. (2006a). Determination of levoglucosan in biomass combustion aerosol by high-performance anion-exchange chromatography with pulsed amperometric detection. *Atmospheric Environment* 40, pp. S299-S311, doi: Doi: 10.1016/j.atmosenv.2005.12.069.

Engling, G., Herckes, P., Kreidenweis, S. M., Malm, W. C. and Collett, J. J. L. (2006b). Composition of the fine organic aerosol in Yosemite National Park during the 2002 Yosemite Aerosol Characterization Study. *Atmospheric Environment* 40, pp. 2959-2972.

Engling, G., Lee, J. J., Tsai, Y.-W., Lung, S. C. C., Chou, C. C. K. and Chan, C. Y. (2009). Size-Resolved Anhydrosugar Composition in Smoke Aerosol from Controlled Field Burning of Rice Straw. *Aerosol Science and Technology* 43, pp. 662-672.

Engling, G., Zhang, Y.-N., Chan, C.-Y., Sang, X.-F., Lin, M., Ho, K.-F., Li, Y.-S., Lin, C.-Y. and Lee, J. J. (2011). Characterization and sources of aerosol particles over the southeastern Tibetan Plateau during the Southeast Asia biomass-burning season. *Tellus B* 63, pp. 117-128, doi: 10.1111/j.1600-0889.2010.00512.x.

Fabbri, D., Torri, C., Simonei, B. R. T., Marynowski, L., Rushdi, A. I. and Fabianska, M. J. (2009). Levoglucosan and other cellulose and lignin markers in emissions from burning of Miocene lignites. *Atmospheric Environment* 43, pp. 2286-2295, doi: DOI 10.1016/j.atmosenv.2009.01.030.

Fine, P. M., Cass, G. R. and Simoneit, B. R. T. (2001). Chemical characterization of fine particle emissions from fireplace combustion of woods grown in the northeastern United States. *Environmental Science & Technology* 35, pp. 2665-2675.

Fine, P. M., Cass, G. R. and Simoneit, B. R. T. (2002). Chemical Characterization of Fine Particle Emissions from the Fireplace Combustion of Woods Grown in the Southern United States. *Environmental Science & Technology* 36, pp. 1442-1451, doi: 10.1021/es0108988.

Fine, P. M., Cass, G. R. and Simoneit, B. R. T. (2004a). Chemical characterization of fine particle emissions from the wood stove combustion of prevalent United States tree species. *Environmental Engineering Science* 21, pp. 705-721.

Fine, P. M., Cass, G. R. and Simoneit, B. R. T. (2004b). Chemical characterization of fine particle emissions from the fireplace combustion of wood types grown in the Midwestern and Western United States. *Environmental Engineering Science* 21, pp. 387-409.

Fisseha, R., Saurer, M., Jaggi, M., Siegwolf, R. T. W., Dommen, J., Szidat, S., Samburova, V. and Baltensperger, U. (2009a). Determination of primary and secondary sources of organic acids and carbonaceous aerosols using stable carbon isotopes. *Atmospheric Environment* 43, pp. 431-437, doi: 10.1016/j.atmosenv.2008.08.041.

Fisseha, R., Spahn, H., Wegener, R., Hohaus, T., Brasse, G., Wissel, H., Tillmann, R., Wahner, A., Koppmann, R. and Kiendler-Scharr, A. (2009b). Stable Carbon Isotope Composition of Secondary Organic Aerosol from b-Pinene Oxidation. *Journal of Geophysical Research* 114, Doi: 10.1029/2008jd011326.

Flanagan, L. B., Ehleringer, J. R. and Pataki, D. E., (2005). *Stable isotopes and biosphere-atmosphere interactions*. Elsevier Academic Press, 0-12-088447-x, California.

Fraser, M. P. and Lakshmanan, K. (2000). Using levoglucosan as a molecular marker for the long-range transport of biomass combustion aerosols. *Environmental Science & Technology* 34, pp. 4560-4564, doi: Doi 10.1021/Es991229l.

Garcia, C. D., Engling, G., Herckes, P., Collett, J. L., Jr. and Henry, C. S. (2005). Determination of Levoglucosan from Smoke Samples Using Microchip Capillary Electrophoresis with Pulsed Amperometric Detection. *Environmental Science & Technology* 39, pp. 618-623.

Goldstein, A. H. and Shaw, S. L. (2003). Isotopes of Volatile Organic Compounds: An Emerging Approach for Studying Atmospheric Budgets and Chemistry. *Chemical Reviews* 103, pp. 5025-5048, doi: 10.1021/cr0206566.

Graham, B., Mayol-Bracero, O. L., Guyon, P., Roberts, G. C., Decesari, S., Facchini, M. C., Artaxo, P., Maenhaut, W., Koll, P. and Andreae, M. O. (2002). Water-soluble organic compounds in biomass burning aerosols over Amazonia - 1. Characterization by NMR and GC-MS. *Journal of Geophysical Research-Atmospheres* 107, Doi 10.1029/2001jd000336.

Haberle, S. G., Hope, G. S. and van der Kaars, S. (2001). Biomass burning in Indonesia and Papua New Guinea: natural and human induced fire events in the fossil record. *Palaeogeography Palaeoclimatology Palaeoecology* 171, pp. 259-268.

Hasan, M., Salam, A. and Alam, A. M. S. (2009). Identification and characterization of trace metals in black solid materials deposited from biomass burning at the cooking stoves in Bangladesh. *Biomass and Bioenergy* 33, pp. 1376-1380.

Hays, M. D., Geron, C. D., Linna, K. J., Smith, N. D. and Schauer, J. J. (2002). Speciation of gas-phase and fine particle emissions from burning of foliar fuels. *Environmental Science & Technology* 36, pp. 2281-2295, doi.

Hennigan, C. J., Sullivan, A. P., Collett, J. L. and Robinson, A. L. (2010). Levoglucosan stability in biomass burning particles exposed to hydroxyl radicals. *Geophysical Research Letters* 37, Doi: 10.1029/2010gl043088.

Herckes, P., Engling, G., Kreidenweis, S. M. and Collett, J. L. (2006). Particle size distributions of organic aerosol constituents during the 2002 Yosemite Aerosol Characterization Study. *Environmental Science & Technology* 40, pp. 4554-4562, Doi: 10.1021/Es0515396.

Hoefs, J., (3rd ed). (1987). *Stable Isotope Geochemistry*. Springer-Verlag, Berlin.

Hoffmann, D., Tilgner, A., Iinuma, Y. and Herrmann, H. (2010). Atmospheric Stability of Levoglucosan: A Detailed Laboratory and Modeling Study. *Environmental Science & Technology* 44, pp. 694-699, doi: Doi 10.1021/Es902476f.

Huang, L., Brook, J. R., Zhang, W., Li, S. M., Graham, L., Ernst, D., Chivulescu, A. and Lu, G. (2006). Stable isotope measurements of carbon fractions (OC/EC) in airborne particulate: A new dimension for source characterization and apportionment. *Atmospheric Environment* 40, pp. 2690-2705.

Iannone, R., Koppmann, R. and Rudolph, J. (2007). A technique for atmospheric measurements of stable carbon isotope ratios of isoprene, methacrolein, and methyl vinyl ketone. *Journal of Atmospheric Chemistry* 58, pp. 181-202, doi: 10.1007/s10874-007-9087-5.

Iinuma, Y., Bruggemann, E., Gnauk, T., Muller, K., Andreae, M. O., Helas, G., Parmar, R. and Herrmann, H. (2007). Source characterization of biomass burning particles: The combustion of selected European conifers, African hardwood, savanna grass, and

German and Indonesian peat. *Journal of Geophysical Research-Atmospheres* 112, Doi: 10.1029/2006jd007120.

Jacob, D. J., Crawford, J. H., Kleb, M. M., Connors, V. S., Bendura, R. J., Raper, J. L., Sachse, G. W., Gille, J. C., Emmons, L. and Heald, C. L. (2003). Transport and Chemical Evolution over the Pacific (TRACE-P) aircraft mission: Design, execution, and first results. *J. Geophys. Res.* 108, Doi: 10.1029/2002jd003276.

Krumal, K., Mikuska, P., Vojtesek, M. and Vecera, Z. (2010). Seasonal variations of monosaccharide anhydrides in PM1 and PM2.5 aerosol in urban areas. *Atmospheric Environment* 44, pp. 5148-5155.

Lau, K. M., Kim, M. K. and Kim, K. M. (2006). Asian summer monsoon anomalies induced by aerosol direct forcing: the role of the Tibetan Plateau. *Climate Dynamics* 26, pp. 855-864, doi: 10.1007/s00382-006-0114-z.

Lawson, D. R. and Winchester, J. W. (1979). Sulfur, Potassium, and Phosphorus Associations in Aerosols from South-American Tropical Rain Forests. *Journal of Geophysical Research - Oceans and Atmospheres* 84, pp. 3723-3727.

Lee, J. J., Engling, G., Candice Lung, S.-C. and Lee, K.-Y. (2008). Particle size characteristics of levoglucosan in ambient aerosols from rice straw burning. *Atmospheric Environment* 42, pp. 8300-8308.

Lin, Y. C., Lin, C. Y. and Hsu, W. T. (2010). Observations of carbon monoxide mixing ratios at a mountain site in central Taiwan during the Asian biomass burning season. *Atmospheric Research* 95, pp. 270-278, doi: 10.1016/j.atmosres.2009.10.006.

Liu, H., Chang, W. L., Oltmans, S. J., Chan, L. Y. and Harris, J. M. (1999). On springtime high ozone events in the lower troposphere from Southeast Asian biomass burning. *Atmospheric Environment* 33, pp. 2403-2410.

Ma, Y., Weber, R. J., Lee, Y. N., Orsini, D. A., Maxwell-Meier, K., Thornton, D. C., Bandy, A. R., Clarke, A. D., Blake, D. R., Sachse, G. W., Fuelberg, H. E., Kiley, C. M., Woo, J. H., Streets, D. G. and Carmichael, G. R. (2003a). Characteristics and influence of biosmoke on the fine-particle ionic composition measured in Asian outflow during the Transport and Chemical Evolution Over the Pacific (TRACE-P) experiment. *J. Geophys. Res.* 108, p. Doi: 10.1029/2002jd003128, doi: 10.1029/2002jd003128.

Ma, Y., Weber, R. J., Lee, Y. N., Orsini, D. A., Maxwell-Meier, K., Thornton, D. C., Bandy, A. R., Clarke, A. D., Blake, D. R., Sachse, G. W., Fuelberg, H. E., Kiley, C. M., Woo, J. H., Streets, D. G. and Carmichael, G. R. (2003b). Characteristics and influence of biosmoke on the fine-particle ionic composition measured in Asian outflow during the Transport and Chemical Evolution Over the Pacific (TRACE-P) experiment. *J. Geophys. Res.* 108, p. 8816, doi: 10.1029/2002jd003128.

Ma, Y., Hays, M. D., Geron, C. D., Walker, J. T. and Gichuru, M. J. G. (2010). Technical Note: Fast two-dimensional GC-MS with thermal extraction for anhydro-sugars in fine aerosols. *Atmospheric Chemistry and Physics* 10, pp. 4331-4341, doi: DOI 10.5194/acp-10-4331-2010.

Martinelli, L. A., Camargo, P. B., Lara, L. B. L. S., Victoria, R. L. and Artaxo, P. (2002). Stable carbon and nitrogen isotopic composition of bulk aerosol particles in a C4 plant landscape of southeast Brazil. *Atmospheric Environment* 36, pp. 2427-2432.

Morales, J. A., Pirela, D. and Durban, J. (1996). Determination of the levels of Na, K, Ca, Mg, Fe, Zn and Cu in aerosols of the western Venezuelan savannah region. *Science of the Total Environment* 180, pp. 155-164.

O'Malley, V. P., Burke, R. A. and Schlotzhauer, W. S. (1997). Using GC-MS/Combustion/IRMS to determine the 13C/12C ratios of individual hydrocarbons produced from the combustion of biomass materials--application to biomass burning. *Organic Geochemistry* 27, pp. 567-581.

Oliveira, C., Pio, C., Alves, C., Evtyugina, M., Santos, P., Goncalves, V., Nunes, T., Silvestre, A. J. D., Palmgren, F., Wahlin, P. and Harrad, S. (2007). Seasonal distribution of polar organic compounds in the urban atmosphere of two large cities from the North and South of Europe. *Atmospheric Environment* 41, pp. 5555-5570.

Oros, D. R., bin Abas, M. R., Omar, N. Y. M. J., Rahman, N. A. and Simoneit, B. R. T. (2006). Identification and emission factors of molecular tracers in organic aerosols from biomass burning: Part 3. Grasses. *Applied Geochemistry* 21, pp. 919-940, doi: DOI 10.1016/j.apgeochem.2006.01.008.

Otto, A., Gondokusumo, R. and Simpson, M. J. (2006). Characterization and quantification of biomarkers from biomass burning at a recent wildfire site in Northern Alberta, Canada. *Applied Geochemistry* 21, pp. 166-183.

Pashynska, V., Vermeylen, R., Vas, G., Maenhaut, W. and Claeys, M. (2002). Development of a gas chromatographic/ion trap mass spectrometric method for the determination of levoglucosan and saccharidic compounds in atmospheric aerosols. Application to urban aerosols. *Journal of Mass Spectrometry* 37, pp. 1249-1257.

Petterson, R. C. (1984). The chemical compostion of wood, In: *Chemistry of solid wood*, Rowell, R. (pp. 57-126), American Chemical Society, Washington, DC.

Pharham, R. A. and Gray, R. L. (1984). Formation and structure of wood, In: *Chemistry of solid wood*, Rowell, R. (pp. 3-56), American Chemical Society, Washington, DC.

Pio, C. A., Legrand, M., Alves, C. A., Oliveira, T., Afonso, J., Caseiro, A., Puxbaum, H., Sanchez-Ochoa, A. and Gelencser, A. (2008). Chemical composition of atmospheric aerosols during the 2003 summer intense forest fire period. *Atmospheric Environment* 42, pp. 7530-7543, doi: DOI 10.1016/j.atmosenv.2008.05.032.

Pochanart, P., Akimoto, H., Kajii, Y. and Sukasem, P. (2003). Carbon monoxide, regional-scale transport, and biomass burning in tropical continental Southeast Asia: Observations in rural Thailand. *Journal of Geophysical Research* 108, Doi: 10.1029/2002jd003360.

Puxbaum, H., Caseiro, A., Sanchez-Ochoa, A., Kasper-Giebl, A., Claeys, M., Gelencser, A., Legrand, M., Preunkert, S. and Pio, C. (2007). Levoglucosan levels at background sites in Europe for assessing the impact of biomass combustion on the European aerosol background. *Journal of Geophysical Research-Atmospheres* 112, Doi 10.1029/2006jd008114.

Radojevic, M. (2003). Chemistry of forest fires and regional haze with emphasis on Southeast Asia. *Pure and Applied Geophysics* 160, pp. 157-187.

Ram, K. and Sarin, M. M. (2010). Spatio-temporal variability in atmospheric abundances of EC, OC and WSOC over Northern India. *Journal of Aerosol Science* 41, pp. 88-98, doi: DOI 10.1016/j.jaerosci.2009.11.004.

Roberts, A. F. (1970). A review of kinetics data for the pyrolysis of wood and related substances. *Combustion and Flame* 14, pp. 261-272.

Rudolph, J., Lowe, D. C., Martin, R. J. and Clarkson, T. S. (1997). A novel method for the compound specific determination of d 13 C in volatile organic compounds at ppt levels in ambient air. *Geophys. Res. Lett* 24, pp. 659-662.

Rudolph, J. (2007). Gas chromatography-isotope ratio mass spectrometry, In: *Volatile Organic Compounds in the Atmosphere,* Koppmann, R. (pp. 388–466), Blackwell Publisher, Oxford.

Sang, X. F., Chan, C. Y., Engling, G., Chan, L. Y., Wang, X. M., Zhang, Y. N., Shi, S., Zhang, Z. S., Zhang, T. and Hu, M. (2011). Levoglucosan enhancement in ambient aerosol during springtime transport events of biomass burning smoke to Southeast China. *Tellus B* 63, pp. 129-139, doi: 10.1111/j.1600-0889.2010.00515.x.

Schauer, J. J., Kleeman, M. J., Cass, G. R. and Simoneit, B. R. T. (2002). Measurement of emissions from air pollution sources. 4. C1-C27 organic compounds from cooking with seed oils. *Environmental Science & Technology* 36, pp. 567-575.

Schkolnik, G., Falkovich, A. H., Rudich, Y., Maenhaut, W. and Artaxo, P. (2005). New Analytical Method for the Determination of Levoglucosan, Polyhydroxy Compounds, and 2-Methylerythritol and Its Application to Smoke and Rainwater Samples. *Environmental Science & Technology* 39, pp. 2744-2752, doi: 10.1021/es048363c.

Schkolnik, G. and Rudich, Y. (2006). Detection and quantification of levoglucosan in atmospheric aerosols: a review. *Analytical and Bioanalytical Chemistry* 385, pp. 26-33, doi: 10.1007/s00216-005-0168-5.

Schmidl, C., Bauer, H., Dattler, A., Hitzenberger, R., Weissenboeck, G., Marr, I. L. and Puxbaum, H. (2008a). Chemical characterisation of particle emissions from burning leaves. *Atmospheric Environment* 42, pp. 9070-9079, doi: DOI 10.1016/j.atmosenv.2008.09.010.

Schmidl, C., Marr, L. L., Caseiro, A., Kotianova, P., Berner, A., Bauer, H., Kasper-Giebl, A. and Puxbaum, H. (2008b). Chemical characterisation of fine particle emissions from wood stove combustion of common woods growing in mid-European Alpine regions. *Atmospheric Environment* 42, pp. 126-141, doi: DOI 10.1016/j.atmosenv.2007.09.028.

Sergejewa, A. S., (1959). *Chemie des holzes und der cellulose.* Theodor Steinkopff Verlag, Dresden.

Shafizaden, F. (1984). The chemistry of pyrolysis and combustion, In: *Chemistry of solid wood,* Rowell, R. (pp. 489-529), American Chemical Society, Washington, DC.

Sheesley, R. J., Schauer, J. J., Chowdhury, Z., Cass, G. R. and Simoneit, B. R. T. (2003). Characterization of organic aerosols emitted from the combustion of biomass indigenous to South Asia. *Journal of Geophysical Research* 108, Doi: 10.1029/2002jd002981.

Simoneit, B. R. T., Schauer, J. J., Nolte, C. G., Oros, D. R., Elias, V. O., Fraser, M. P., Rogge, W. F. and Cass, G. R. (1999). Levoglucosan, a tracer for cellulose in biomass burning and atmospheric particles. *Atmospheric Environment* 33, pp. 173-182.

Simoneit, B. R. T., Kobayashi, M., Mochida, M., Kawamura, K., Lee, M., Lim, H. J., Turpin, B. J. and Komazaki, Y. (2004). Composition and major sources of organic compounds of aerosol particulate matter sampled during the ACE-Asia campaign. *Journal of Geophysical Research* 109, Doi 10.1029/2004jd004598.

Stohl, A., Forster, C., Huntrieser, H., Mannstein, H., McMillan, W. W., Petzold, A., Schlager, H. and Weinzierl, B. (2007). Aircraft measurements over Europe of an air pollution plume from Southeast Asia - aerosol and chemical characterization. *Atmospheric Chemistry and Physics* 7, pp. 913-937.

Stott, P. (1988). The forest as Phoenix: towards a biogeography of fire in mainland South East Asia. *The Geographical Journal* 154, pp. 337-350.

Streets, D., Yarber, K., Woo, J. and Carmichael, G. (2003). Biomass burning in Asia: Annual and seasonal estimates and atmospheric emissions. *Global Biogeochem. Cycles* 17, p. 1099.

Thompson, A. M., Witte, J. C., Hudson, R. D., Guo, H., Herman, J. R. and Fujiwara, M. (2001). Tropical Tropospheric Ozone and Biomass Burning. *Science* 291, pp. 2128-2132, doi: 10.1126/science.291.5511.2128.

Venkataraman, C., Habib, G., Eiguren-Fernandez, A., Miguel, A. H. and Friedlander, S. K. (2005). Residential biofuels in south Asia: Carbonaceous aerosol emissions and climate impacts. *Science* 307, pp. 1454-1456, doi: DOI 10.1126/science.1104359.

Wan, E. C. H. and Yu, J. Z. (2006). Determination of sugar compounds in atmospheric aerosols by liquid chromatography combined with positive electrospray ionization mass spectrometry. *Journal of Chromatography A* 1107, pp. 175-181.

Wang, G. H., Kawamura, K., Xie, M. J., Hu, S. Y., Cao, J. J., An, Z. S., Waston, J. G. and Chow, J. C. (2009). Organic Molecular Compositions and Size Distributions of Chinese Summer and Autumn Aerosols from Nanjing: Characteristic Haze Event Caused by Wheat Straw Burning. *Environmental Science & Technology* 43, pp. 6493-6499, doi: Doi 10.1021/Es803086g.

Wang, Q., Shao, M., Liu, Y., William, K., Paul, G., Li, X., Liu, Y. and Lu, S. (2007a). Impact of biomass burning on urban air quality estimated by organic tracers: Guangzhou and Beijing as cases. *Atmospheric Environment* 41, pp. 8380-8390.

Wang, S. H., Lin, N. H., Chou, M. D. and Woo, J. H. (2007b). Estimate of radiative forcing of Asian biomass-burning aerosols during the period of TRACE-P. *Journal of Geophysical Research-Atmospheres* 112, Doi: 10.1029/2006jd007564.

Wang, T., Wong, H. L. A., Tang, J., Ding, A., Wu, W. S. and Zhang, X. C. (2006). On the origin of surface ozone and reactive nitrogen observed at a remote mountain site in the northeastern Qinghai-Tibetan Plateau, western China. *Journal of Geophysical Research* 111, Doi: 10.1029/2005jd006527.

Ward, T. J., Hamilton, R. F., Dixon, R. W., Paulsen, M. and Simpson, C. D. (2006). Characterization and evaluation of smoke tracers in PM: Results from the 2003 Montana wildfire season. *Atmospheric Environment* 40, pp. 7005-7017, doi: DOI 10.1016/j.atmosenv.2006.06.034.

Xie, M., Wang, G., Hu, S., Gao, S., Han, Q., Xu, Y. and Feng, J. (2011). Polar organic and inorganic markers in PM10 aerosols from an inland city of China -- Seasonal trends and sources. *Science of the Total Environment* 408, pp. 5452-5460.

Yang, S., He, H., Lu, S., Chen, D. and Zhu, J. (2008). Quantification of crop residue burning in the field and its influence on ambient air quality in Suqian, China. *Atmospheric Environment* 42, pp. 1961-1969, doi.

Yttri, K. E., Dye, C. and Kiss, G. (2007). Ambient aerosol concentrations of sugars and sugar-alcohols at four different sites in Norway. *Atmospheric Chemistry and Physics* 7, pp. 4267-4279.

Zdrahal, Z., Oliveira, J., Vermeylen, R., Claeys, M. and Maenhaut, W. (2002). Improved method for quantifying levoglucosan and related monosaccharide anhydrides in atmospheric aerosols and application to samples from urban and tropical locations. *Environmental Science & Technology* 36, pp. 747-753, doi: Doi 10.1021/Es015619v.

Zhang, T., Claeys, M., Cachier, H., Dong, S., Wang, W., Maenhaut, W. and Liu, X. (2008). Identification and estimation of the biomass burning contribution to Beijing aerosol using levoglucosan as a molecular marker. *Atmospheric Environment* 42, pp. 7013-7021.

Zhang, Y. X., Shao, M., Zhang, Y.-h., Zeng, L.-m., He, L.-y., Zhu, B., Wei, Y.-j. and Zhu, X.-l. (2007). Source profiles of particulate organic matters emitted from cereal straw burnings. *Journal of Environmental Sciences* 19, pp. 167-175.

Zhang, Y. N., Zhang, Z. S., Chan, C. Y., Engling, G., Sang, X. F., Shi, S. and Wang, X. M. (2011). Levoglucosan and carbonaceous species in the background aerosol of coastal southeast China: case study on transport of biomass burning smoke from the Philippines. *Environmental Science and Pollution Research* (Accepted)

Zhang, Z., Engling, G., Lin, C. Y., Chou, C. C. K., Lung, S. C. C., Chang, S. Y., Fan, S., Chan, C. Y. and Zhang, Y. H. (2010). Chemical speciation, transport and contribution of biomass burning smoke to ambient aerosol in Guangzhou, a mega city of China. *Atmospheric Environment* 44, pp. 3187-3195.

7

Groundwater and Health Implications of Biofuels Production

Rosane C.M. Nobre and Manoel M.M. Nobre
Universidade Federal de Alagoas / IGDEMA
Brazil

1. Introduction

This chapter presents an overview of environmental and health problems associated with ethanol production in large scale in Brazil. Brazil and the United States are the leading producers of biofuels, accounting together for almost 90% of the total worldwide production in 2009 (REN21, 2010). The main biofuel in the United States is currently ethanol derived from corn kernels, whereas in Brazil the main biofuel production is ethanol derived from sugarcane crops (approximately 99%). In both countries, biodiesel derived from soybeans and Brazilian vegetable oils comprise a very small fraction of the total production.

Brazil is the world largest producer of sugarcane and its industry is the most energy-efficient producer of ethanol. As the demand for renewable fuel production worldwide increases, concerns exist about potential environmental impacts such as deforestation, biodiversity, soil erosion and water resources contamination. Sugarcane crops are especially suitable to be mostly cultivated in tropical areas, with natural ecosystems such as tropical forests with biodiversity hotspots. These areas may be replaced by feedstock plantations in the near future. The agricultural areas currently in use for sugarcane plantations occupy around 70.000 km^2 or 7 million ha (União da Indústria de Cana-de Açúcar [UNICA], 2011), more than half in the state of Sao Paulo, which retains approximately 60% of the Brazilian harvest and is responsible for 62% of the ethanol production (Goldemberg et al., 2008).

Sugarcane is also grown in the northeastern Brazil, in areas previously occupied by the Atlantic rainforest, mostly deforested, and in the Cerrado, the largest savanna species in South America with high biodiversity. The Cerrado occupies a great area of the state of Goias, the fourth greater producer of Brazilian sugarcane, in which has been observed the greatest expansion of cultivated areas in 2009, approximately 40% within one year (UNICA, 2011). Higher biofuel demands are also responsible for rainforest loss in Indonesia and other countries.

Despite being self-sufficient in petroleum oil production, almost half of the Brazilian energy, 47% in 2009, comes from renewable resources (Ministry of Mines and Energy [MME], 2010). Biofuels from sugarcane represent 18% of our national energy matrix (MME, 2010), and this figure tends to increase in the following years. In Brazil, this may be attributable to the following factors: i) high demand for sugar and ethanol worldwide due to high energy prices; ii) the development of new vehicle models with gasoline-and-ethanol mixtures (flex fuels) representing 90% of Brazilian cars; iii) the Kyoto protocol which demands an increased reduction in CO_2 emissions; iv) lack of regulatory criteria for land use; v) cheap labor and cheap production, with an average cost of US$ 0.20 per kilogram of sugar or US$ 0.15 cents per liter of ethanol.

2. Environmental concerns

It is not an easy task to quantify the numerous environmental threats/impacts associated with ethanol production and use. Kusiima & Powers (2010) recommend the evaluation of these impacts in terms of their monetary values in order to have a unit of measure. The authors quantify external costs associated with ethanol production from various biomass feedstocks especially corn. The indirect impacts, such as global climate change, greenhouse gas emissions, soil erosion, regional eutrophication are usually addressed with many uncertainties.

A study conducted by the environmental protection agency in California/USA indicates that ethanol from sugarcane provides less air pollution when compared with ethanol from corn or gasoline and emits much less greenhouse gases than other biofuels during the whole life-cycle (Coelho et al., 2006). Considering both direct and indirect effects (i.e., deforestation and other effects), the amount of CO_2 emission per megajoule (MJ) of energy produced is around 96 g from gasoline, 99 g from ethanol (corn) and 72 g from ethanol (sugarcane), as indicated in **Figure 1** (USEPA, 2009). Mathematical models were used to simulate these indirect effects.

As currently debated by environmental parties worldwide, the risk of the Amazon deforestation may play a smaller role in the global scenario of environmental implications of biofuels production. Besides, sugarcane cultures are not suitable for production in that area. In fact, sugarcane crops are moving to areas already deforested by soybeans culture and pasture. Sugarcane plantation currently represents only 2% of agricultural areas (UNICA, 2011) and about 0.8% of the total Brazilian territory. Besides, genetic improvements to sugarcane cultures have allowed its increased production without excessive land-use expansion (Coelho et al., 2006).

The threat to food production as a consequence to biofuels increased market is also of less concern. In Brazil, it is possible to amplify cultivated areas without being a threat to food harvest. It is not a matter of land scarcity but the lack of an effective land use management and the urgent need of better policies to encourage best agricultural practices. The transformation of extensive pasture practice to agriculture land should be a possibility. In the present conditions, there is no competition for land with food in Brazil. On the other hand, biofuels production either from corn or soybean may have a negative impact worldwide, in the near future, on the availability of grain for direct consumption (Tirado et al., 2010). In the United States, about 25% of the domestic corn yield went to ethanol production (Stone et al., 2010) and Mexico has already been feeling the effects of rising US corn, since the price of tortillas suddenly doubled in 2006 (Journal of the American Dietetic Association, nov/07). Recent catastrophic weather and geological events around the globe, such as the 2011 tsunami in Japan, could put even more pressure on the cost of food (Reuters, 03/02/11), an issue that has already brought about protests across the Middle East.

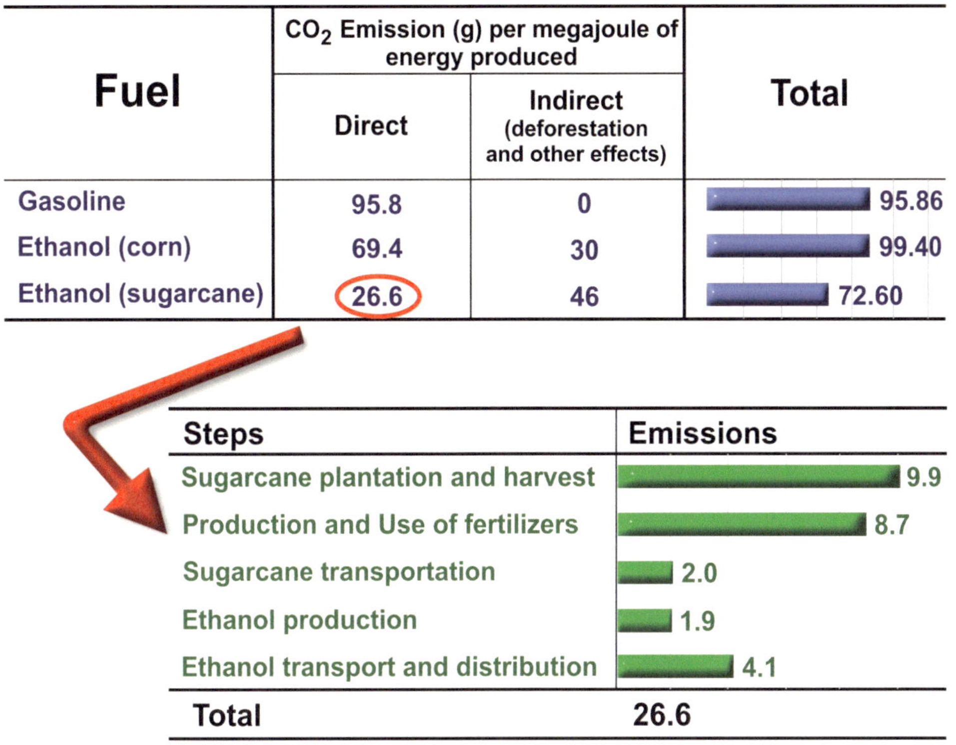

Fig. 1. Ethanol from sugarcane: less air pollution (source: USEPA, 2009)

The major environmental issues related to the sugarcane industry include watershed problems and groundwater contamination due to intensive use of agrochemicals and fertilizers. An additional concern is also deforestation of riparian vegetation and the impacts on streambank stability. Sugarcane plantations demand excess use of water, a precious resource used for many purposes. Despite elevated water availability, some Brazilian groundwater resources are already significantly stressed and vulnerable to contamination (Nobre et al., 2007).

For example, in Sao Paulo, major recharge areas of the Guarani Aquifer, the largest freshwater South American aquifer (1.2 million km^2 or 120 million ha), are located in extensive sugarcane fields (Queiroz et al, 2009). It has been confirmed the presence of groundwater contamination associated with pesticide use in this aquifer in the last few years due to cane plantations. Similarly, large portions of the north American Ogallala aquifer, a vast groundwater reservoir, show water table declines over 30 meters since the 1940s and is disappearing in some areas (Scientific American Earth 3.0, 2009). The Ogallala is three-quarters of the groundwater under the Great Plains region known as the High Plains Aquifer. It is the largest groundwater system in North America and over 90 percent of the extracted water is used for irrigation. In the United States, the demand for corn is such that more land is now being cultivated in drier regions of the Great Plains to the west of the corn belt where intensive irrigation is required, increasing water demand even further (The Economist, Feb 28th 2008).

In this case, the need to expand the cultivated areas to ethanol production is an additional threat, since these cultures require more water than most other crops. Since subsidies for these crops, especially in the United States, are very high (as compared to land conservation), the choice for ethanol production, in much larger scales, has apparently been made. In Brazil, as a consequence to the present efficiency and observed cost reduction, subsidies were fully eliminated by 1997 and the industry relies exclusively on private investments (Goldemberg et al., 2008; Coelho et al., 2006). The good news is that ethanol plants are becoming more efficient and use about half as much water per liter of ethanol as they did a decade ago (The Economist, Feb 28th 2008). Moreover, the residual waste from sugarcane (bagasse) is used to provide electricity heat which results in a very competitive price. In order to be sustainable, biofuels production must preserve natural resources, including water and energy.

3. Water availability and demand

Agriculture is the dominant water user, and increasing the production of biomass feedstocks will certainly compete with food crops, increase water demand and change water resource allocation. Ethanol production requires much more water if compared to other fossil fuels. More than 90% of the required water is used for feedstock irrigation and a very small fraction is used in the processing of bio mass (Berndes, 2002).

Even in Brazil, where water availability is relatively high if compared to other parts of the world, there is a risk that these supplies be further depleted and deteriorated as fuel consumption increases. The United States Department of Agriculture states that about 25% of all irrigation in 2007 was for corn production (United States Department of Agriculture [USDA], 2009). The High Plains Aquifer states are to top corn produces. The natural occurrence of droughts and intensive irrigation to produce corn has caused the dramatic reduction of water levels in most regions of the aquifer (Scientific American Earth 3.0, 2009).

Water use in divided into two parts: crop production and ethanol production. The water requirement for sugarcane production is about 8-12 mm/ton of crop production (Stone et al., 2010). Sugarcane crops are historically grown in areas with total annual precipitation of 1500-2500 mm (Goldemberg et al., 2008) and irrigation is applied only during dry seasons. In Brazil, water usage in irrigation is small, around 3.3 Mha compared to 230 Mha worldwide (Coelho et al., 2006).

The crops, however, do not use all of the rainfall, some infiltrates further and some water evaporates from the soil (and leaves) of the plant during evapotranspiration. As the demand for ethanol production increases, crops are being cultivated in areas not previously suitable for grass growth and irrigation will be essential in these cases. Regardless of all, sugarcane crops, as well as corn crops, require vast amounts of water and sun. University researchers worldwide, however, are investigating new plant varieties, including genetic modified grains, so that they can tolerate lower amounts of water without affecting yields.

The water requirements to produce corn grain are much higher than water required for sugarcane. **Figure 2** depicts these figures, adapted form Stone et al. (2010) comparing three cases: world corn grain, U.S.A. corn (from the state of Nebraska) and world sugarcane production. The graph on the right shows crop water requirements for ethanol production considering conversion of 409 and 334 liters of ethanol per 1 ton of corn grain and sugarcane, respectively, from biomass to ethanol (after Stone et al., 2010). In general, it is needed 4-5 times more water to produce the same amount of ethanol using corn instead of sugarcane.

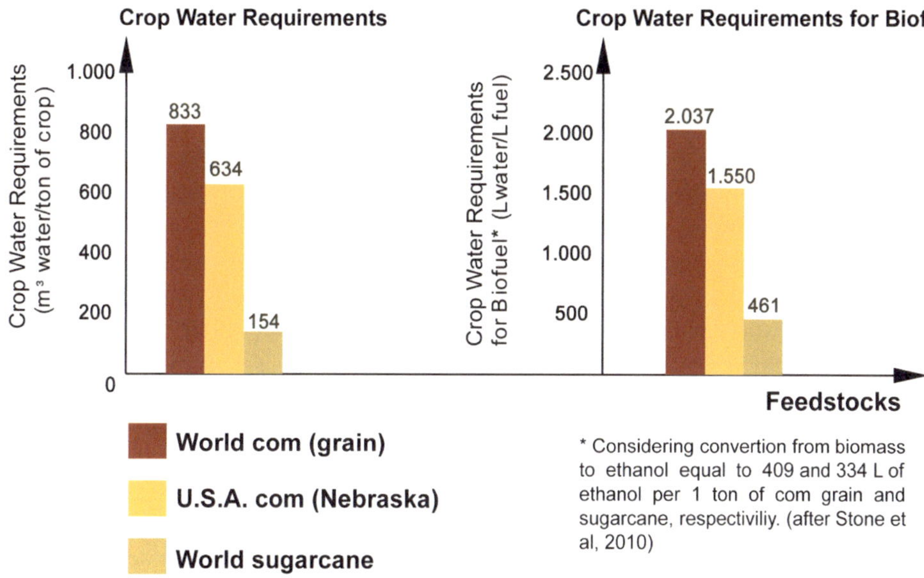

Fig. 2. Crop water requirements for corn and sugarcane cultures

Another advantage from the sugarcane production in Brazil is that farmers that own extensive lands in both northeast and southeast Brazil can concentrate their efforts in planting and harvesting their production in different periods of the year (including the seasonal migration of field workers) due to different weather conditions in both areas. In the southeast, rainy seasons occur during the summer whereas in the northeast, it happen during winter. Moreover, more than 50% of sugar-growing lands are controlled by ethanol refineries. This makes production more optimized and lucrative. However, this model contributes to workers exhaustion and other health and social related-problems. For instance, the Gini coefficient, a measure of social inequality, is very high (0.88) for export oriented crops like sugarcane (Martinelli et al., 2010).

Ethanol production facilities require large amounts of water in processing sugarcane into ethanol even though they have already improved their water efficiency over time. They currently use approximately three liters of water to produce a liter of ethanol. Ten years ago, water consumption was doubled. Some predictions due to emerging technologies simulate a reduction of water usage to two liters within a short timeframe. Water is used for four processes: cane washing, condenser in evaporation and vacuum; fermentation cooling and alcohol condenser cooling (Stone et al., 2010).

4. Water pollution and nitrogen loading

Incorporating biofuel crops into agricultural practices will affect not only water quantity but also water quality. The sugarcane industry is a great pollutant, with serious implications to the environment and human health. These include problems associated with water pollution due to fertilizer and agrochemical loading, inadequate disposal of wastewater from the alcohol and sugar processing plants, soil erosion, among others.

In agricultural areas, groundwater has a distinct water quality signature and is usually composed of nitrate, potassium, chloride, calcium and magnesium. These compounds are originated from fertilizers, animal manure, lime and wastewater/sludge. The presence of these constituents becomes a problem when the amount present is beyond the allowable values that may pose a threat to human health.

The old-established methods of monoculture sugarcane production will probably persist in Brazilian fields, with drastic environmental side effects. Crop rotation is not practiced in these sugarcane fields, causing increased vulnerability to pests, and the need of major inputs of agrochemicals than most crops. In addition, standard agricultural practices demand the use of fertilizers such as nitrogen and phosphorus, which may bring about an increase in the loadings of nitrate to groundwater. Nitrate in contaminated water is known to cause many health problems such as methemoglobinemia in infants as well as stomach cancer in adults (Ward et al., 2005).

In the United Stated, in the "corn belt" of the upper Mississippi river there is the huge problem of nitrogen loss associated with annual corn plantations, due to a shallow rooting system and a short time of active nutrient uptake. This source of nutrient pollution is considered a great contributor to the "dead zone" in the plume of the Mississippi River in the Gulf of Mexico. It is argued that perennial crops such as switch grass add to much less water pollution.

In the Brazilian sugarcane fields, on the other hand, it is a standard practice to apply inorganic fertilizers as well as pesticides at high application rates. Partially treated effluents from sugarcane industry operations are also recycled and re-applied to the fields by sprinkler irrigation techniques or drip irrigation as an effective fertilizer. This waste, denominated vinasse, is nutrient rich, causing eutrophication of ecosystems and polluted runoff when discharged to surface water bodies.

The use of pesticides, which include herbicides and insecticides, is a common practice in the sugarcane industry. The Brazilian agrochemical market is the largest in the world where the major enterprises concentrate 80% of its sales in this country (Bava, 2010). The monocultures of sugarcane alone accounts for about 13% of Brazilian´s herbicide application. Many chemicals already banished in many countries (such as the insecticides endosulfan, malathion, tamaron and gramoxone) are still used in Brazil but is now being under evaluation by the Ministry of Environment and Health.

The use of atrazine, a known endocrine disruptor (ED) compound, is still used in many areas. Low doses of ED can cause developmental harm by interfering with hormonal triggers at key points in the development of an organism. Monteiro et al. (2011), for instance, found high concentrations of herbicides, mostly triazines and hexazinone, in studies of water and sediment toxicity along the Corumbatei river basin in the state of Sao Paulo where sugarcane is cultivated. Table 1 presents a list of all products currently used in Brazilian sugarcane fields, indicating the product names, major active ingredients and related endocrine disrupting effects.

This situation follows "the Circle of Poison" in which pesticides that are banned in industrialized countries continue to be manufactured there but the production is totally exported to developing countries (Galt, 2008). For example, over 25% of the US exports in the 70´s were products that have been prohibited or heavily restricted for use in North American fields. Developing countries, on the other hand, use the imported pesticides mostly on export crops, and return to industrialized countries as pesticide residues on food. More restrictions on pesticide use, however, are being pressured by environmental parties in Brazil but the battle is far from over.

5. Health problems

The individual and environmental factors predisposing workers to illness were identified in this section. Also, there are the indirect health consequences to nearby communities that consume agrochemical contaminated groundwater. The main individual factors related to harvest of the sugarcane industry are the physical effort due to excess work and exhaustion. Among the environmental factors, we can enlist the intense solar radiation, respiratory problems due to smoke breathing and exposure to agrochemicals in air, soil and water. There are also situations of mental suffering and the use of drugs to alleviate pain and stimulate output performance. According to the International Labor Organization, the risk of sugarcane worker deaths at the work place is, at least, two times greater than that of workers in other industries.

Although pesticides have been used in Brazil in large scales over the last 50 years, it is not easy to establish links and connections between human environmental exposure (by different media) and diseases, due to the inherent difficulties in proving the connection. This is mostly the case with carcinogenic compounds and the incidence of cancer, a multifactor disease related to many different risk factors. It is known that either the environmental factors alone or genetic variations alone are not enough to cause cancer (McKelvey et al., 2004).

The indiscriminate use of agrochemicals in Brazil, however, has certainly contributed to the environmental impact and elevated incidence of intoxication, mostly occupational. Levels of contamination have also been detected in the living environment. In many rural areas, the plantation areas are mostly close to workers houses. Another problem is that, in some cases, pesticides are stored at home and contaminated clothing is used indoors exposing the whole family with volatile toxic compounds (Jacobson et al, 2009). This makes occupation exposure a challenging problem. **Figure 3** presents a chart with the increased incidence in cases of intoxication due to pesticides in Brazil (Sistema Nacional de Informações Toxico Farmacologicas [SINITOX], 2011). These figures are underestimated because of sub-notification of cases in the rural areas.

The Brazilian Federal Constitution of 1988 establishes that the government is responsible to carry out actions of sanitary and epidemiological vigilance as well as those related to worker health (Banco Nacional de Desenvolvimento Econômico e Social [BNDES], 2008). This is the responsibility of the government Sole System of Healthcare (SUS). In the present scenario, this system is also responsible for preventive measures to avoid the impacts on the environment created by the intensive production of ethanol. Brazilian government is being successful in many actions related to basic healthcare actions but there is a great challenge to be faced ahead as the ethanol industry will continue to grow in the coming years, creating a substantial demand for new cultivated areas. **Figure 4** below summarizes part of the challenge faced by SUS in the regions where the sugarcane industry is more intense (adapted from BNDES, 2008).

Compound	Products	Class	Toxicological Class	Endocrine Disrupting (ED) Effects [1]
2,4-D	Deferon; Tento 867 SC; U 46 D- Fluid	Selective/ Hormonal Herbicide	I	Synergistic androgenic effects when combined with testosterone.
Ametrine	Ametrina Agripec; Ametron SC; Simetrex SC; Topeze SC	Selective Herbicide / Herbicide (triazines)	III	
Atrazine	Extrazin SC; Gesaprim 500; Siptran 500 SC	Herbicide (triazines) / Selective Herbicide	III	Androgen inhibitor with a weak oestrogenic effect. Disrupts the hypothalamic control of lutenising hormone and prolactin levels. Induces aromatase activity, increasing oestrogen production.
Carbamaryl	Sevin 480 SC	Insecticide	II	
Carbaryl	Carbaryl Fersol 480 Sc	Insecticide (carbamato)	II	Weak oestrogen mimic.
Carbofuran	Furadan 350 SC / TS	Insecticide	I	Acute doses increase levels of progesterone, cortisol and oestradiol whilst decreasing testosterone levels.
Clomazone	Gamit	Herbicide	II	
Cyanazine	Bladex 500	Herbicide (triazines)	III	
Cyclanilide	Finish	Regulator of vegetable growth	I	
Diuron	Advance; Ametron SC; Cention SC; Diuron 500 SC; Fortex SC; Velpar k Grda	Herbicide / Selective Herbicide	II	Inhibits the actions of androgens.
Endosulfan	Disulfan CE, Endosulfan 350 EC; Thiodan CE	Insecticide	I	Antagonises the action of androgens via binding competitively to their receptors and inhibiting the genetic transcription they induce. Mimics the actions of oestrogens indirectly by stimulating the production of their receptors.
Ethephon	Ethephon 480; Ethrel 720; Finish	Regulator of vegetable growth	II	
Ethoxysulfuron	Gladium	Selective Herbicide	III	
Fipronil	Regent 20 g; Regent 800 WG	Insecticide	II	Disrupts the production of thyroid hormones.
Glyphosate	Glifosato 480 Agripec	Systemic Herbicide		
Hexazinone	Advance; Velpar k Grda	Herbicide / Selective Herbicide	III	
Isoxaflutole	Merlin / Karmex DF	Herbicide	II	
Methanearsonic Acid	Volcane	Herbicide	II	
Metribuzin	Sencor	Herbicide	II	Causes hyperthyroidism, alters somatotrophin levels.
MSMA	Daconate 480; Fortex 00	Herbicide / Selective Herbicide	II	
Simazine	Extrazin SC; Simetrex SC; Topeze SC	Herbicide (triazines) / Selective Herbicide	III	Induces aromatase activity, increasing oestrogen production.
Trifluralin	Premerlin 600 CE	Selective Herbicide	II	Interacts with the pregnane X cellular receptor, interfering with themanufacture of enzymes responsible for steroid hormone metabolism.
Trinexapac-ethyl	Moddus	Regulator of vegetable growth	III	

I - Very High Toxicity II- High Toxicity III- Medium Toxicity 1 – McKinlay et al (2008)

Table 1. Pesticides currently used in sugarcane crops in Brazil

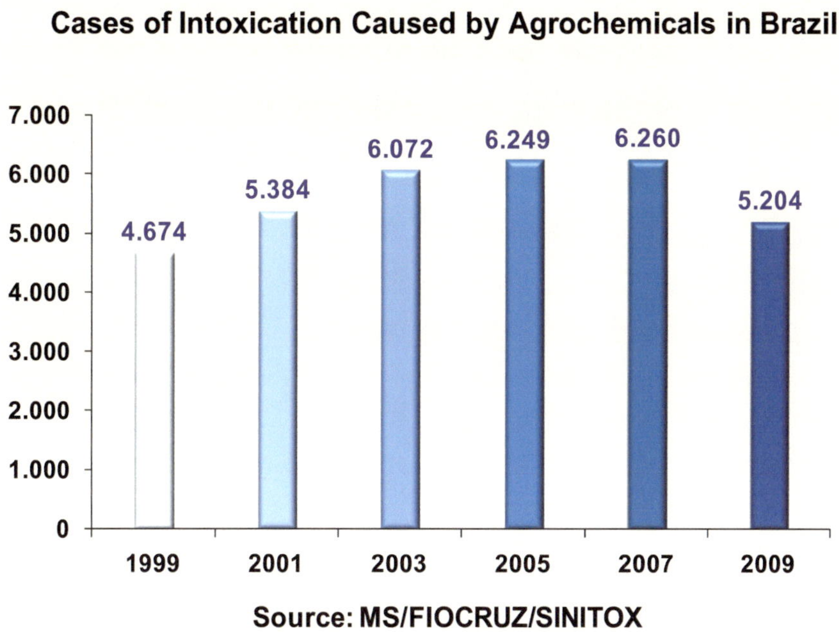

Fig. 3. Incidence in cases of intoxication caused by agrochemicals in Brazil

5.1 Respiratory problems

In many situations of sugarcane production in Brazil, burning the crops (and the sugarcane leaves) is a common practice that precedes harvesting in order to facilitate its cutting by hand. It also increases the sugar content by weight due to water evaporation. Atmospheric pollution due to the presence of particulates and gases from the burning causes many respiratory problems to workers, cardiovascular diseases and lung cancer. It also contributes to acid rain and high nitrogen deposition in soil. An increased number in hospital admissions has been observed due to asthma and other respiratory problems in the last few years. In the state of Sao Paulo, a law has been established in order to stop the burning practice completely (Goldemberg et al., 2008) and it is still under regulation. There is a great pressure to extend this law to other sugarcane producing states in the country.

When biomass burns, incomplete combustion results in the formation of toxic compounds, such as PAH (polycyclical aromatic hydrocarbons) emission, methane and fine particulates. As a consequence, food products derived form sugarcane may contain traces of pesticides, their metabolites and even the presence of PAHs.

The PAHs, such as benzo(a)pyrene, are the most harmful to health and are considered endocrine disruptor compounds. These compounds are found in a variety of food items including sugarcane juice, a common beverage commercialized in many Brazilian cities. Tfouni et al. (2009) investigated levels of PAHs in sugarcane juices for different cities and periods of time and verified that higher concentration levels of these contaminants were registered in juices collected in the harvest period. Also, Bosso et al. (2006) confirmed the present of the substance 1-OHP in the urine of sugarcane workers, a secondary indicator of the presence of PAHs in the organism. Higher concentrations were detected during the harvesting season.

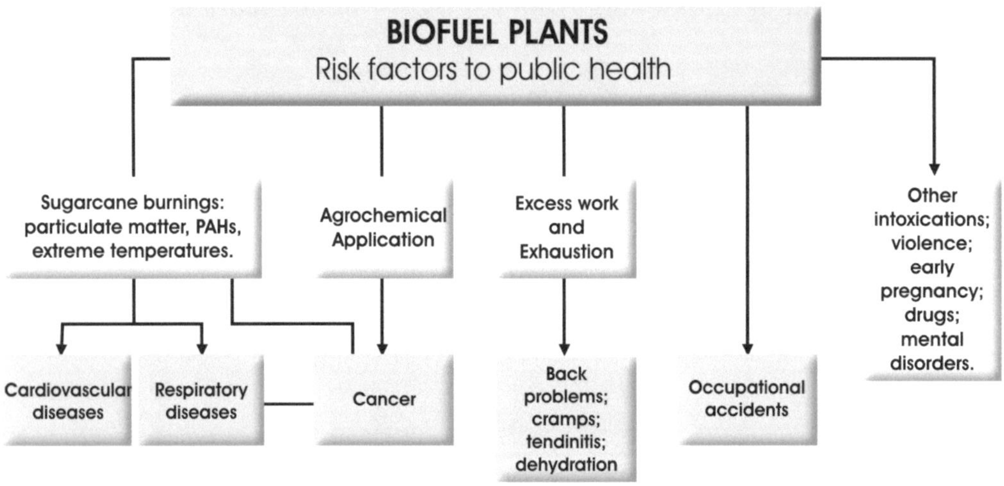

Fig. 4. Health problems related to the sugarcane industry in Brazil

5.2 Health problems due to groundwater contamination

In this section, emphasis is given to groundwater contamination due to nitrate and agrochemicals. Pesticides are generally over used in the sugarcane fields, presenting a serious risk to the environment. Many pesticides have already been confirmed as endocrine disruptors (ED). These compounds have estrogenic activity that may disrupt the hormonal system of mammals, causing birth defects and infertility, diabetes, cancer and even changes in behavior. The Brazilian Ministry of Health and the Environment are currently re-evaluating the use of these compounds.

Potential sources of diffuse contamination are common in agricultural areas and usually in close proximity to the population. Chlorinated organics pesticides can cause cancer by co-carcinogenic process (Vieira et al., 2005). For example, DDT and its metabolites (DDD, DDE) are the substances most cited in the literature for their roles as endocrine disruptors and impacts on human health and the environment (Wolff & Toniolo, 1995). For persistent compounds like DDT, human milk is the most contaminated of all human foods. Although these compounds have been prohibited in many countries, they still have an important role in many hormone-dependent cancers such as breast and prostate. This is possible due to high recalcitrance in soils and groundwater that may persist for many decades. This is also true to other organochlorine pesticides and triazine herbicides.

The herbicide 2,4-dichlorophenoxyacetic acid (2,4-D), still used in sugarcane plantations in Brazil (see **Table 1**), is an endocrine disruptor organophosphate pesticide. Human epidemiological studies have already linked this compound to endocrine related cancers (McKinlay, 2008). The compound diuron, an herbicide commonly present in many pesticides formulas used in sugarcane fields, is known to inhibit the actions of androgens. The insecticide carbaryl, on the other hand, is a weak oestrogen mimic. **Table 1** also includes the known endocrine disrupting effects related to many other pesticide contaminants currently used for sugarcane production in many parts of Brazil such as atrazine, carbofuran, endosulfan, fipronil, metribuzin, simazine and others.

There are studies that indicate that nitrate, derived from nitrogen, a plant nutrient supplied by inorganic fertilizer and animal manure, raises the risk of several types of cancer, especially colon and stomach (Ward et al., 2005; Irigaray et al., 2007). Beneath agricultural lands, nitrate is the primary form of nitrogen. It is soluble in water and can easily pass through soil to the groundwater table. Nitrate can persist in groundwater for decades and accumulate to high levels, as it is very stable in its oxidative form. Infants under six months of age are susceptible to nitrate poisoning in water. The resulting condition is referred to as methemoglobinemia, commonly called "*blue baby syndrome.*" High concentrations of nitrate are a risk factor in developing gastric and intestinal cancer. Due to these health risks, great efforts are made on treatment processes to reduce nitrate concentrations to safe levels. Prevention measures should be applied to avoid the leaching of nitrate from the soil. Some suggest that reducing the amount of fertilizers used in agriculture will help alleviate the problem.

O'Leary et al. (2004) investigated a site contaminated by pesticides on the island of Long Island (NY) and its association with breast cancer incidence. Brody et al. (2006) conducted a similar study with women diagnosed with cancer in the peninsula of Cape Cod (Massachusetts) and the correlation between the etiology of cancer and the exposure to pesticides contaminated groundwater. Nitrate-N was used as the main tracer of contamination levels. The same database was used by Vieira et al., (2008), considering the use of statistical techniques and geographic information system for the visualization of spatial trends of breast cancer, aiming to identify the possible environmental exposure pathways.

The incidence of skin and digestive cancers among a group of rural workers in the central part of Sao Paulo State has also been verified to be correlated with the intensive use of agrochemicals in sugarcane plantations (Stoppelli & Crestana, 2005). The study indicated an almost two fold increase in the probability of cancer incidence among rural workers. Nobre et al., (2011), on the other hand, conducted a quantitative risk analysis related to groundwater contamination in a city located in northeastern Brazil that has a long history of sugarcane monoculture and a high incidence rate of breast cancer. For the last 40 years, the community consumed groundwater as the sole water source. The intensive use of fertilizers and inadequate solid and waste water disposal were considered the main environmental risk factors. The results presented high values for the carcinogenic and non-carcinogenic risk indices.

6. Final remarks

Biofuels are becoming widely used as a viable alternative to petroleum-based fuels. Higher demands for ethanol worldwide are compelling some countries, both developed and developing, to revise their plans in terms of increasing production in order to avoid future shortcomings related to food shortage, threat to biodiversity and environmental degradation.

Although Brazil is the biofuel industry leader, and the most successful and energy-efficient producer of ethanol, many concerns exist in terms of potential environmental impacts including water quality and depletion, health associated problems and social inequity as discussed earlier in this chapter. These are the major restrictions for the sustainable and certified sugarcane production in Brazil, considering the increase in sugarcane industry (and ethanol production) in the following years. These concerns must be addressed by independent parties and better understood based on current scientific knowledge.

Since the first release of the bestselling Silent Spring from Rachel Carson in 1962, there is a consensus that chemical substances in the environment may pose profound effects in animals and that the environmental preservation is inexplicable associated to human health. In her book, chapter 3 (Elixirs of Death), Rachel says "For the first time in the history of the world, every human being is now subjected to contact with dangerous chemicals ...residues of these chemicals linger in soil to which they may have been applied a dozen years before... they have been found in fish in remote mountain lakes, in earthworms burrowing in soil, in the eggs of birds and in man himself.... All this has come about because of the sudden rise and prodigious growth of an industry for the production of manmade or synthetic chemicals with insecticidal properties. This industry is a child of the Second World War." (Carson, 1962). It is hoped that the new generation industry of biofuels production does not cause new environmental impacts as those predicted by Rachel Carson 50 years ago.

7. References

Bava, S.C. Alimentos Contaminados. Le Monde Diplomatique Brasil 2010, ed.33.

Berndes, G. Bioenergy and Water - The implications of large scale bioenergy production for water use and supply. *Global Environmental Change* 12 (2002): 253-271.

BNDES. Impactos da indústria canavieira no Brasil. November 2008. Accessed on February 2011. Available at <http://www.plataformabndes.org.br/>

Bosso, R.M.V., Amorim, L.M.F., Andrade, S.J., Rossini, A., Marchi, M.R.R., Leon, A.P., Carareto, C.M.A., Froes, N.D.T.C. Effects of genetic polymorphisms CYP1A1, GSTM1, GSTT1 and GSTP1 on urinary 1-hydroxypyryne levels in sugarcane workers. *Science of the Total Environment* 370 (2006): 382-390.

Brody, J.G., Aschengrau, A., McKelvey, W., Swartz, C.H., Kennedy, T., Rudel, R.A. Breast cancer risk and drinking water contaminated by wastewater: a case control study. *Environmental Health: A Global Access Science Source* (2006) 5:28.

Carson, R. Silent Spring (1962). Crest Book, 1992, 155p.

Coelho, S.T., Goldemberg, J., Lucon, O., Guardabassi, P. Brazilian sugarcane ethanol: lessons learned. *Energy for Sustainable Development* X-2 (2006): 26-29.

Galt, R.E. Beyond the circle of poison: Significant shifts in the global pesticide complex. 1976-2008. *Global Environmental Change* 18 (2008): 786-799.

Goldemberg, J., Coelho, S.T., Guardabassi, P. The sustainability of ethanol production from sugarcane. *Energy Policy* 36 (2008): 2086-2097.

Irigaray P., Newby J.A., Clapp R., Hardell L., Howard V., Montagnier L., Epstein S., Belpomme D. Life style-related factors and environmental agents causing cancer: an overview. *Biomedicine & Pharmacotherap* 61 (2007):640-658.

Kusiima, J.M., Powers, S.E. Monetary value of the environmental and health externalities associated with production of ethanol from biomass feedstocks. *Energy Policy* 38 (2010): 2785-2796.

Jacobson, L.S.V., Hacon, S.S., Alvarenga, L., Goldstein, R.A., Gums, C., Buss, D.F., Leda, L.R. Comunidade pomerana e uso de agrotóxicos: uma realidade pouco conhecida. *Ciência e Saúde Coletiva* 14-6(2009): 2239-2249.

Martinelli, L.A., Naylor, R., Vitousek, P.M., Moutinho, P. Agriculture in Brazil: impacts, costs, and opportunities for a sustainable future. *Current Opinion in Environmental Sustainability* 2 (2010):431-438.

McKelvey, W., Brody, J.G., Aschengrau, A., Swartz, C.H. Association between Residence on Cape Cod, Massachusetts, and Breast Cancer. *Ann Epidemiol* 14 (2004): 89-94.

McKinlay. R., Plant, J.A., Bell, J.N.B., Voulvoulis, N. Endocrine disrupting pesticides: Implications for risk assessment. *Environment International* 34 (2008): 168-183.

Monteiro, R.T.R., Armas, E.D., Messias, T.G., Falqueto, M.A., Santos, M.A.P.F., Abreu Jr., C.H., Queiroz, S.C.N. Evaluation of herbicides and chemical elements and its relationships with bioessay toxicity of water and sedimento of Corumbatei river, SP, Brazil. *Environmental Health Conference*, Elsevier, 05-09 February 2011, Salvador, Brazil.

Nobre, R.C.M., Rotunno Filho, O.C., Mansur, W.J., Nobre, M.M.M., Cosensa, C.A.N. Groundwater vulnerability and risk mapping using GIS, modeling and a fuzzy logic tool. *Journal of Contaminant Hydrology* 94(2007): 277-292.

Nobre, G.C.M, Nobre, R.C.M., Araújo, M.M.V., Amorim, H.J.C.A.L., Andrade, A.C.M. Breast cancer as an environmental disease in the city of Maceió-AL. *Environmental Health Conference*, Elsevier, 05-09 February 2011, Salvador, Brazil.

O`Leary, E.S. *et al.* Pesticide exposure and risk of breast cancer: a nested case-control study of residentially stable women living on Long Island. *Environmental Research* 94 (2004): 134-144.

Queiroz, S.C.N., Ferracini, V.L., Gomes, M.A.F., Rosa, M.A. Comportamento de herbicida hexazinone em área de recarga do aqüífero Guarani cultivada com cana-de-açúcar. *Quimica Nova* 32:2 (2009): 378-381.

REN21, 2010. Renewables 2010: Global Status Report (Paris: REN21 Secretariat). Available at http://www.ren21.net/REN21Activities/Publications/GlobalStatusReport/

Scientific American Earth. Saving the Ogallala Aquifer. *Scientific American Earth* 3.0, 19-1 (2009): 32-39.

SINITOX, 2011. Sistema Nacional de Informações Tóxico Farmacológicas. Registros de Intoxicações. Available at <http://www.fiocruz.br/sinitox_novo/>

Stone, K.C.; Hunt, P.G., Cantrell. K.B., Ro, K.S. The potential impacts of biomass feedstock production on water resource availability. *Bioresource Technology* 101 (2010): 2014-2025.

Stoppelli, I.M.B.S., Crestana, S. Pesticide exposure and cancer among rural workers from Bariri, São Paulo State, Brazil. *Environment International* 31 (2005): 731-738.

Tfouni, S.A.V., Souza, N.G., Neto, M.B., Loredo, I.S.D., Leme, F.M., Furlani, R.P.Z. Polycyclic aromatic hydrocarbons (PAHs) in sugarcane juice. *Food Chemistry* 116 (2009): 391-394.

Tirado, M.C., Cohen, M.J., Aberman, N., Meerman, J.; Thompson, B. Addressing the challenges of climate change and biofuel production for food and nutrition security. *Food Research International* 43 (2010): 1729-1744.

UNICA, 2011. Dados e Cotações Estatísticas. Accessed on February 2011. Available at <http://www.unica.com.br/dadosCotacao/estatistica>.

USDA, 2009. 2007 Census of Agriculture, Unites States Summary and State Data. Available at http://www.agcensus.usda.gov/Publications/2007/Full_Report/index.asp

USEPA, 2009. California Environmental Protection Agency. Air Resources Board. Accessed on February 2009. Available at http://www.arb.ca.gov/homepage.htm

Vieira, V.M., Aschengrau A., Ozonoff, D. Impact of tetrachloroethylene-contaminated drinking water on the risk of breast cancer: using a dose model to assess exposure in a case-control study. *Environ Health* 4:1 (2005): 3-13.

Vieira, V.M. *et al.* Spatial-temporal analysis of breast cancer in upper Cape-Cod, MA. International Journal of Health Geographics 7 (2008): p.46.

Ward M.H. *et al.* Drinking-water nitrate and health: recent findings and research needs. *Environ Health Perspectives* 113 - 11 (2005): 1607-1614.

Wolff M.S., Toniolo, P.G. Environmental organochlorine exposure as a potential etiologic factor in breast cancer. *Environ Health Perspectives* 103 – 7 (1995):141-145.

8

Biobased Economy – Sustainable Use of Agricultural Resources

S. Kulshreshtha[1], B. G. McConkey[2], T. T. Liu[2], J. A. Dyer[3], X. P. C. Vergé[4] and R. L. Desjardins[5]
[1]*University of Saskatchewan, Saskatoon,*
[2]*Agriculture & Agri-Food Canada, Swift Current,*
[3]*Agro-environmental Consultant, Cambridge, Ontario,*
[4]*Consultant to AAFC, Ottawa, Ontario,*
[5]*Agriculture & Agri-Food Ottawa, Ontario*
Canada

1. Introduction

The biobased economy can be to the 21st century what the fossil-based economy was to the 20th century. Agriculture has the potential to be central to this economy, providing source materials for commodity items such as liquid fuels and value-added products (chemicals and materials). At the same time, agriculture will continue to provide food and feed that are healthful and safe, which may give rise to some situations of trade-offs.

The use of agricultural raw material in a biobased economy is not new. However, now agriculture has to compete with alternative land uses in order to claim the status of socially responsible entrepreneurship. Conservation of valuable landscapes, habitats, biodiversity have come to the forefront of some policy makers' agenda. The public-good benefits that could accrue from the biobased economy are compelling. They include increased security in some countries (such as USA), economic advantages to farmers, industry, rural communities, and society, environmental benefits at the global, regional, and local levels, and other benefits to society in terms of human health and safety.

How should this economy develop so that whatever is done is done well? This question requires examining some of the issues related to sustainability of this economy. Such an investigation has not taken place and thus, there is a need to explore this aspect of the biobased economy. In this chapter, opportunities and challenges facing the bioeconomy are introduced, primarily through a review of the literature. Major concentration of this study is on the agricultural feedstocks for use in the production of liquid transportation fuels, and related products. Some attention is also paid to production of biogas for electricity and heating purposes.

2. Definition of biobased economy

As an alternative, researchers working in the agriculture, forestry, and fisheries sectors recognize the use of biobased products for competing with the fossil-based industry (CARC, 2003), commonly referred to as the 'biobased economy'. This economy uses renewable bio-resources, biological tools, eco-efficient processes that contribute to GHG emission reductions to produce sustainable bioproducts for medical treatments, diagnostics, and more-nutritional foods, energy, chemicals and materials while improving the quality of the environment and standard of living (OECD, 2001). Biobased resources are materials derived from a range of plant systems, and may include starch, sugar, wood, cellulose, lignin, proteins etc. These resources are produced from different sources such as, biomass, crop residue, dedicated crops and crop processing by-product.

The major commodity produced in the biobased economy is energy, in the form of liquid fuels (ethanol and biodiesel) and biogas (Hardy, 2002). The types of energy generated from these products include uses in transportation, heating, electric appliances etc. Agricultural and forest products are generally used in the production of the above biofuels.

Generally, agricultural activity generates a variety of feedstocks for the production of bio-products, particularly bioenergy. Main feedstocks of agricultural activity are from crop biomass including crop residues and livestock waste. Canada, possessing about 67.5 M ha of agricultural farmland, has the potential to offer feedstocks for bioenergy (including biofuels). Of this area, 31.87 M ha are planted each year to grow starch (wheat, barley, corn and oat), oil (rapeseed, soybean and flaxseed) and forage crops (Rye, fodder corn and tame hay), with a total carbon content of about 33.5 Mt C/yr, and an energy content of about 2 exajoules (EJ) yr^{-1} or 2 times 10^{18} J yr^{-1} (Wood & Layzel, 2003). Additionally, agricultural crop residues were estimated to contain about 56 Mt C/year. Although some of this residue may be incorporated into the soil to maintain soil fertility and carbon content, the recoverable portion contains 14.6 Mt C/yr and has an energy potential of 0.52 EJ/yr. To this estimate, one can add livestock wastes in Canada, which could produce over 3 billion m^3 of biogas which is equivalent to energy of 0.065 EJ/yr (Wood & Layzel, 2003).

3. Definition of sustainability

3.1 What is sustainability?

Sustainability is inherently about durability and endurance. The World Commission on Environment and Development defines it as "the capacity to meet the needs of the present without compromising the ability of future generations to meet their own needs" (UNGA, 1987). It emphasizes strategies that promote economic and social development to meet human needs in ways that avoid environmental degradation, overexploitation or pollution (Khanna et al., 2009). At the 2005 World Summit it was noted that this requires the reconciliation of economic, environmental and social demands - the "three pillars" of sustainability (UNGA, 2005). The concept of sustainability is shown in Fig. 1.

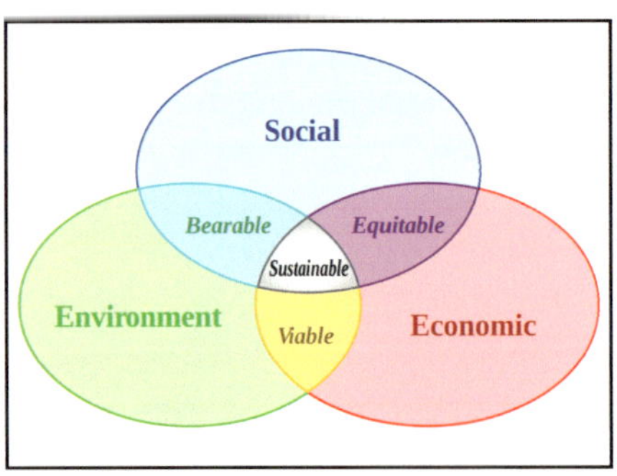

Fig. 1. Framework for Assessment of Sustainability

Figure 1 shows that an economy would be sustainable if it is: (1) Economically viable (uses natural, financial and human capital to create value, wealth and profits); (2) Environmentally compatible (uses cleaner, more eco-efficient products and processes to prevent pollution, depletion of natural resources as well as loss of biodiversity and wildlife habitat), and minimizes damage to the ecosystem services that provide many ecological goods and services to the society; and (3) Socially responsible (behaves in an ethical manner and manages the various impacts of its production through initiatives).

3.2 Sustainability in the context of biobased economy

The biobased economy can contribute to a more sustainable society, not only because it leads to an economy no longer primarily dependent on fossil fuels for energy and industrial raw materials, but also by generating less waste, by a lower energy consumption and by using less water. In addition, the biobased economy provides also for the established industries the opportunity for further growth in a sustainable way (Albrecht et al., 2010). However, does it mean that the production and use of bioenergy is intrinsically sustainable? The Environmental Audit Committee (EAC) found that although biofuels can reduce GHG emissions from road transport, most first generation biofuels have a detrimental impact on the environment overall. In addition, most biofuels are often not an effective use of bioenergy resources, in terms either of cutting GHG emissions or value-for-money (EAC, 2008). Stoeglehner & Narodoslawsky (2009) answered this question from an ecological footprint perspective. They found, by comparing different technologies, that biofuels are considerably more sustainable than fossil options presently in use. Yet, to what extent biofuel use is sustainable remains open as this can only be answered in a regional context taking other land use demands, visions and values into account (Stoeglehner & Narodoslawsky, 2009).
Major utilitarian frameworks define and identify sustainable choices as those that maximize per capita utility subject to an ethical constraint that per capita utility will not decline over time. The utilitarian framework can be applied to derive sustainable outcomes in the context of biofuels, and in particular to identify which biofuels to produce and to what extent, by assuming that utility is derived from the consumption of food, fuel (fossil fuel and biofuel) and other private goods and is maximized subject to budget constraints, land availability and various sustainability constraints. Biofuels would be considered a sustainable substitute if they can compete with fossil fuels in a free market setting at prices that internalize all environmental costs of production, minimize damages to the environment and allow food and other goods and services to be available such that overall utility is non-decreasing over time (Khanna et al., 2009). The production of any type of biofuel is likely to involve trade-offs among these multi-dimensional aspects of sustainability. The degree to which biofuels can accommodate the three pillars of sustainability, taking account of potential tradeoffs among these pillars, needs to be evaluated

3.2.1 Economic sustainability

The economic sustainability of biofuels depends on the costs of production and market price of supply. The sustainability of the corn ethanol industry depends on its ability to deal with volatility in both gasoline and corn prices. Variability in the price of corn could lead to cycles of boom and bust for the biofuel industry with the impact of supply shocks being exacerbated when inventories are low (Hochman et al., 2008). The oil price, commercially viable technology to produce cellulosic biofuels, and trade barriers also affect economic viability of the biofuel industry. The rising oil price has contributed to higher corn prices because of increased cost of production of corn, in addition to its demand. Besides the supply-side considerations, the demand for ethanol and the availability of infrastructure to deliver the ethanol produced to the blenders are the driving forces behind the biofuel industry sustain expansion.

3.2.2 Environmental sustainability

Biofuels are occasionally claimed as being carbon neutral and fossil-fuel free, but serious concerns about the carbon benefits of current biofuels have been raised. Actually, biofuels consume a significant amount of energy that is derived from fossil fuels. Equally important is the fact that production of biofuels has other environmental impacts, such as soil erosion due to tilling, eutrophication due to fertilizer runoffs, impacts of exposure to pesticides, habitat, and biodiversity loss due to land-use change, etc., which have not received the same attention as GHG emissions (Rajagopal & Zilberman, 2007). Conversely, the grain used for ethanol feedstock production is often the poor quality, impure grains which are mostly unsuitable for either human or livestock, and which also do not require as much pesticide (Dyer et al., 2011). In contrast to grain-based ethanol, cellulosic biofuels from perennial grasses (such as switchgrass) have the potential to produce more biofuel per hectare of land and thus have smaller indirect land use effects. While, the environmental benefits of cellulosic biofuels depend on the mix of feedstocks use, the location and management practices used to grow them are equally important. There might also be some trade-offs between environmental benefits and most profitable methods of producing cellulosic feedstocks (Khanna et al., 2009).

3.2.3 Social sustainability

Khanna et al. (2009) consider that the social sustainability of biofuel depends on the distribution of biofuel costs and benefits across countries, income groups, and rural and urban areas. One should keep in mind that human rights, health and equity are also important issues that are related to social sustainability. Higher crop prices in response to increased demand of biofuel will improve farm incomes. However, the higher commodity price may be capitalized into land rent and prices of inputs, which will reduce the future benefit to farmers. Cost of food to consumers may also increase, which may create a heavy burden on the urban poors. The development of biofuel production may also bring to the forefront equity and gender-related issues, such as labour conditions on plantations, constraints faced by small holders and the disadvantaged position of female farmers (FAO, 2008). All of these could affect the welfare of the society and sustainability.

3.3 The criteria and indicators for assessing the sustainability of bioenergy development

An indicator can be used to quantify a specific impact of bioenergy production (e.g. the rate of soil erosion) (Smeets, 2008). Ideally, to evaluate the sustainability of bioenergy use, the impacts of bioenergy production, conversion and trade must be analysed using an integrated approach, taking account of the three dimensions of sustainable development: people (social well-being; the social impacts), planet (maintaining environmental quality; the environmental impact), and profit (economic viability of bioenergy production and its welfare impacts; and other economic impacts). The production and use of bioenergy can only be deemed sustainable if the net impact is positive (Smeets, 2008). Practically applicable criteria and/or indicators are required to monitor and assess the sustainability of bioenergy production and use.

Various ongoing initiatives aim to ensure the sustainability of bioenergy production and use through certification, a form of communication that assures the buyer of bioenergy that the supplier complies with specific sustainability criteria. The European Union and several individual countries, most notably the UK and The Netherlands, are currently developing certification systems. Other countries, for example Brazil, are linking biofuel certification with tax reductions and other incentives to stimulate sustainable bioenergy use. Also, various non-governmental organisations are formulating sustainability criteria.

	Area of concern	Loose set of criteria	Strict set of criteria
Social-economic	Food supply	Energy crop production must not endanger the food supply.	
	Child labour	(Child labour is prohibited.)	Child labour is prohibited
	Wages	Fair wages must be paid to avoid poverty as defined by (inter)national standards.	Fair wages must be paid to avoid poverty as defined by (inter)national standards and to ensure that wages are fair compared to national average.
	Employment	Energy crop production must contribute to employment.	Energy crop production must contribute to employment, including all indirect and induced effect.
	Education	(Education must be provided for workers' children).	Education must be provided for the workers' children by the energy crop producer.
	Healthcare	(Healthcare services must be provided for the all workers' family members).	Healthcare services must be provided for all workers' family members by the energy crop producer.
Environ-mental	Deforestation	Energy crop production must not result in deforestation.	
	Soil erosion	Soil erosion rates must not exceed those due to conventional agriculture land use	Soil erosion rates must not exceed those due to conventional agricultural land use; they must be reduced to match the natural soil-regeneration capacity.
	Depletion of fresh water resources	(Energy crop production must not deplete ground water).	
	Nutrient losses and soil nutrient depletion	Soil nutrient depletion must be prevented as far as reasonably achievable.	Soil nutrient depletion and nutrient leaching must be prevented as far as reasonably achievable.
	Pollution	Agrochemical pollution must be avoided as far as reasonably achievable	
	Biodiversity	Biodiversity must be protected.	

Table 1. Areas of concern and sustainability criteria in Smeets's study, criteria in parentheses are not translated into cost

Smeets (2008) analysed to what extent implementing a sustainability certification system affects the management system (costs) of bioenergy production and availability (quantity) of land for energy plantations. The certification system takes account of twelve sustainability criteria and accompanying indicators (Table 1). However, this certification system lacks the important criterion of "GHG emissions". A project group "Sustainable Production of Biomass" was established in 2006 by the Interdepartmental Programme Management Energy Transition to develop a system for biomass sustainability criteria for the Netherlands for the production and conversion of biomass for energy, fuels and chemistry. A set of generic sustainability criteria and corresponding sustainability indicators was formulated (Table 2) (Cramer et al., 2006).

The need to secure the sustainability of biomass production and trade in a fast growing market is widely acknowledged by many stakeholder groups and setting standards and establishing certification schemes are recognized as possible strategies that help ensure sustainable biomass production and trade (Dam & Junginger, 2008). McBridge et al. (2011) have developed a selection criteria framework for bioenergy sustainability (Fig. 2).

There seems to be a general agreement that it is important to include economic, social and environmental criteria in the development of a biomass certification system. However, mutual differences are also visible in the strictness, extent and level of detail of these criteria, due to various interests and priorities (WWF, 2006) and geographic constraints. The development of biomass certification systems is still in its infancy and largely in development. Therefore, it is worthwhile to consider in this preliminary phase which ways can be followed if the strategy to be taken in the development of a reliable and efficient biomass certification system (Dam & Junginger, 2008).

4. Environmental impacts of biobased economy

Agriculture involves a large human manipulation of the biosphere that impacts the environment. For all the impacts considered, Engstrom et al., (2007) noted that agriculture affects the environment through: eutrophication of water resources, GHG emissions, and loss of biodiversity. On a life cycle analysis basis the impacts are even larger but much of that environmental harm is associated with fossil fuel use. In addition to direct fossil fuel use for agriculture, agriculture production involves further fossil fuel use for energy-intensive inputs like N fertilizers and for transportation of inputs to the farm and products from farm to market (Dyer and Desjardins, 2009).

Bioenergy production is an important existing bioeconomy initiative whose current and potential environmental impacts have been studied extensively. Bioenergy production may cause eutrophication of water, increases ecosystem and human exposure to toxins, causes loss of biodiversity, degrades air quality, and increases acidification of the ecosystem (Bai et al., 2010).

Informed decisions by society require comparative studies of environmental impact of alternatives. For agriculture, the most useful information for decision-makers is not the damage from agriculture to the environment but the comparative measures of environmental harm between food types, production practices, and/or geographical situations. This information facilitates making choices that best balance food need with acceptable environment damage (Brentrup et al., 2004). A similar situation exists for bioenergy. The comparative values of environmental impact between energy sources are required to make sound choices in bionergy (de Vries et al., 2010). Thus, the problem becomes a multi-objective, albeit limited, optimization across the considered alternate energy sources or across considered alternative ways to provide energy-related functions, such as km of passenger travel (European Environment Agency, 2008).

Criterion	Level	Indicator/procedure
1. GHG balance	Net emission reduction ≥50%.	• Testing with the aid of calculation methods. • Use of standard values for different steps in standard chains.
For all the themes below a dialogue with local and national stakeholders is required		
2.Competition with food, local energy supply, medicines and building material	Availability of biomass for food, local energy supply, building materials or medicines must not decrease.	• Comply with minimum requirements testable by means of performance indicators[a].
3. Biodiversity	No deterioration of protected areas or valuable ecosystems. Insight into active protection of the local ecosystem.	• Comply with minimum requirements testable by means of performance indicators[a]. • Reporting obligation on a "management plan for active protection of the local ecosystem".
4. Economic prosperity	No negative effects on the local and regional economy. Insight into the active contribution to the increase of local prosperity.	• Comply with minimum requirements testable by means of performance indicators[a]. • Reporting obligation on the way in which active contribution is made to local prosperity.
5. Well-being 5a Working conditions of workers	No negative effects on the social well-being of the workers and local population	• Comply with Social Accountability 8000 and with the Tripartite Declaration of Principles concerning Multinational Enterprises and Social Policy compiled by the International Labour Organisation.
5b Human Rights		• Comply with the Universal Declaration of Human Rights (concerning: non-discrimination; freedom of association; child labor; forced and compulsory labor; disciplinary practices; security practices and indigenous rights).
5c Property rights and rights of use	Insight into the active contribution to improvement of	• Comply with the following requirements: • No land use without the consent of sufficiently informed original users. Land use is carefully described and officially laid down.

5d Insight into the social circumstances of local population 5e Integrity	social circumstances of local population.	• Official property and use, and customary law of the indigenous population is recognized and respected. • Comply with minimum requirements testable by means of performance indicators[a]. • Reporting obligation in which is described how an active contribution to the social circumstances of the local population is made. Here an open and transparent communication is expected with and, in consultation with, the local population. • Companies in the supply chain comply with the Business Principles for Countering Bribery.
6. The environment (6a) Waste Management (6b) Use of agro-chemicals. (6c) Prevention of erosion and soil exhaustion (6d) Insight into the conservation of quality and quantity of surface and ground water. (6e) Emission to air	No negative effects on the environment.	• Comply with local and national legislation and regulations. • Apply Good Agricultural Practice guidelines on integrated crop management. • Comply with the strictest local, international and EU rules and regulations • Comply with minimum requirements testable by means of performance indicators[a]. • Comply with EU regulations.

Note: [a] These have been developed on the basis of obligatory reports from period 2007-2010.

Table 2. Criteria and indicators for sustainable biomass production for 2011 (Cramer et al., 2006)

4.1 Greenhouse gas emissions

Reducing GHG emissions compared to fossil-fuel alternative is often considered the environmental value of biofuels. Several standards require that biofuels provides GHG emission reductions at least 60% (Zahniser, 2010) lower than those for competing fossil fuel. The estimated GHG benefits of bioenergy are complex, variable, and controversial. Most biofuel production systems provide GHG benefits, typically at least 30% less than fossil fuels (Scharlemann & Laurance, 2008). Some favourable systems such as biodiesel from palm oil and ethanol from sugarcane in Brazil can achieve life-cycle reduction of 50% to 90% (FAO, 2008). Second generation biofuels using biomass crops and crop residues have been estimated to achieve GHG reductions greater than 50%. (Bai et al., 2010) However, some studies argue that the GHG emissions associated with bioenergy production are underestimated and that there is no net GHG savings for many biofuels (Crutzen et al., 2008).

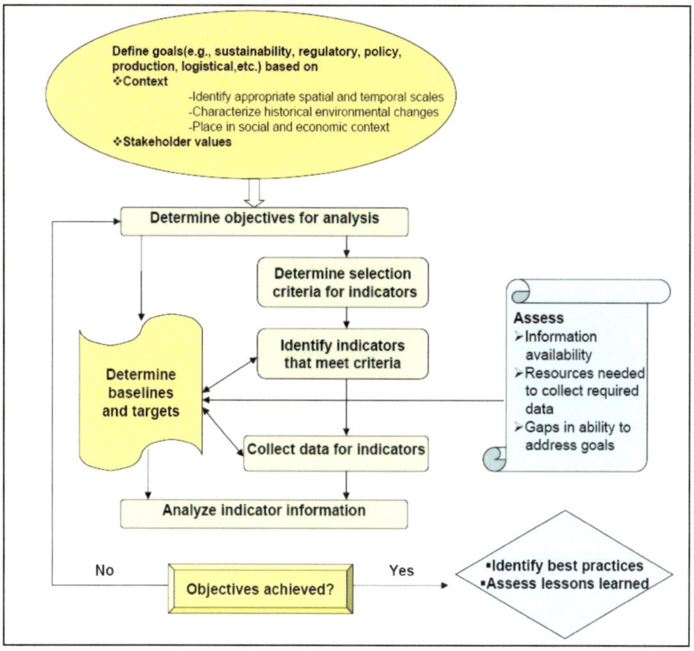

Fig. 2. Framework for Selecting Sustainability Indicators for Bioenergy (adapted from McBridge et al., 2011)

Considering changes in soil carbon associated with crop production can reduce GHG emissions. Where there is an increase in land carbon stocks this reduces net GHG emissions (Adler et al., 2007) and, if the carbon stock change is sufficient, GHG emission can become negative, i.e. a net removal (Brandão et al., 2010).

Searchinger et al. (2008) included indirect land-use change (ILUC) from major increases in ethanol production from US corn. There are large GHG emissions from the land use change, particularly from clearing of forests. They calculated that it would take 150 years of biofuel production before the aggregate GHG emission reductions from ethanol compared to fossil-fuel gasoline are larger than the GHG emission from biofuel-induced ILUC. Fargione et al. (2008) estimated that the GHG effects of ILUC increases the GHG emission for ethanol from US corn by 17 to 420 times. However, the analysis of Searchinger et al. (2008) has attracted criticism that it oversimplifies trade effects, neglects the effect of increases in yield over time, and the use of alternatives pathways to ethanol from feedstock other than corn (Mathews and Tan, 2009).

Kløverpris et al. (2010) used a global trade model to show that land use impact is complex and depends on where feedstock production is taking place. Gains in productivity are more feasible in some regions than others. For example, Denmark has high yield and restrictions on use of fertilizer and pesticides so opportunity for increased production is lower than countries with lower initial yield and fewer restrictions on farming activities. Feasible increases in yield of crops can overcome the ILUC associated with bioenergy. Schmidt et al. (2009) determined that selection of location for sourcing food to replace that lost from bioenergy is important to ILUC effects. For example, exports of Canadian rapeseed oil to Europe would displace palm oil from tropical countries where palm plantations threaten the rain forests in those countries (Klein and LeRoy, 2007). Similarly, by strengthening the market demand for field crops in the Canadian Prairies, the demand for biofuel feedstock will increase the area seeded to crops, rather than left fallow, a practice that is known to increase wind erosion (Dyer et al., 2011).

4.2 Land use and biodiversity

Gomiero et al. (2010) have argued that agreed limits to human appropriation of ecosystem services and global net primary productivity are needed. The world will not be able to support biofuels and food production when loss of agricultural land for transportation, industry, and settlements are considered. Appropriation of net primary productivity beyond the current 50% is unsustainable. They point out that the area impact of biofuel is already much larger than that of fossil fuels considering their relative impacts on energy supply. Fibre and bioenergy needs will exacerbate the pressure on global biodiversity from conventional food production. Bioenergy is a tradeoff between GHG reductions and biodiversity (Schmidt et al., 2009).

Land use impact is not only how much land but also what land and how land is used. Dale et al. (2010) present a potential scenario of increases in biofuel production with increases in biodiversity, mostly through increase production of perennial biomass crops included vegetation mixtures more similar to natural prairies. Solid biofuels for commercial and industrial applications could be an effective and sustainable way to grow the bioeconomy. The use of biomass pellets - which can be produced from wood, switchgrass or straw, would not only create new market oppourtunities for the forest and agricultural industries, it would reduce dependence on coal as well as the GHG emissions associated with coal use. Sophisticated geographical analysis involving land use, habitats, and sensitive ecosystems allows for design of bioenergy production that minimizes potential biodiversity impact (Dragisic et al., 2010). However, Gomiero et al. (2010) note that efficient biofuel production requires monoculture and mechanization for land near the biofuel plants to achieve maximum efficiency. Such production practices could be detrimental to biodiversity.

Bioenergy feedstock production will affect land use which can impact biodiversity to varying degrees, depending on the crop type and the region. Growing grain crops probably has the greatest detrimental impact on biodiversity if these crops are managed more intensively, with increased inputs and fewer rotations (Dyer et al., 2011). Growing perennial herbaceous crops on marginal land can often reduce biodiversity loss compared to using the land for row crops such as corn (Williams et al., 2009). However, Dyer et al. (2011) found that if the marginal land is natural grassland, such as much of the rangeland in Western Canada, rather than the result of land degradation, even a perennial feedstock crop (such as switchgrass) could result in the loss of extensive areas of natural habitat. When cattle are displaced by feedstock crops (**ILUC**), they may be grazed at unsustainable stocking rates or in rangeland not previously used for grazing (Dyer et al., 2011). Good geographic planning of bioenergy development can protect high-carbon high-biodiversity compared to letting market forces determine land use (Schmidt et al., 2009).

4.3 Sustaining land productivity

Crop residues are an attractive feedstock for bioenergy since they do not reduce food production, are available in large quantities, and are relatively low cost. However, crop residue protects the soil from erosion and maintains soil organic matter.

The removal of 20-30% of crop residue is probably sustainable (Gomiero et al., 2010) although residue removal will eventually require additional fertilizer to replace nutrients removed (Wilhelm et al., 2010). The balance between the residue removal rate and long-term soil health is a challenge (Williams et al., 2009).

Soil erosion is affected by crop type and its production practices. Generally, increased bioenergy production increases erosion risk (de Vries et al., 2010). The choice of crops is important, especially if maize replaces grass and forages (Searchinger & Heimlich, 2009). Production practices, such as winter cover crops where appropriate, can mitigate erosion risk (Kim & Dale, 2005).

4.4 Eutrophication

Nutrient loss through runoff leads to eutrophication of water bodies. This is largely a consequence of fertilizing crops for bioenergy feedstock (Dale et al., 2010). Consequently, bioenergy can increase eutrophication compared to fossil fuels even in highly optimized production systems (Cherubini & Jungmeier, 2010). The use of perennial biomass crops for bioenergy feedstocks can decrease contamination of water with nutrients compared to annual crops (Williams et al., 2009). Similarly, removal of crop residue can increase nutrient contamination from surface runoff (Blanco-Canqui et al., 2009).

5. Economic impacts of biobased economy

The economics of biofuels critically depend on the price of fossil fuels, price of feedstocks, the cost of conversion (including investment needs) and the revenues generated by the by-products. Storage, transport and logistic costs also need to be included (Vermeulen & Vorley, 2007). Two major sources of revenue from biofuel production are sale of the fuel, and sale of by-products, which may include dry distiller's grain and sollubles (DDGS), glycerine and carbon dioxide, as well as rapeseed or soybean meal.

Investigations by $(S+T)^2$ & Edna Lam Consulting (2005) for ethanol and biodiesel production suggest that these products cannot compete with fossil-based products without a subsidy. The impact of biofuel production on various sectors of the society is also very different. Benefits are realized by the ethanol industry, but at the cost of state revenues, and consumer expenditures. But with new markets that respond differently than conventional food markets, the rural economy is enhanced (Klein and LeRoy, 2007). Society as a whole benefits from the country's reduced reliance on crude oil imports and reduced economic costs for mitigating GHG emissions (Hardy, 2002; Domac et al., 2005).

5.1 Job creation and rural development

Brazil is one of the examples of successful job creation from bioenergy industry. The bioenergy industry offers direct or indirect employment opportunities[1]. Employment generation from a biofuel plant differs between the two stages: construction stage and operations stage. During the construction phase, employment impacts are large but

[1] Direct employment refers to the creation of employment opportunities from increased biofuel feedstocks production, transportation and construction and operation, maintenance of conversion processing plants. Indirect employment is jobs created through the supporting industries, for example, marketing and distribution of end products from biofuel industries (Domac et al., 2005).

temporary in nature. Plant operation generates fewer but permanent jobs. For example, Haig (2006) estimated that the impact of producing 2 billion litres of ethanol on the rural economy would generate 6,645 jobs in rural Canada.

Urbanchuk (2006) has found that local ownership of biofuel plants maximizes the rural development potential. He estimates that the full contribution to the local economy of a farmer-owned co-operative ethanol plant is likely to be as much as 56 percent higher than the impact of an absentee-owned corporate plant. This is attributed to two main factors unique to farmer owned plants: (1) A larger share of operational expenditures is made in the local community; and (2) The distribution of dividend payments to farmer-owners of a co-operative ethanol plant represents additional income to farmers and their families.

Meanwhile, if a market for selling carbon credits could be established, this would provide another source of revenue to farmers.

5.2 Improved trade balance

The activities associated with the biobased economy such as the expansion of biofuel would cause, in some cases, substantial increase in exports of agriculture commodities (Timilsina et al., 2010) due to a diversified set of agricultural products. In addition, a biobased economy is economically viable in a longer term perspective. In a study of Thailand, although the costs of biofuel production may exceed the cost of importing equivalent petroleum, domestic production of biofuels allows virtually all of the money to stay within the country's economy, and thus, adds to the balance of payment for the country (Bell et al., 2011).

5.3 Establishment of new industries

An increase in feedstock production for biobased industry results in an increased production of by-product and residues that are in turn utilized as raw materials for several other sectors, such as livestock production, cosmetics and pharmaceutical industries, among others (IEA-Bioenergy, 2009). Input providing industries, such as agricultural equipment manufacturing firms and fertilizer industries, will expand to supply additional goods and services to support the increased biomass production activity (Han et al., 2011). Byproducts and inputs can be important criteria for feedstock crop choices. For example, soybean-based biodiesel was shown to have a lower carbon footprint than rapeseed-based biodiesel due to both providing more livestock feed byproduct than rapeseed oil and being a legume that does not require N-fertilizer input (Dyer et al., 2010).

The oil price plays an important role in determining the economics of biofuels (Baker and Zahniser, 2007). If the world oil price remain high, biofuels will be more financially viable even without government support. The remote areas (or countries) usually have the comparative advantage of labor, but due to poor facility and transportation system, prices of oil may be markedly higher than the international prices. In these cases, if biofuel production and processing are located near consumption centers or can be transported to them at relatively low costs, they can be competitive against imported fossil fuels (Vermeulen & Vorley, 2007).

5.4 Fiscal effects of biofuel development

Biofuel development can affect several levels of governments through one or a combination of three pathways: (1) Provision of public subsidies; (2) Generation of new and different sources of government revenues; and (3) Change in government expenditures. Under current fossil based fuel prices, biofuels are not competitive. Many jurisdictions have accepted the need for public subsidies to enhance the public cause. However, biofuel support programs can act as a substitute for other agricultural program subsidies. For example, the U.S. ethanol tax credit, according to Gardner (2003), has served to displace some of the government deficiency payments related to corn. The financial impact on government is likely to include both positive and negative components. There is a cost to government for any incentives provided to the biofuel industry, but there will also be tax revenues that flow to government from the income generated by these operations. Intuitively, if subsidies are retired at some point in time, the benefits from the program would exceed costs to government.

In the case of an energy importing country, impact on the government would be through replacement of petroleum imports. However, this cost should be weighed against government spending to develop the biofuel industry. In some countries such as Brazil, development of the biofuel industry has resulted in a net benefit even after all government support expenditures are included.

6. Social implications

There are mainly two major social benefits of biobased industry: increased standard of living and increased social cohesion and stability (Domac et al., 2005). While the biobased industries help create income generation and other positive impacts, their effectiveness depends on a number of other factors, as shown in Fig. 3. These may include: whether the industry can provide full-time jobs or part-time and night shift jobs; total employment created per energy unit or per amount of land; number of households or people employed in a region; whether skilled or unskilled labour are required, etc (Domac et al., 2005). Some of the identifiable social benefits and social costs are discussed below.

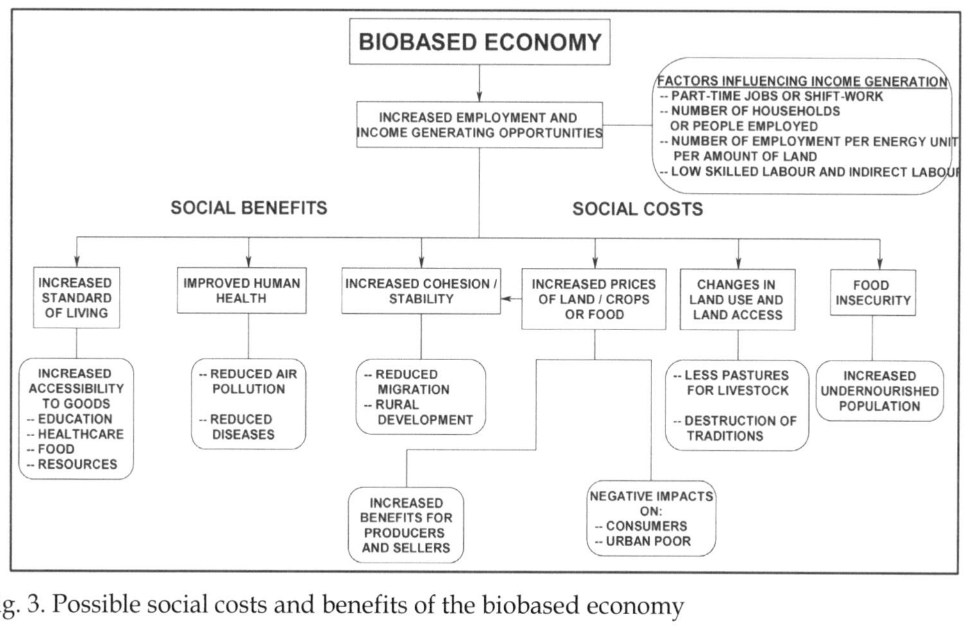

Fig. 3. Possible social costs and benefits of the biobased economy

6.1 Social benefits

6.1.1 Improved quality of life in rural areas

The increased income in a household or community would further help increase a community's or individual's accessibility to good education, health care, resources (e.g. water, land), food products and employment opportunities etc. Biobased industry, being located in rural areas, may provide many of these benefits by establishing livelihood opportunities for the local people. In addition, increased income may help strengthen the cohesion or stability of a community.

6.1.2 Improved human health

The biobased economy may also play an important role in improving human health and safety. For example, sugarcane bagasse used for making paper and fiberboard would otherwise be burnt in the field releasing harmful air pollutants (Phalan, 2009). In addition, improved air quality will reduce diseases such as asthma, and biodegradability characteristics of biobased products, compared to petroleum-based alternatives, are an added advantage (Hardy, 2002). Finally, the local energy security created by bioenergy sector especially biogas will help replace the use of firewood which otherwise would cause air pollution creating negative impact on health of people. In poor countries, increased family incomes would make health care more affordable.

6.1.3 Poverty alleviation

Although liquid fuels are currently being developed for transportation, modern technologies to convert biomass into energy promises to be a more directed way to alleviate poverty, especially in remote oil-dependent regions (Federal Ministry of Food, Agriculture & Consumer Protection, 2006). Some of this would happen through providing employment opportunities in regions where alternatives are scarce or non-existent.

6.1.4 Economic and social impacts on indigenous people

Well-planned biofuel projects could allow indigenous communities to generate capital and maintain or rebuild livelihoods based on the sustainable use of natural resources. In Canada, there is evidence that aboriginal communities and organizations have seldom been incorporated into rural/regional economic development planning, and biobased economy could offer them this opportunity.

6.2 Social costs associated with biobased economy

Some of the social challenges that may arise from biobased industry include changes in land-use rights, food insecurity, and destruction of traditions, among others. Selected social costs are shown in Fig. 3.

6.2.1 Land-use change and impacts on land access

Changes in land use due to increased expansion of agricultural lands for the cultivation of biofuel crops may affect land access and rights of local people (Cotula et al., 2008). In addition, increased economic value created for agricultural biomass may attract agricultural producers to shift from food or cash crops to feedstock. This change would indirectly affect many others whose livelihoods are partially or completely dependent on food crops (Cotula et al., 2008). Further, land values tend to rise when policies and market incentives are provided to convert lands for biofuels production. This increased land value may displace poor people from their land (Cotula et al., 2008).

6.2.2 Food security and cultural impacts

Several studies have argued that increasing demand for biofuel feedstock will pose some serious threats to the food security of people (Yang et al., 2009; Pimentel et al., 2009). In general, development of biobased industry could affect food security in two ways. One, higher food prices caused from the demand of feedstock for biofuel production will limit the purchasing power of the poor or marginalized people (Yang et al., 2009). Two, higher land-use change, such as diverting crop lands to biofuel feedstock production, can have major negative effects on local food security and on the social and cultural dimensions of land use (Cotula et al., 2008). Increased livelihood opportunities from biobased industry would lead to destruction of traditional economic or cultural activities, such as hunting, fishing and trapping. Additionally, using food and feed crops for ethanol production would increase the prices of other food items which are derived directly (e.g. breads, cereals) or indirectly (e.g. chicken, eggs, milk) from these biofuel crops. Although higher food prices represent higher income for farmers, they will affect those whose livelihoods are not linked to agriculture (e.g. urban poor).

6.2.3 Social impacts of rapid growth

Biofuel development could occur over a very short period of time and could change the social fabric of communities. New industrial developments always bring about some costs to communities. According to Finsterbusch (1980) some of these costs include: (i) new residents are frustrated by crowded housing (mainly trailers) and lack of amenities - especially recreational opportunities; (ii) These conditions aggravate family relations and lead to family tension, child abuse and neglect, and delinquency; and (iii) Reported cases of depression, alcoholism, and attempted suicide greatly increase, as do mental health cases. Researchers have provided documentation of a general increase in crime, drug abuse, mental illness, child abuse, and related problems in communities among both new and long-time resident (Gartrell et al., 1984) resulting from a rapid growth over a very short period of time.

7. Providing the balance to sustainability – trade-offs to be made

Biobased economy cannot provide all of society's material and energy needs. One therefore, needs to look at the value of displaced food production in social-economic context to know if trade-offs are worthwhile. Other possible trade-offs that may exist are: (i) Between economic and environmental goals of the society; (ii) Between environmental and social objectives of the society; and (iii) Between economic and social objectives.

7.1 Environment and economy

Traditionally, there has been a view that investments for mitigation of environmental damage (environmental protection) is a cost that takes resources away from investments that would increase production efficiency. Consequently, there are trade-offs between environment and the economy. Many countries have developed (or proposed) policies for reducing GHG emissions, such as subsidies, carbon tax, import tariffs for biofuels, and mandates for quantities to be produced or blended. These policies may promote investments in environmental protection and related technology development, while they can also distort markets and are subject to political decisions that may make them unsustainable. At the same time, some policies strive at maximizing the economic benefit, but will cause environment degradation. An example of this is the U.S. volumetric tax credit for cellulosic biofuels, that does not differentiate across feedstocks and rewards monocultures of high-yielding biofuels per unit of land and are therefore unlikely to create incentives for maintaining biodiversity (Khanna et al., 2009).

7.1.1 Climate change mitigation vs. energy security

Biofuels are attractive to governments which can diversify energy budget and reduce their exposure to international oil market to maintain economic sustainability. Corn-based ethanol in the United States and sugarcane-based ethanol in the Brazil have been built successfully with this objective in mind. While the well-to-wheel environmental benefits are different, such as sugarcane-based ethanol and cellulosic biofuels may achieve significant reduction of GHG, the corn-based ethanol performs poorly due to intensive fossil fuel input (Vermeulen et al., 2008).

7.1.2 GHG vs. other environmental goods

Besides GHG emission reduction, there are many other environmental benefits associated with a biobased economy, such as decreasing soil erosion, water eutrophication, loss of biodiversity, that should be considered. Treating GHG emissions as the only environmental cost, with no concern for other environment threats, can probably result in the other environmental goods and services, such as soil, water and biodiversity, becoming the unintended casualties. Decision makers need to include the full range of desired environmental outcomes in the design of appropriate and robust biofuel policies.

7.2 Environment and society

Emphasis on biofuels as renewable energy sources has developed globally. The use of food crops for biofuel production raises major nutritional and ethical concerns (Pimentel et al., 2009). As a result some trade-offs may exist. One such trade-offs is use of agricultural commodities for food vs. for fuel production.

The food versus fuel debate arises because increased use of land and water for bioenergy production reduces the availability of these resources to produce food for human consumption. The competition is direct in terms of first generation biofuel production that uses feedstocks of cereal grains (e.g. corn, wheat, etc.), oilseeds (e.g. rapeseed, soybean, palm oil), or other crops (e.g. sugar cane) that are conventionally used for food. However, even if the bioenergy feedstock crop is not suitable for food directly, it uses land that could be used for food production.

Secure and affordable food is basic to social sustainability. However, bioenergy may be at the origin of social benefits in providing better quality of life for rural population. It also has great potentials to mitigate environmental impacts. Therefore, if bioenergy is seen as a net environmental benefit, then the extent to which bioenergy production threatens the supply of secure and affordable food becomes an environment and society trade-off. However, if bioenergy is seen as environmental benefit, then the trade-off becomes between society and environment.

7.3 Economy and society

Usually, it is hard to clearly distinguish between economic and social issues. While economic sustainability emphasizes the economic feasibility and viability, society sustainability focuses more on distribution, human health, human rights and equity. Some social conflicts hide behind the economic benefit maximization. For example, the smaller scale operations generally have higher cost. However, the social sustainability policy goals for biofuels include promotion rural development and inclusion of small farmers. This trade off is important as many commodity dependent developing countries are characterised by a high proportion of small producers (Vermeulen & Vorley, 2007).

If an industrialized form of bioenergy crop cultivation is practiced, then the land required will most probably be controlled by large land owners or national companies (WWF, 2006). From maximization of the economic profits, crop cultivation tends to be industrialized which in turn will affect small landowners and poor people's right and welfare. Land ownership should be equitable, and land-tenure conflicts should be avoided. This requires clearly defined, documented and legally established tenure rights. To avoid leakage effects, poor people should not be excluded from the land. Customary land-use rights and disputes should be identified. A conflict register might be useful in this context (WWF, 2006).

7.4 SWOT analysis of biobased economy development

A Strength-Weakness-Opportunities-Threats (SWOT) analysis of the biobased economy is developed which would help decision makers understand strengths and need for developing appropriate policies to overcome limitations for such developments in the future. This analysis is presented in Table 3. One can see whether taking an action or building a project based on biobased economy depends on consideration of many positive and negative factors.

	Internal	External
Positive	**Strengths** • Energy security • Job creation and rural development • Improved trade activities • Establishment of new industries • Reduce GHG emissions	**Opportunities** • Renewable energy requirement • Policy encouragement and technology development
Negative	**Weakness** • Food security • Economic viability • Environmental impact uncertainty • Equity concerns	**Threats** • Rise in fuel and food price • Natural hazards and Crisis on financial market

Table 3. Relevant factors identified in each SWOT category

How to get win-win outcomes from biobased economy development? A map and related policies are urgently needed for the global biofuels industry that supports sustainability. Preventing environmental degradation and social-economic disruption from activities associated with bioenergy supply is seen as a basic principle of sustainability (WWF, 2006). Vermeulen et al. (2008) mentioned that it may be better for the EU to miss its target of reaching 10 per cent biofuel content in road fuels by 2020 than to compromise the environment and human wellbeing. The "decision tree" outlined in Fig. 4, which is developed by Vermeulen et al. (2008), can guide the interdependent processes of deliberation and analysis needed for making tough choices in biofuels to balance the tradeoffs between environment, economy and society.

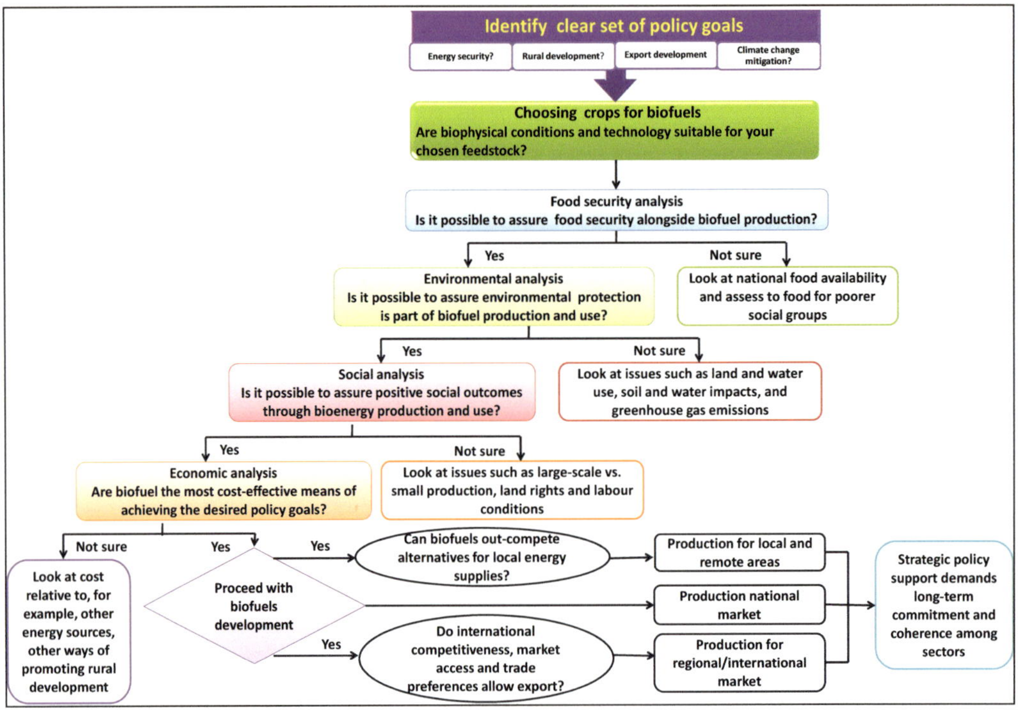

Fig. 4. A decision tree for sustainable strategic national choices on biofuel development (Vermeulen et al., 2008)

8. Conclusions

There exist significant opportunities and challenges with biobased economy. If done correctly, such developments can provide important environmental, economic, and social benefits. The challenge is to have desired outcomes well defined and then develop structures and policies to make those outcomes a reality.

The biobased economy is a major new opportunity for agriculture, which could enable to take it from its recurring overproduction for limited food, feed, and fiber markets to a more sustainable and profitable productions. But the benefits of this biobased economy will extend beyond agriculture to society as a whole, necessitating broad-based support in terms of public policy and investment.

Biobased economy, being located in rural areas, may provide many social benefits, including: (i) Increased employment opportunities in rural areas, resulting in reduced out-migration of local people; (ii) Health and sustainable rural communities; and (iii) Emergence of new investment opportunities for local entrepreneurs (e.g. trucking). Many new challenges would also emerge as a result. Among these are included some of the economic challenges, such as: (i) biomass crops have only one local market, making the local economy more sensitive to its price; (ii) Cost of infrastructure improvement and maintenance; (iii) Increased specialization; (iv) Lack of local control (since heavily capitalized portions of business are less likely to be locally owned -- such as biorefineries to process corn into ethanol); (v) GHG mitigation could cause agricultural activities to be reduced (e.g. through decreases in livestock population which currently provide important incomes and employment); (vi) Higher priced food (local, national, and international); (vii) seasonal employment; (xi) Many low-skill jobs, e.g. machinery operator, truck driver, etc.; (x) Road congestion, less safe highways due to truck traffic to transport biomass; (xi) Potential competition for water between population and industry, affecting some social functions in the communities; and (xii) Destruction of traditions, e.g. displacement of livestock, farmers into forest plantation managers, pastures into biomass grass.

To develop a sustainable biobased economy, two important needs must be addressed. First, it is essential to identify and implement mechanisms for the sustainable production of biomass as current practice of agriculture already facing challenges related to environment degradation and food security due to unsustainable practices. Policy incentives to adopt sustainable agriculture methods that help maintain soil cover, increase water use efficiency and reduce soil erosion are critical (Langeveld et al., 2010) and, research focus on ecosystem services to provide the necessary information to make appropriate land management decisions is also required. Second, developing technologies in order to improve the efficiency of conversion of biomass to biofuels is essential. This not only improves the energy yield of bio-fuels but also reduces the overall environmental and economic burden and hopefully could provide sufficient quantities to satisfy the energy needs of the society.

Ultimately, in a short to medium term, the success of biofuels market completely dependent on the economic factors and not ecological aspects (Festel, 2008). However, Coelho (2005) argues that the full potential of biofuel industry is hindered currently because the fossil fuels do not reflect their real costs and risks. The externalities associated with fossil fuels, such as additional health and environmental costs, are not taken into consideration and the policies of biofuels are mostly focus on side effects, such as local agricultural and food effects.

9. Acknowledgements

Authors would like to thanks Mrs. Poornima Sheelenere for assistance provided in searching the literature and providing its critical assessment. Financial assistance provided by Agriculture and Agri-Food Canada is gratefully acknowledged.

10. References

Adler, P.R., Del Grosso, S.J. & Parton, W.J. (2007). Life-cycle assessment of net greenhouse-gas flux for bioenergy cropping systems. *Ecological Applications*, Vol.17, No.3, (April 2007), pp.675-691, ISSN 1051-0761

Albercht J.; Carrez D.; Cunningham P.; Mancia R.; Máthé L.; Raschka M. & Piotrowski S. (2010). The Knowledge Based Bio-Economy (KBBE) in Europe: Achievements and Challenges, available from: http://www.bio-economy.net/reports/files/KBBE_2020_BE_presidency.pdf

Bai, Y., Luo, L. & Van Der Voet, E. (2010). Life cycle assessment of switchgrass-derived ethanol as transport fuel. *International Journal of Life Cycle Assessment,* Vol.15, No.5, (June 2010), pp.468-477, ISSN 0948-3349

Baker, A. & Zahniser, S. (2007). Ethanol Reshapes the Corn Market. AmberWaves May 2007 Special Issue. http://www.ers.usda.gov/AmberWaves/May07SpecialIssue/Features/Ethanol.htm

Blanco-Canqui, H., Stephenson, R.J., Nelson, N.O. & Presley, D.R. (2009). Wheat and sorghum residue removal for expanded uses increases sediment and nutrient loss in runoff. *Journal of Environmental Quality,* Vol.38, No.6, (November 2009), pp.2365-2372, ISSN 0047-2425

Bell, D. R., Silalertruksa, T., Gheewala, S. H. & Kamens, R. (2011). The net cost of biofuels in Thailand-An economic analysis. *Energy Policy,* Vol.39, No.2, (February 2011), pp. 834-843, ISSN 0301-4215

Brandão, M., Milà i Canals, L. & Clift, R. (2010). Soil organic carbon changes in the cultivation of energy crops: Implications for GHG balances and soil quality for use in LCA. *Biomass and Bioenergy,* article in press, ISSN 0961-9534, available from: http://www.sciencedirect.com/science/article/B6V22-4Y4XCMW-1/2/26051898140dc493640a7c3fe71df0fb

Brentrup, F., Kusters, J., Lammel, J., Barraclough, P. & Kuhlmann, H. (2004). Environmental impact assessment of agricultural production systems using the life cycle assessment (LCA) methodology - II. The application to N fertilizer use in winter wheat production systems. *European Journal of Agronomy,* Vol.20, No.3, (February 2004), pp.265-279, ISSN 1161-0301

CARC. (2003). An Assessment of the Opportunities and Challenges of a Bio-Based Economy for Agriculture and Food Research in Canada, Canadian Agri-Food Research Council (CARC), Ottawa

Cherubini, F. & Jungmeier, G. (2010). LCA of a biorefinery concept producing bioethanol, bioenergy, and chemicals from switchgrass. *International Journal of Life Cycle Assessment,* Vol.15, No.1, (January 2010), pp.53-66, ISSN 0948-3349

Coelho, S.T. (2005). Biofuels - Advantages and Trade Barriers, United Nations Conference on Trade and Development

Cotula, L., Dyer, N. & Vermeulen, S. (2008) Fuelling exclusion? The biofuels boom and poor people's access to land, International Institute for Environment and Development, London

Cramer, J., Hamelinck, C., Heuvel, E., Bergsma, G., Junginger, M., Smeets, E. & Faaij, A. (2006). Criteria for sustainable biomass production, available from: http://www.globalproblems-globalsolutions-files.org/unf_website/PDF/criteria_sustainable_biomass_prod.pdf

Crutzen, P.J., Mosier, A.R., Smith, K.A. & Winiwarter, W. (2008). N_2O release from agro-biofuel production negates global warming reduction by replacing fossil fuels. *Atmospheric Chemistry and Physics,* Vol.8, No.2, (January 2008), pp.389-395, ISSN 1680-7316

Dale, B.E., Bals, B.D., Kim, S. & Eranki, P. (2010). Biofuels done right: Land efficient animal feeds enable large environmental and energy benefits. *Environmental Science and Technology,* Vol.44, No.22, (November 2010), pp.8385-8389, ISSN 0013-936X

Dam, J. & Junginger, J. (2008). Overview of recent developments in sustainable biomass certification. *Biomass and Bioenergy,* Vol.32, No.8, (August 2008), pp.749-780, ISSN 0961-9534

de Vries, S.C., van de Ven, G.W.J., van Ittersum, M.K. & Giller, K.E. (2010). Resource use efficiency and environmental performance of nine major biofuel crops, processed by first-generation conversion techniques. *Biomass and Bioenergy,* Vol.34, No.5, (May 2010), pp.588-601, ISSN 0961-9534

Domac, J., Richards, K. & Risovic S. (2005). Socio-economic drivers in implementing bioenergy projects. *Biomass and Bioenergy,* Vol.*28,* No.2, (February 2005), pp. 97-106, ISSN 0961-9534

Dragisic, C., Ashkenazi, E., Bede, L., Honzák, M., Killeen, T., Paglia, A., Semroc B. & Savy C. (2010). Tools and methodologies to support more sustainable biofuel feedstock production. *Journal of Industrial Microbiology and Biotechnology,* article in press, pp.1-4, ISSN 1367-5435

Dyer, J.A. & Desjardins, R. L. (2009). A review and evaluation of fossil energy and carbon dioxide emissions in Canadian agriculture. Journal of Sustainable Agriculture. 33(2):210-228.

Dyer, J.A., Vergé, X.P.C., Desjardins, R.L., Worth, D.E., McConkey, B.G. (2010). The impact of increased biodiesel production on the greenhouse gas emissions from field crops in Canada. Energy for Sustainable Development 14(2):73–82.

Dyer, J.A., Hendrickson, O.Q., Desjardins, R. L., & Andrachuk, H.L. (2011). An Environmental Impact Assessment of Biofuel Feedstock Production on Agro-Ecosystem Biodiversity in Canada. In: Agricultural Policies: New Developments. Chapter 3. Editor: Laura M. Contreras, ISBN 978-1-61209-630-8. Nova Science Publishers Inc. Hauppauge, NY 11788. 29 pp. (In press)

EAC (Environmental Audit Committee). (2008). *Are Biofuels Sustainable?* the House of Commons London: The Stationery Office Limited, retrieved from: http://image.guardian.co.uk/sys-files/Environment/documents/2008/01/18/EACbiofuelsreport.pdf

Engstrom, R., Wadeskog, A. & Finnveden, G. (2007). Environmental assessment of Swedish agriculture. *Ecological Economics,* 60(3), (January 2007), pp.550-563

European Environment Agency. (2008). *Maximising the environmental benefits of Europe's bioenergy potential.* European Environment Agency, Copenhagen, Denmark, retrieved from: http://www.eea.europa.eu/publications/technical_report_2008_10

Fargione, J., Hill, J., Tilman, D., Polasky, S. & Hawthorne, P. (2008). Land clearing and the biofuel carbon debt. *Science,* Vol.319, No.5867, (February 2008), pp.1235–1238.

Federal Ministry of Food, Agriculture and Consumer Protection. (2006). *Biofuels for Transportation: Global Potential and Implications for Sustainable Agriculture and Energy in the 21st Century.* Prepared by the Worldwatch Institute, Washington D.C., in cooperation with the Agency for Technical Cooperation (GTZ) and the Agency for Renewable Resources (FNR).

FAO (Food and Agricultural Organization). (2008). *The state of food and agriculture-Biofuels: prospects, risks and opportunities,* retrieved from: http://www.fao.org/docrep/011/i0100e/i0100e00.htm

Festel, G. W. (2008). Biofuels - Economic Aspects. *Chemical Engineering & Technology*, Vol.31, No.5, (May 2008), pp. 715-720, ISSN 1521-4125

Finsterbusch, K. (1980). *Understanding Social Impacts: Assessing the Effects of Public Projects*. London: Sage Publications.

Gardner, B. (2003). *Fuel Ethanol Subsidies and farm price support: Boon or Boondoggle?* College park, MD: University of Maryland, Department of Agricultural and Resource Economics.

Gartrell, J. M., Krahn, H. & Trytten, T. (1984). Boom Towns: The Social Consequences of Rapid Growth, in D. Detomasi and J. Gartrell (eds.). *Resource Communities: A Decade of Disruption*. Boulder: The Westview Press.

Gomiero, T., Paoletti, M.G. & Pimentel, D. (2010). Biofuels: Efficiency, Ethics, and Limits to Human Appropriation of Ecosystem Services. *Journal of Agricultural and Environmental Ethics*, Vol.23, No.5, (2010), pp.403-434, ISSN 1187-7863

Haig, T. (2006). Ethanol and Biodiesel - Growing a New Economy. Power Point Presentation made in Alberta Bio-energy Information Meeting. Red Deer Alberta. Accessed on line at:
http://www1.agric.gov.ab.ca/$department/deptdocs.nsf/all/bdv11191/$FILE/tr h_reddeer_03_oct_2006.pdf

Han, F. X., King, R. L., Lindner, J. S., Yu, T.-Y., Durbha, S. S., Younan, N. H., Monts, D.L., Su, Y., Luthe, J.C. & Plodinec, M.J. (2011). Nutrient fertilizer requirements for sustainable biomass supply to meet U.S. bioenergy goal. *Biomass and Bioenergy*, Vol.35, No.1, (January 2011), pp. 253-262, ISSN 0961-9534

Hardy, R. W. F. (2002). The Bio-based Economy. In: *Trends in new crops and new uses*, Janick, J. & A. Whipkey (Eds.), pp. 11–16, ASHS Press, ISBN 9-970756-5-5, Alexandria, VA.:.

Hochman, G., Sexton, SE. & Zilberman, D. (2008). The economics of biofuel policy and biotechnology. *Journal of Agricultural and Food Industrial Organization*, Vol.6, No.8, (November 2008), pp.1-34.

IEA-Bioenergy. (2009). *Bioenergy – A Sustainable and Reliable Energy Source, Main Report*: IEA Bioenergy.

Khanna, M., Hochman, G., Rajagopal, D., Sexton, S. & Ziberman, D. (2009). Sustainability of food, energy and environment with biofuels. *CAB Reviews: Perspectives in Agriculture, Veterinary Science, Nutrition and Natural Resources*, Vol.4, No.028, (April 2009), pp. 1-10, ISSN 1749-8848.

Kim, S. & Dale, B.E. (2005). Life cycle assessment of various cropping systems utilized for producing biofuels: Bioethanol and biodiesel. *Biomass & Bioenergy*, Vol.29, No.6, (December 2005), pp.426-439, ISSN 0961-9534

Klein, K.K. & LeRoy, D.G. (2007). The Biofuels Frenzy: What's in it for Canadian Agriculture? Green Paper Prepared for the Alberta Institute of Agrologists. Presented at the Annual Conference of Alberta Institute of Agrologists. Banff, Alberta, March 28, 2007. Department of Economics, University of Lethbridge. 46 pp.

Kløverpris, J.H., Baltzer, K. & Nielsen, P.H. (2010). Life cycle inventory modelling of land use induced by crop consumption: Part 2: Example of wheat consumption in Brazil, China, Denmark and the USA. International *Journal of Life Cycle Assessment*, Vol.15, No.1, (January 2010), pp.90-103, ISSN 0948-3349

Langeveld, J. W. A., Dixon, J. & Jaworski, J. F. (2010). Development Perspectives of the Biobased Economy: A Review. *Crop Science,* Vol. 50, No. 2, (March-April, 2010), pp. S142-S151, ISSN 0011-183X

Mathews, J.A. & Tan, H. (2009). Biofuel and indirect land use emissions in the life cycle of biofuels: The debate continues. *Biofuels, Bioproducts and Biorefining,* Vol.3, No.3, (May/June 2009), pp.305-317

McBridge, A., Dale, V. & Baskaran, L., Downing M., Eaton L., Efroymson R., Garten Jr. L., Kline K., Jager H., Mulhooland P., Parish E., Schweizer P. & Storey J. (2011). Indicators to support Environmental Sustainability of Bioenergy Systems, Jaunary 2011, available from: http://www.ornl.gov/sci/ees/cbes/forums/Slides_Jan11.pdf

OECD (Organization for Economic Cooperation and Development). (2001). *The Application of Biotechnology to Industrial Sustainability - A Primer*: Paris.

Phalan, B. (2009). The social and environmental impacts of biofuels in Asia: An overview. *Applied Energy,* Vol. *86*(Supplement 1), (November 2009), pp. S21-S29. ISSN 0306-2619

Pimentel, D., Marklein, A., Toth, M., Karpoff, M., Paul, G., McCormack, R., Kyriazis, J. & Krueger, T. (2009). Food Versus Biofuels: Environmental and Economic Costs. *Human Ecology,* Vol.*37*, No.1, (January 2009) pp.1-12, ISSN 1520-7439

Rajagopal D. & Ziberman D. (2007). *Review of environmental, economic and policy aspects of biofuels,* retrieved from: http://elibrary.worldbank.org/content/workingpaper/10.1596/1813-9450-4341

Scharlemann, J.P.W. & Laurance, W.F. (2008). Environmental science: How green are biofuels? *Science,* Vol.319, No.5859, (January 2008), pp.43-44, ISSN 0036-8075

Schmidt, J.H., Christensen, P. & Christensen, T.S. (2009). Assessing the land use implications of biodiesel use from an LCA perspective. *Journal of Land Use Science,* Vol.4, No.1-2, (2009), pp.35-52, ISSN 1747-423X

Searchinger, T. & Heimlich, R. (2009). Likely impacts of biofuel expansion on midwest land and water resources. *International Journal of Biotechnology,* Vol.11, No.1-2, (September 2009), pp.127-149, ISSN 0963-6048

Searchinger, T., Heimlich, R., Houghton, R.A., Dong, F., Elobeid, A., Fabiosa, J., Tokgoz, S., Hayes, D. & Yu, T.H. (2008). Use of U.S. Croplands for Biofuels Increases Greenhouse Gases Through Emissions from Land Use Change. *Science,* Vol.319, No.5867, (February 2008), pp.1238-1240.

(S&T)² Consultants Inc. & Edna Lam Consulting. (2005). *Economic, Financial, Social Analysis and Public Policies for Biofuels. Phase 2.* Prepared For Natural Resources Canada, Ottawa.

Smeets, E. (2008). Possibilities and limitations for sustainable bioenrgy production systems. *Utrecht University,* ISBN 978-90-8672-029-3.

Stoeglehner, G. & Narodoslawsky, M. (2009). How sustainable? Answers and further questions arising from ecological footprint perspective. *Bioresource Technology,* Vol.100, No.16, (August 2009), pp.3825-3830.

Timilsina, G. R., Beghin, J. C., Mensbrugghe, D. v. d., & Mevel, S. (2010). *The Impacts of Biofuel Targets on Land-Use Change and Food Supply - A Global CGE Assessment*: Environment and Energy Team The World Bank.

UNGA (United Nations General Assembly). (1987). Towards Sustainable Development, *Report of the World Commission on Environment and Development: Our Common Future,* retrieved from: http://www.un-documents.net/ocf-02.htm.

UNGA (United Nations General Assembly). (2005). 2005 World Summit Outcome, *Resolution adopted by the General Assembly,* retrieved from: http://data.unaids.org/Topics/UniversalAccess/worldsummitoutcome _resolution_24oct2005_en.pdf.

Urbanchuk, J. M. (2006). Economic Impacts on the Farm Community of Cooperative Ownership of Ethanol Production. LECG LLC. Accessed on line at: http://www.agobservatory.org/library.cfm?refID=88963

Vermeulen, S. & Vorley, B. (2007). Biofuels: Strategic Choices for Commodity Development Countries. In: *Common Fund for Commodities*, available from: http://www.globalbioenergy.org/uploads/media/0711_CFC__Strategic_Choices_ on_Biofuels.pdf

Vermeulen, S., Dufey, A. & Vorley, B. (2008). Biofuels: making tough choices. In: *International Institute for Environment and Development,* 05.03.2011, available from: http://pubs.iied.org/17032IIED.html

Wilhelm, W.W., Hess, J.R., Karlen, D.L., Johnson, J.M.F., Muth, D.J., Baker, J.M., Gollany, H.T., Novak, J.M., Stott, D.E. & Varvel, G.E. (2010). Balancing limiting factors and economic drivers for sustainable Midwestern US agricultural residue feedstock supplies. *Industrial Biotechnology,* Vol.6, No.5, (October 2010), pp.271-287, ISSN 1550-9087

Williams, P.R.D., Inman, D., Aden, A. & Heath, G.A. (2009). Environmental and sustainability factors associated with next-generation biofuels in the U.S.: What do we really know? *Environmental Science and Technology,* Vol.43, No.13, (July 2009), pp.4763-4775, ISSN 0013-936X

Wood, S. M. & Layzell, D. B. (2003). *A Canadian Biomass Inventory: Feedstocks for a Bio-based Economy.* Kingston: BIOCAP Canada Foundation.

WWF (World Wildlife Federation). (2006). Sustainability Standards for Bioenergy, available from: http://www.wwf.de/fileadmin/fm-wwf/pdf_neu/Sustainability_Standards_for_Bioenergy.pdf

Yang, H., Zhou, Y. & Liu, J. (2009). Land and water requirements of biofuel and implications for food supply and the environment in China. *Energy Policy,* Vol.37, No.5, (May 2009), pp. 1876-1885, ISSN 0301-4215

Zahniser, S. (2010). EU Environmental Sustainability Requirements and Brazilian Biofuel Exports. *EuroChoices*, Vol. 9, No.2, (August 2010), pp.28-34, ISSN 1478-0917

9

Implications of Biofuel Feedstock Crops for the Livestock Feed Industry in Canada

J. A. Dyer[1], X. P. C. Vergé[2], R. L. Desjardins[3] and B. G. McConkey[4]
[1]*Agro-environmental Consultant, Cambridge, Ontario,*
[2]*Consultant to AAFC, Ottawa, Ontario,*
[3]*Agriculture & Agri-Food Canada, Ottawa,*
[4]*Agriculture & Agri-Food Canada, Swift Current*
Canada

1. Introduction

The rapid growth of liquid biofuel production could eventually require three or four times the amount of land currently used to supply the feedstock for biofuels (FAO, 2008). The 2007 US Energy Independence and Security Act set the target for 2022 for national ethanol production at nearly four times the present production. It is predicted that this goal would result in the largest and most rapid changes in land use in history (Sinclair and Sinclair, 2010), especially when combined with the similar changes that can be expected in Canada (Klein and LeRoy, 2007).

In spite of the major impact on agriculture that can be expected from such change in land use, biofuels will satisfy a relatively small share of the fuels needed for transportation (FAO, 2008; Karman et al., 2008). Consequently, small increases in the addition of ethanol to gasoline (from 5% to 10%) have meant very large changes in crop distributions (Dufey, 2007; Fritshe et al, 2009). The adoption of 5% biodiesel in Canada could have a similar impact on land use (Dyer et al., 2010a). The increased demand for biofuel may, in turn, lead to higher retail prices for meat and dairy products because of higher livestock feed costs (Zhang and Wetzstein, 2008). Agricultural policy must take the growth of biofuels into account as part of planning for future food security.

Since anthropogenic global warming/climate change will likely be the greatest challenge to mankind in the 21st century (thanks to our addiction to oil), renewable energy supply and Greenhouse Gas (GHG) emissions are the prime justification for biofuel production (Karman et al., 2008). If properly developed, biofuels can potentially help to reduce fossil CO_2 emissions from transport (IEA, 2004; Klein and LeRoy, 2007; Murphy, 2008). Because of the sensitivity of the agricultural resource base to the expansion of biofuel feedstock production, the real potential reduction in GHG emissions from biofuel should take into account any related changes in land use. Such changes should include both the use of the actual land on which the biofuel feedstock was grown and any secondary, or indirect, shifts in land use (Dyer et al., 2011). In addition, land use effects may end up being as important in altering weather as changes in climate patterns associated with GHG buildup (Pielke, 2005).

While it is not clear whether the impacts on food production from increased biofuel feedstock production will always be negative, some shrinkage of resources available to produce livestock feed is expected (Auld, 2008; Klein and LeRoy, 2007). The objective of this chapter was to assess the impact from a shift in land use on the GHG emissions from the Canadian livestock industries. To achieve this goal, the actual area changes will first be identified. While the purely ecological concerns are beyond the scope of this chapter, we recognize that the reallocation of land from livestock feed to feedstock production may re-align several of Canada's agro-ecosystems. The integrity of these agro-ecosystems, particularly those that involve livestock production, will involve a range of environmental considerations, including biodiversity, soil structure or the water cycle (Vergé et al., 2011).

2. Background

In order to decrease dependence on foreign oil in the USA, the Bush administration introduced incentives in 2005 to stimulate the ethanol industry (Whyte, 2008). The result has been rapid growth in the grain ethanol and biodiesel industries over the last five years in both Canada and the USA. Historical trends prior to this period, therefore, provide the only realistic baseline for this assessment. Although Canada does not have the same energy security concerns as the USA, the Canadian biofuel industries are still growing (Klein et al., 2004). The growth of the US biofuel industries, particularly grain ethanol, will have inescapable economic consequences for Canadian livestock producers, regardless of how these industries develop in Canada.

An important spinoff from replacing livestock feed crops with biofuel feedstock crops is the expanded market opportunities for crop producers (IEA, 2004). Whereas most field crop producers should gain economically from the increase in grain prices, livestock farmers are expected to suffer from the rising costs of feed (FAO, 2008; Khanna et al., 2009). From 2006 to 2008, livestock feed prices nearly doubled, in part because of increasing use of corn for ethanol (GAO, 2009). Almost one-third of the US corn crop in 2008 was used for ethanol production. The amount of land available for grazing cattle has also been declining. In 2007 corn used for ethanol production in Canada increased by about 34% while corn grown for feed increased only slightly (Sawyer, 2007).

2.1 Biofuel industry profiles

An environmental impact assessment of biofuel feedstock production on Canadian agro-ecosystem biodiversity used case study scenarios from canola biodiesel, cellulosic ethanol, and corn ethanol (Dyer et al., 2011). Several other possible scenarios were identified in that assessment, including wheat-based ethanol in western Canada and soybean-based biodiesel in eastern Canada. Dyer et al. (2011) predicted only minor impacts from the latter two biofuel industries. Wheat used as a feedstock in western Canada is a small share of the wheat that goes into the food market and should result in very little shrinkage in the land available to support livestock in that region. Since this diversion to biofuel feedstock provides a market for low quality wheat (EIC, 2010), there should be minimal environmental impacts from the production of wheat for ethanol feedstock.

Some use of soybeans for biodiesel feedstock is already in operation in eastern Canada (McKague, 2009). But high corn prices have still tempted many Ontario farmers to stray from their usual corn/soybean crop rotation in order to raise more corn (Sawyer, 2007). A stronger market for biodiesel made from soy oil would stimulate soybean production in the corn growing regions of Canada and displace some of the expanding popularity of corn in central Canada, and thus slow the trend towards a corn monoculture. Therefore, the net impact from soy-biodiesel on the environment should be positive. Since soybean meal, the biggest fraction of this crop (Halliday, 2003; Yacentiuk, 2001), is still available as feed, the impact from soy-biodiesel on livestock feed supply would be minimal.

When cellulosic ethanol facilities become commercially viable, they could replace older grain ethanol facilities, creating more demand for biomass (Simpson, 2009). This certainly would be the case in the US with their national ethanol production target for 2022 of 86.4 billion liters of ethanol per year from non-grain feedstock (Sinclair and Sinclair, 2010). However, the quantitative changes resulting from biomass feedstock for cellulosic ethanol are highly speculative at this stage because this industry is still in its infancy. Since biomass can be produced on almost any class of land, the only land use shift would likely involve moving cattle from higher to lower quality grazing land (Sawyer, 2008). The changing use of rangelands have not attracted as much interest with respect to GHG emissions as have impacts from cattle displaced into forested areas (Baker, 2010). However, if rangeland was used to either support biomass production or to graze too many displaced cattle, biodiversity loss from those previously-undisturbed rangeland habitats would be a greater concern than increased GHG emissions (Dyer et al., 2011).

2.2 Livestock GHG emissions in Canada

Agriculture and Agri-Food Canada (AAFC) researchers undertook to make an inventory of GHG emissions from livestock farms in Canada (Dyer et al., 2010b). This inventory procedure recognized that farm animal populations are limited by the area available to grow the feed grains and forage they consume. Consequently, animal-based production cannot be effectively assessed without first determining the GHG emissions from growing those crops. The land base on which those crops are grown was defined as the Livestock Crop Complex (LCC). The cost of feedstock crop production must include N_2O emissions, farm inputs and farm fossil energy use (Reijnders, 2008). Therefore, manure and enteric methane emissions, nitrous oxide from nitrogen fertilizer and manure, and fossil carbon dioxide emissions associated with feed grain and forage production in the LCC were part of the AAFC methodology for the livestock GHG emissions assessment (Vergé et al., 2007).

Commodity-specific crop complexes were defined for the Canadian beef, dairy, pork and poultry industries (Vergé et al., 2007; 2008; 2009a,b). For each livestock industry, the crop type composition and amount of each crop in the respective diet defined the total crop area in each respective crop complex. This methodology also exploited the differences in diet among age-gender categories of each type of livestock (Elward et al., 2003). Historical GHG emission trends were generated from the statistical assessments for the four livestock industries (Dyer et al., 2008; Vergé et al., 2008; 2009a,b) over the 1981 to 2006 census years (5-year intervals). The whole set of required computations were assembled together in one unified spreadsheet model that can be driven by agricultural census records of livestock populations. This unified model has been used to estimate protein-based GHG emission intensities (Dyer et al., 2010c).

3. Methodology

Simplistic approaches are unlikely to deliver a sustainable biofuel industry or contribute to the climate change challenge (Otto, 2009). Estimating GHG emissions from livestock requires a detailed and deterministic set of estimates for those emissions prior to, or in the absence of, the growth of the biofuel industries. The same methodology must be applicable to altered livestock industries under a range of scenarios for those expected biofuel crops. The unified spreadsheet model for livestock GHG emissions in Canada (mentioned above) provided the GHG estimates used in this chapter. The 2001 livestock GHG emission estimates from this model were used as the baseline GHG emissions for the pre-Bush Administration incentives in this chapter.

The environmental impacts from livestock feed production are specific to agro-ecosystems (Vergé et al., 2011). Therefore, the effects of expanding biofuel feedstock production into areas that had previously been used to grow livestock feed will also vary by region. The only areas of the two feedstock crops (corn and canola) that will be considered are those areas that will encroach on the land dedicated to producing feed grains for livestock. Six hypothetical scenarios involving canola biodiesel and corn ethanol used in this chapter to demonstrate the biofuel feedstock and livestock feed interactions in Canada are summarized in Table 1. The expected or required volumes of ethanol or biodiesel were used to estimate the required weights of grain corn or canola to be diverted to feedstock and away from livestock. Any corresponding shrinkage in the respective livestock GHG emissions were then added to the fossil fuel savings from each respective biofuel type.

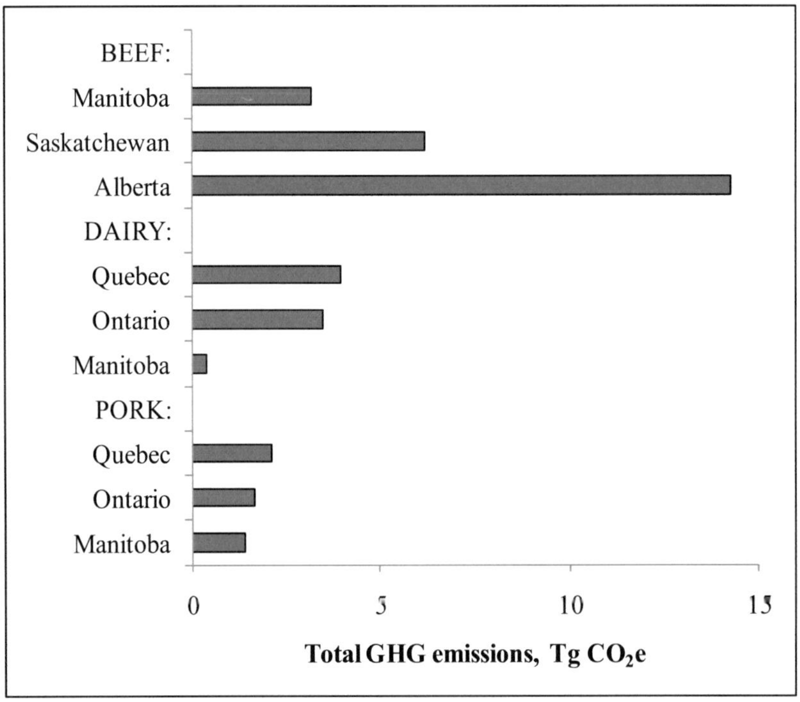

Fig. 1. Total GHG emissions from beef farms in the three Prairie Provinces and from dairy and hog (pork) farms in the three central provinces of Canada in 2001

The 2001 GHG emissions from dairy, beef and hog farms as estimated by Vergé et al. (2007; 2008; 2009) were used as the baseline for the livestock-related GHG emissions in this analysis. Those GHG emission calculations were re-run for this analysis with the virtual age-gender category and total population changes required to test each of the three livestock types. Since the goal of this chapter was to compare the total CO_2e emissions of GHG with the avoided fossil CO_2 from biofuels, only the total GHG emissions are shown in Figure 1, rather than specific types of GHGs. In this application, avoided emissions refer to the net amount of fossil fuel that would not be burned as a result of the increase in biofuel energy assumed in this analysis.

3.1 Biofuel feedstock area and avoided fossil fuel

The starting point for the conversion of biofuel to both the feedstock area and avoided CO_2 emissions from fossil fuel was an assumed target energy quantity of 8 PJ. For equivalent fossil CO_2 emissions, energy was converted to the equivalent volumes of diesel at 36 MJ/litre and gasoline at 32 MJ/litre (Karman et al., 2008). With CO_2 emissions per volume of liquid fossil fuel of 2.73 and 2.36 kg/litre for diesel and gasoline, respectively (Neitzert et al., 1999), the weights of CO_2 emissions from the initial quantities of bioenergy from these two fuels could then be calculated.

With CO_2 emissions per unit of energy given by Jaques (1992) as 70.69 t/TJ for diesel and 67.98 t/TJ for gasoline, the weights of CO_2 from these fossil fuels could also be calculated (as a cross-check) directly from the assumed energy. The weights of CO_2 emissions to produce and consume a litre of fuel (Peña, 2008), expressed as an index of gasoline, provided a basis by which to derive the net avoided fossil CO_2 as a result of using biofuels. This index gave the fossil CO_2 emission cost of corn ethanol produced with natural gas as 68% of gasoline, whereas biodiesel is given as 52% of gasoline and 47% of petro-diesel. Hence the substitution value of corn ethanol for gasoline was 32% of the imbedded CO_2 emissions and the substitution value of biodiesel for petro-diesel was 53%.

The assumed target energy quantities were converted to the equivalent volumes of canola oil at 34 MJ/litre and ethanol at 21 MJ/litre (Karman et al., 2008). The volumetric energy of ethanol reflects the relatively low energy content per unit volume compared to gasoline (Karman et al., 2008). An average estimate of 377.5 litres of ethanol per t of grain corn was derived from three literature sources (AAFC, 2009; Bonnardeaux, 2007; Hardin, 1996). The tons of feedstock crop (F) of grain corn (*gc*) was computed as:

$$F_{gc} = V_{\text{ethanol}} / 377.5 \quad (1)$$

Since canola loses 39% of its weight during oil extraction (Vergé et al., 2007), and the density for canola oil is 0.915 kg/litre (Elert, 2000), the weight in tons of feedstock crop (F) of canola seed (*cs*) was computed from the volume in litres of canola oil as:

$$F_{cs} = 0.915 \times V_{\text{canola oil}} / 0.39 \quad (2)$$

The two biofuel byproducts, dry distillers grain (DDG) and canola meal, were added back into the respective livestock diets to offset some of the expected shrinkage from these LCC area losses. Both of these byproducts were treated as high energy grain substitutes, rather than as extra roughage for ruminants. The DDG byproduct from the ethanol processing was 31.9% of the grain corn feedstock weight (Bonnardeaux, 2007). The canola meal byproduct from the biodiesel processing was 61% of the canola feedstock weight (Vergé et al., 2007). While they are both high in protein (McKague, 2009; EIC, 2010), the dietary benefits of this protein were ignored in this analysis. These feedstock weights were factored by provincial crop yields to estimate the crop areas needed to produce these fuel volumes. The scenario tests involved the subtraction of these estimated net feedstock crop areas from the respective LCC areas.

3.2 Livestock scenarios for biofuel expansion

For cattle, producers may respond to less available feed grain by feeding more forage, a system that has proven to be economically viable in some countries (Casey and Holden, 2005, 2006). This strategy was the basis of Scenarios B1 to B4 for beef (described below). In Ontario and Quebec, however, virtually all arable land is in cultivation and so no land would be available to expand forage production to compensate for reduced grain corn supply (Whyte, 2008). The two Central Canada scenarios are as follows.

- Scenario D: given the lack of land for expanding forage production, no attempt was made to redefine the balance between grain and roughages (forage) in dairy cow diets to accommodate the changing crop distribution in the LCC. When the supply of feed grain in the dairy cattle diet was reallocated to feedstock, reduction of the entire population was assumed, rather than adjusting the herd for possible increased roughage consumption.
- Scenario P: no forage crops are involved in the non-ruminant hog diet. The Canadian hog population includes either breeding stock or animals destined for slaughter, with almost no differences in diet between the two categories. Therefore, reductions in the total populations were assumed for the pork industry, in response to reallocation of land in annual crops to feedstock production.

Scenario	Required action	Animal type	Feedstock	Biofuel	Region
B		Beef	Canola	Biodiesel	Prairie Provinces[1]
	B1 Send the calves and yearling slaughter animals in feedlots for slaughter.				
	B2 Transfer calves and yearling slaughter animals in feedlots from a grain diet to the predominantly forage-based diet of replacement heifers.				
	B3 Feed all slaughter and replacement animals the same forage-based diet as the grazing, breeding cattle.				
	B4 Reduce the whole beef population across all age-gender categories.				
D		Dairy	Grain corn	Ethanol	Central Canada[2]
	Reduce the whole dairy population across all age-gender categories.				
P		Pork	Grain corn	Ethanol	Central Canada[2]
	Reduce the whole hog population across all age-gender categories.				

[1] Manitoba, Saskatchewan and Alberta
[2] Quebec, Ontario and Manitoba

Table 1. Scenarios used to test the effect of reallocating farmland from feed grains used in the Canadian livestock industry to feedstock crop production for biofuel

3.3 Scenarios for western Canadian beef

Because the Canadian beef industry is a mix of grain-based and grazing-based production systems, several farm level responses are possible from the expansion of canola feedstock areas into the beef crop complex (BCC). The Canadian beef industry is also unique in that these different production systems are typically managed independently (ranches and feedlots under different ownership), with different decision processes (Vergé et al., 2008). The four possible scenarios specific to beef (B) production (Table 1) were ranked in order of the number of beef animal categories they affected. Because of the complexity of the western Canadian beef industry, the age-gender category populations (as defined by Vergé et al. (2008)) and the mean live weights are summarized in Table 2. The grain-based differences in diet among the age-gender beef categories are illustrated in Figure 2.

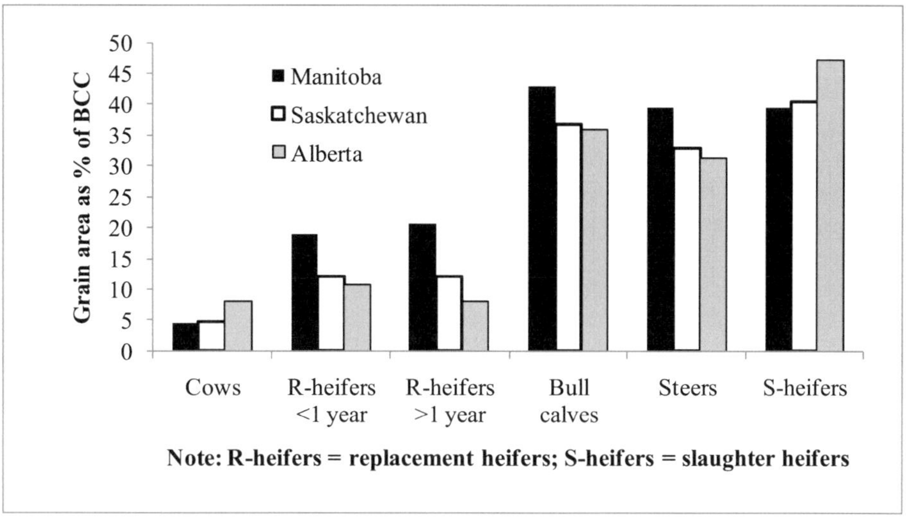

Fig. 2. The areas in feed grains as % of the Beef Crop Complex (BCC) for six age-gender categories in each prairie province of Canada in 2001

Two beef scenarios are similar to the dairy (D) and pork (P) scenarios. In scenarios B1 and B4 changes were limited to the outright removal of animals from the system, rather than reallocation of animals from one livestock category to another within the same industry. The other two scenarios (B2 and B3) were based on shifting the diet of one or more age-gender livestock categories to the diet of other categories that consume less grain. From a feed supply perspective, a number of yearling steers would be re-designated as range-fed breeding cows, for example, taking into account the difference in their respective live weights. These two scenarios required more grain area to be reallocated than the area needed to produce the desired biodiesel energy. This was because some additional area of grain is needed to meet the grain dietary components of the expanded population of the new category.

- Scenario B1: the impact would be limited to the reduction of the slaughter cattle (slaughter calves, steers and non-replacement heifers), mostly in feedlots. The assumption behind this scenario was that, with less grain or high energy feed, there would be no value in keeping these animals alive during the period they would normally be in the feedlot. Hence, they were slaughtered straight away and thus eliminated from the industry (and its carbon footprint).
- Scenario B2: instead of immediate slaughter of these animals (from Scenario 1), they would be kept on a diet equivalent to that of the replacement heifers, which is based on more forage and less grain than that of slaughter animals. Hence the impact would be broadened to include the population expansion of the replacement category by the slaughter animals. In this assumption, these animals become mainly grass-fed, rather than mainly grain-fed, beef. With respect to diet-based GHG emission calculations

(Vergé et al., 2008) they became virtual replacement heifers, with allowance for the live weight differences (Table 2).

- Scenario B3: both slaughter and replacement beef cattle are transferred to a predominantly forage-based diet. The components of this grazing-based diet would be defined by the prairie beef ranch where the breeding cows are maintained. Hence the cattle being transferred (slaughter calves, steers, replacement and non-replacement heifers, and bull calves) are treated as virtual grazing, breeding beef cows, with corrections for live weight differences (Table 2).
- Scenario B4: the impact of less high energy feed being available to the feedlot industry would be felt throughout the whole beef production system. The assumption behind this scenario was that beef producers have become sufficiently dependent on marketing their product through a high feed energy finishing process (the feedlot) that, without sufficient feedlot capacity, the unfinished beef would not be economically viable, and so the impact would be felt throughout the entire industry.

	Breeding stock		Replacemant heifers		Bull	For slaughter		Calves
	Bulls	Cows	> 1 year	< 1 year	calves	Steers	Heifers	< 3 months
Provinces	Head of beef cattle x 10^3							
Manitoba	25	545	114	132	85	120	89	116
Saskatchewan	60	1,200	254	284	174	216	68	234
Alberta	104	2,028	1,083	621	372	438	760	497
	Live weight, kg/head							
Manitoba	765	671	490	319	319	356	451	153
Saskatchewan	712	601	467	317	317	371	443	150
Alberta	666	609	539	315	315	386	505	142

Table 2. Populations and live weights of beef cattle by age-gender categories in the three Prarie provinces of Canada in 2001

In Scenarios B2 and B3 it was assumed that the expansion of the grass-fed slaughter animals would be based on land capable of growing perennial forage but not annual grains or oilseeds. While this is typically marginal land, it is not necessarily publically-owned rangeland. In Canada, the only significant quantities of such land would be in the western provinces. This assumption brings new land into production (albeit under permanent cover) and potentially raises the net GHG emissions from beef production. It also raises the possibility of non-GHG related impacts on the land being brought into production (IRGC, 2008; Vergé et al., 2011). Because this land would probably be managed as improved pasture or hay, the chemical inputs and introduced forage crops could threaten biodiversity (Dyer et al., 2011).

3.4 Area reallocation calculations

Adjustment of the category populations called for in the respective scenarios was achieved through the ratio of the net feedstock area (A_{nc}) to the baseline areas of annual crops. The net converted (*nc*) area for feedstock (*fs*) was adjusted for the land freed from feed production by the biofuel byproduct (*bf*) (IEA, 2004) as follows:

$$A_{nc} = A_{fs} - A_{bf} \quad (3)$$

where

A_{fs} = area to grow the biofuel feedstock crop

A_{bf} = area required to grow the feed equivalent to the weight of feed byproduct

The ratio of the net converted feedstock area to the BCC feed grain areas (AR) of the beef categories being displaced was calculated as a fraction of the BCC:

$$AR = A_{nc} / \sum^{c} A_{beef,c} \quad (4)$$

Thus the reduced (*r*) beef population ($P_{r,c}$) for each age-gender category (*c*) was computed as:

$$P_{r,c} = P_{bl,c} \times AR \quad (5)$$

where

$P_{bl,c}$ = the baseline (*bl*) population (P) in each beef animal category (*c*).

Based on the changed beef populations, the areas in forage (mainly perennial grass, alfalfa and hay) were recalculated by re-running the unified livestock GHG emissions model with these re-aligned beef cattle populations.

4. Results and discussion

4.1 Overview of Canadian agricultural land use

Table 3 shows three levels of area data on an east-west basis. The crop areas needed to feed Canadian livestock (beef and dairy cattle, swine and poultry) are shown as the first level in Table 3. To put these LCC areas into context, they are compared to the national crop areas as reported in the 2001 agricultural census. Any areas in each crop type that do not supply livestock feed were excluded from the LCC (Dyer et al., 2010b). In the second level, only the types of crops used in animal diets (as identified in the LCC) are included, but the entire areas planted for those crops in Canada are given, regardless of whether they are used to feed livestock. In the third level, all types of field crops were taken into account, and the entire area planted to each of those crops is included. The crop types were grouped as either grains (including oilseeds) or forages.

On a national basis, forages represented more than 60% of the LCC. The largest portion of the LCC was in the western provinces (70% of the total). In eastern Canada, areas were evenly distributed between grains and forages. This was very different in the west, where the area for forages (9.3 Mha) was twice as high as the area for grain (4.4 Mha). Table 3 illustrated that most of the cultivated crop types correspond to those used for animal feed. The difference between Level 2 and Level 3 was only 3.25 Mha. Since forages were grown exclusively for animal feed, all of this difference was accounted for by the grain crops. In the west, grains and oilseeds represented about 70% of all crop lands, whereas there was almost no difference between grains and forages in the east.

Table 3 also illustrates that the LCC represented almost half the total Canadian crop land. The grain portion of the LCC represented about one fourth of the total grain and oilseed areas. The small difference in forage areas between Levels 2 and 3 was due to sheep and horses not being included in the LCC (Dyer et al., 2010b). About 80% of grain areas in the east were used for animal feed (2.84 Mha compared to 3.53 Mha). In the west, feed grains only accounted for 17% (4.38 Mha compared to 25.04 Mha) of the western grain areas.

Regions	Grains & oilseeds	Forages	Total
	Mha		
	Crop areas included in the LCC[1]		
East[2]	2.8	2.6	5.5
West[3]	4.4	9.3	13.7
Canada	7.2	11.9	19.2
	All areas for only those crops in the LCC		
East	3.3	2.8	6.0
West	22.0	9.6	31.6
Canada	25.3	12.4	37.7
	All areas for all Canadian crops		
East	3.5	2.8	6.3
West	25.0	9.6	34.6
Canada	28.6	12.4	40.9

1 Livestoc Crop Complex for beef, dairy, hogs and poultry
2 Atlantic Provinces, Quebec and Ontario
3 Manitoba, Saskatchewan, Alberta and British Columbia

Table 3. Overview of the use of arable land in Canada, as recorded in the 2001 agricultural census in relation livestock

4.2 Reallocations of livestock areas to biofuel feedstock

Table 2 shows the complexity of the Canadian beef industry, particularly when the differences in the way replacement stock and animals destined for slaughter are taken into account. This complexity was a critical factor in the response by beef farmers to changes in feed grain areas and the need for four test scenarios for this industry. The largest share of the provincial beef populations was in the breeding categories (replacement heifers and cows). These animals were also the heaviest. The Alberta beef cattle population is almost twice as high as the Saskatchewan population which is more than twice as high as the Manitoba population. There were appreciable differences among the age-gender categories with respect to both population and live weights.

Only the six categories that were involved in the four beef cattle scenarios are shown in Figure 2. All grains, pulses and oilseeds in the beef diet were grouped together as feed grains. The dependence on grain consumption shown in Figure 2 varies noticeably among the age-gender categories. The percent of the total area supporting the cattle (the BCC) on which feed grain was grown in 2001 demonstrates that, among the cows and replacement heifers, only a small share of their diet was in grains. In comparison, the diet for the cattle destined for slaughter required that almost half of the areas that feed slaughter animals be in grain production. Grain consumption by the breeding cows was much less than by the replacement stock. The differences in diet among the age-gender categories were quite consistent across the three provinces.

Table 4 shows that the baseline area (in the LCC prior to any area reallocation for ethanol feedstock) for hog and dairy farms was much higher than the areas being reallocated to corn ethanol feedstock production. The small area for corn for feedstock use in Manitoba reflects the relatively low acreages of this crop in Manitoba in 2001. The changes in area shown for forage are due to a reduction in areas of forage required for dairy cattle as a result of reduced feed grain and shrinkage of the dairy population. In the Ontario and Manitoba dairy industries, the areas in forage only moderately exceeded the areas in feed grains, but in Quebec the forage area was more than twice the area in feed grains. The areas in feed grain to support the hog farms was about the same in all three central Canadian provinces, while feed grain areas for dairy suggest a much smaller dairy industry in Manitoba than in the other two provinces.

	Corn ethanol[1]	Feed grain baseline		Forage for dairy	
		hogs	dairy	baseline	changes
Provinces		Mha			
Quebec	42	582	238	597	48
Ontario	66	574	421	551	57
Manitoba	5	605	57	87	11

[1] are also equal to changes in feed grain areas for hog and dairy farms

Table 4. Areas required to support the Canadian pork (hog) and dairy industries and the changes in those areas in order to increase corn production for 8 PJ of bio-ethanol energy supply

For the beef industry, the feed grain areas shown for scenarios B1 and B4 in Table 5 were equal to the areas reallocated to canola feedstock areas for this analysis. This was a necessary condition of this analysis for these two scenarios. For the two scenarios in which beef cattle were transferred to other age-gender categories (B2 and B3), the area changes shown in Table 5 for categories of the animals being transferred were appreciably greater than the areas of expanded canola. However, the net change in feed grain area was still equal to the area reallocated to canola expansion. The difference between the feed grain area changes shown for B2 and B3 (Table 5), and the expanded canola area reflected the consumption of at least some feed grain by cattle transferred to the replacement or breeding stock diets (Figure 2).

	Beef diet baseline	Scenarios for beef production			
		B1	B2	B3	B4
Provinces		Mha, feed grains			
Manitoba	223	56	89	62	56
Saskatchewan	380	144	187	191	144
Alberta	1,106	99	115	141	99
		Mha, forages			
Manitoba	1,108	86	0	0	279
Saskatchewan	2,652	266	20	-356	925
Alberta	4,371	130	1	-134	1,113

Table 5. Areas required to support the Canadian beef production and the changes in those areas in order to increase canola production for 8 TJ of biodiesel energy suply

For scenarios B1 and B4, Table 5 showed reductions in the forage areas that were greater than the feed grain areas for expanded canola. Scenario B2 showed no appreciable reduction in forage areas, whereas for B3, the transfer of the slaughter animals to the diet of the breeding animals in Saskatchewan and Alberta actually required increases. In this case, the area changes are shown as negative quantities because they represent forage area that had to be taken from other uses instead of being freed for other uses. The decreased forage areas in B1 were from one and a half to twice as high as the expanded canola areas, while the decreased forage areas in B4 were 4 to 10 times as high as the expanded canola areas.

4.3 Changes in GHG emissions from feedstock expansion

Figure 1 presents the total provincial GHG emissions from the three livestock industries considered in this analysis. The largest GHG emitters were the Alberta and Saskatchewan beef industries, followed by the Manitoba beef industry and the Quebec and Ontario dairy industries. The Manitoba dairy industry was the lowest GHG source. These GHG emissions are primarily N_2O and CH_4 (Desjardins et al., 2010).

	Tg CO_2e					Gg CO_2e/PJ{biofuel}		
	Corn	Farm-related		Ethanol plus farm		Corn	Ethanol plus farm	
Provinces	ethanol	Dairy	Pork	Dairy	Pork	Ethanol	Dairy	Pork
Quebec	0.069	0.635	0.149	0.704	0.218	24	240	74
Ontario	0.114	0.517	0.188	0.631	0.302	24	130	62
Manitoba	0.005	0.029	0.008	0.035	0.013	24	153	58

Table 6. Avoided CO_2 and farm-related greenhouse gas (GHG) emissions, and the intensities of avoided emissions as a result of displacing dairy and pork production with corn for bio-ethanol feedstock in the three central provinces of Canada in 2001

The results for hog and dairy farms are both shown in Table 6 because the only scenario involved in the two ethanol feedstock expansion tests was a decrease in the entire population. The avoided GHG emissions from the changes in both the pork and dairy production systems far exceeded the avoided fossil CO_2 emissions resulting directly from the corn ethanol energy. This difference was most evident in Quebec where the dairy diet was more heavily dependent on forages. The last three columns of Table 6 use the intensity of avoided GHG emissions to put these comparisons on a basis that can be extrapolated to larger quantities of biofuel energy.

Table 7 shows that the enhancement of avoided GHG emissions was much less certain for the beef industry than for the pork and dairy industries. In the B4 scenario (5th column) where the whole population was reduced (just as with pork and dairy), the savings in emissions were overwhelming in comparison to the directly avoided CO_2 emissions by bio-ethanol. This was because of the greater dependence of beef over dairy on forages. Under Scenario B1 (2nd column of Table 7), feedlots would be the most affected activity of the beef industry since most of the cattle in these two age-gender categories are finished for market in feedlots in Canada. Even in this scenario, which involved the elimination of the high feed grain based finishing of slaughter animals without any increase in grazing, the avoided on-farm GHG emissions exceeded the directly avoided CO_2 emissions by bio-ethanol by several times.

In scenarios B2 and B3 (the 3rd and 4th columns of Table 7), the opposite trend is evident. This was because the transfer of beef cattle into more forage based diets meant that the consumption of forages by the beef cattle population increased more than the grain consumption was decreased. The effect of dietary changes from one age-gender category to another on crop distributions in the BCC was evident in Figure 2. These dietary differences meant that, under scenarios B2 and B3, total cattle numbers would have to undergo little change. With greater use of forage (and a higher roughage share in the diet) enteric methane emissions would increase rapidly (Desjardins et al., 2010). Although the B1, B2 and B3 scenarios were considered much more realistic than B4, the latter scenario provided a useful perspective and boundary condition on the set of possible responses by the beef industry.

	Canola biodiesel	Scenarios for beef production			
		B1	B2	B3	B4
Provinces	Tg of avoided fossil CO_2	Tg CO_2e Farm-related GHG emissions			
Manitoba	0.067	0.245	-0.080	0.138	1.574
Saskatchewan	0.143	0.538	-0.098	-0.565	4.219
Alberta	0.111	0.315	-0.151	-0.358	7.118
		Total GHG emissions			
Manitoba	-	0.312	-0.012	0.206	1.642
Saskatchewan	-	0.681	0.045	-0.422	4.363
Alberta	-	0.426	-0.040	-0.247	7.229
	Gg CO_2e/PJ{biodiesel}				
Manitoba	40	186	-7	123	980
Saskatchewan	40	191	13	-118	1,224
Alberta	40	154	-15	-90	2,620

Table 7. Avoided CO_2 and farm-related greenhouse gas (GHG) emissions, and the intensity of avoided emissions as a result of displacing beef production with canola for biodiesel feedstock in the Prarie Provinces of Canada in 2001

5. Summary and conclusions

This analysis provides a good understanding of the interaction between livestock farming and feedstock production for biofuels in Canada. It has shown that target levels of liquid biofuel energy translate directly into cropland reallocations. It demonstrated that where dislocation of livestock is a possible outcome of the expansion of biofuel feedstock production, the carbon footprint will extend beyond the cultivation of the feedstock crop. Given how much of Canada's arable land is in the LCC (Table 3), this extended carbon footprint should be a major consideration in the Canadian biofuel development strategy.

This analysis also revealed the dependence of the ultimate value of biofuels as a GHG reduction tool on previous or alternative uses of the land targeted for feedstock production. For the expansion of feedstock crops into land that supports non-ruminant livestock (poultry or pork), the impact would be straight forward since there is no significant fall-back on grazing. For ruminants however, these interactions are highly complex, even when considered on the one-dimensional basis of GHG emissions taken in this analysis.

It is also important to understand what livestock-feedstock interactions will mean to other environmental issues (Dufey, 2007; Karman et al., 2008; Vergé et al., 2011). The environmental impact assessment of biofuel feedstock production on habitat and biodiversity in Canada raised several issues that are relevant to biofuel-livestock interactions addressed in this chapter (Dyer et al., 2011). That study found that many of the impacts on biodiversity will be the result of decisions made by farmers that are not profiting directly from feedstock crops, but wish to continue farming livestock. This is particularly true of the so-called cow-calf, or ranch, operations and how they respond to any reductions in the grain-based feedlot operations.

What this set of tests came down to for ruminants is that farmers can respond to reduced feed grain supply in two ways: by reducing their livestock numbers or by returning to a more roughage-based diet with more forage and less grain. The general case for eastern dairy farmers was for farm land on which to expand forage production to be a limiting factor (Whyte, 2008). In this case, simply reducing the herd size was the most plausible option, given the limited land resources. The type of beef operations most likely to be affected are the feedlots because, with a limited land base, they are the most vulnerable to feed grain price increases. The greater availability of land on which to expand forage production in the Prairie Provinces, along with the complexity of the beef population (Table 2) and large feedlot industry makes it difficult to predict how beef producers will react to expanded canola production.

Displacement of ruminants by biofuel feedstock is an effective GHG reduction strategy if the populations of those displaced animals are actually reduced. However, when they are simply transferred to the more forage-based diet, the enhanced benefit from reduced enteric methane emissions is either cancelled out or reversed (Table 7). Feeding beef cattle more forage and less grain in response to expanded canola is more likely if the canola biodiesel industry opts for vertical integration (ownership of the feedstock production) and exclusion of the beef farmers. The numbers of beef producers who would choose to reduce their herds to grow canola for biodiesel, compared to the numbers that would feed their cattle more forage, depends on giving them the opportunity to sell their canola to the biodiesel processing plants as an alternative income to cattle. Although this only applies on an appreciable scale to the beef industry, beef is Canada's largest livestock commodity and is the largest source of livestock GHG emissions (Figure 1).

Increased canola production in western Canada can displace wheat as well as feed grains. If the byproduct from the entire western Canadian canola industry were to be used as livestock feed, the canola meal byproduct may be sufficient to support an increased livestock population (cattle or hogs). However, since the market for canola as a source of healthy cooking oil is competitive with food quality wheat, only part of the expansion of canola area in western Canada should be attributed to biodiesel feedstock. To the extent that canola expansion would be into food-quality wheat, rather than into the LCC, the canola meal byproduct would be available to livestock. However, none of the reductions in GHG emissions from the existing cattle populations could be credited to the expanded canola production unless the cattle transferred to a more canola meal-based diet (with less forage) were displaced, or came, from the existing cattle populations.

This assessment was critically dependent on the set of livestock GHG emission inventory models developed by Vergé et al. (2007; 2008; 2009a,b). Given the magnitude of GHG emissions from the Canadian livestock industries (Figure 1), any future assessments of biofuel feedstock production in Canada should also make use of this methodology. Caution is needed in interpreting or applying these test results because the responses to the conversion of crop land to feedstock production were based on assumed decisions by the farm operators. The ultimate value of biofuels as a GHG reduction tool depended on previous or alternative uses of that land that were beyond the scope of these livestock GHG emission models. What is really critical from a policy perspective is that those farmers operate independently from the decision makers who purchase the biofuel feedstock crops. It would therefore be useful to assess the social and economic pressures that drive these decisions.

This chapter has not dealt with the changes in soil carbon as a result of land use changes. This term would depend on the use to which the land removed from forage production was put. If it was seeded with other feed grains or annual crops, then some soil carbon would be lost (Davidson and Ackerman, 1993). If, however, it was used for grazing, then this may serve to reduce pasture stocking rates, and lower the dependence on rangeland for grazing beef cattle. Lower stocking rates will mean healthier turf, whether in improved pasture or rangeland, which is likely to result in an overall increase in soil carbon. Another looming possibility is the developing cellulosic ethanol industry which could exert pressure on ruminant livestock farming from the forage supply side (rather than feed grains) while at the same time, maintaining perennial ground cover, and soil carbon levels. This is not to say that changes in soil carbon will not make a difference in this extended carbon footprint for biofuels. But it is equally unlikely that those changes would always fully compensate for changes in enteric methane. Therefore, even without taking soil carbon into account, the implications of including livestock industries in biofuel GHG calculations should not be ignored. However, incorporating soil carbon sequestration is a future challenge for the set of livestock GHG emission models used in this chapter.

The final caveat to the GHG mitigation benefits of the livestock displacement described in this chapter is that Canadian agriculture would produce less meat. In North America and Europe, the loss of some meat is not a major threat to the human diet. Nutritionally, there might be health benefits for many consumers if they were encouraged by higher meat prices to consume more vegetables and whole grains, and less red meat. In the developing world, however, dietary protein is often a limitation to improved health, and will be more so as human populations continue to grow. As many of these countries achieve higher incomes, the demand for meat will increase and other sources will be sought. Nevertheless, the assumption that displaced livestock will mean lower GHG emissions attributed to biofuel production may not apply to countries that are protein deficient or where the demand for meat is growing.

6. References

AAFC (Agriculture and Agri-Food Canada) (2009). Corn: Situation and Outlook. Market Outlook Report. AAFC No. 10918E, ISSN 1920-20082X, Vol.1(2), June 19, 2009, www.agr.gc.ca or http://www.agr.gc.ca/pol/mad-dam/index_e.php?s1=pubs&s2=rmar&s3=php&page=rmar_01_02_2009-06-19

Auld, D. (2008). The Ethanol Trap: Why Policies to Promote Ethanol as Fuel Need Rethinking. C.D. Howe Institute Commentary 268, July 2008. ISSN 1703-0765 (online). 20pp. www.cdhowe.org

Baker, P. (2010). Recent developments in the world of biofuels: Spring 2010. CABI's Biofuels Information Xchange. http://biofuelexperts.ning.com/ Accessed 2 June 2010.

Bonnardeaux, J. (2007). Potential Uses for Distillers Grains. Department of Agriculture and Food Western Australia, State of Western Australia, 3 Baron-Hay Court South Perth WA 6151. 2007.
www.agric.wa.gov.au/objtwr/imported_assets/content/sust/biofuel/potentialusesgrains

Casey, J.W. & Holden, N.M. (2005). Analysis of greenhouse gas emissions from the average Irish milk production system. *Agricultural System*, Vol.86, pp.97–114.

Casey, J.W. & Holden, N.M. (2006). Quantification of GHG emissions from sucker-beef production in Ireland. *Agricultural Systems*, Vol.90, pp.79–98.

Davidson, E.A. & Ackerman, I.L. (1993). Changes in soil carbon inventories following cultivation of previously untilled soils. *Biogeochemistry* Vol.20, pp.161–93.

Desjardins, R.L., Worth, D.E., Vergé, X.P.C., McConkey, B.G., Dyer, J.A. & Cerkowniak, D. (2010). Agricultural Greenhouse Gases. Chapter 16, Pages 110-117. In: *Environmental Sustainability of Canadian Agriculture. Agri-Environmental Indicator Report Series: Report* #3. Eilers, W., R. MacKay, L. Graham and A. Lefebvre (editors), Agriculture and Agri-Food Canada. ISBN 978-1-100-15576-0, Ottawa, Ontario K1A 0C5.

Dufey, A. (2007). Biofuels production, trade and sustainable development: emerging issues. The International Institute for Environment and Development (IIED). Environmental Economics Programme/Sustainable Markets Group. ISBN: 978-1--84369-643-8. 62 pp.

Dyer, J.A., Vergé, X., Desjardins, R.L. & Worth, D. (2008). Long term trends in the GHG emissions from the Canadian Dairy industry. *Canadian Journal of Soil Science*, Vol.88, pp.629-639 (Special issue).

Dyer, J.A., Vergé, X.P.C. Desjardins, R.L. Worth, D.E. & McConkey, B.G. (2010a). The impact of increased biodiesel production on the greenhouse gas emissions from field crops in Canada. *Energy for Sustainable Development*. Vol. 14, No.2, pp.73–82. doi: 10.1016/j.esd.2010.03.001.

Dyer, J.A., Vergé, X.P.C. Desjardins, R.L. Worth, D.E. & McConkey, B.G. (2010b). Understanding, Quantifying and Reporting Greenhouse Gas Emissions from Canadian Farmland. *Sustainable Futures*. Fall issue, pages 10-12. http://www.aic.ca/sustainable/pdf/Sustainable_Futures_Fall_2010.pdf.

Dyer, J.A., Vergé, X.P.C. Desjardins, R.L. & Worth, D.E. (2010c). The protein-based GHG emission intensity for livestock products in Canada. *Journal of Sustainable Agriculture*, Vol.34, No.6, pp.618-629.

Dyer, J.A., Hendrickson, O.Q. Desjardins, R.L. & Andrachuk, H.L. (2011). An Environmental Impact Assessment of Biofuel Feedstock Production on Agro-Ecosystem Biodiversity in Canada. In: *Agricultural Policies: New Developments. Chapter 3*. Editor: Laura M. Contreras, ISBN 978-1-61209-630-8. Nova Science Publishers Inc. Hauppauge, NY 11788. 29 pp. (In press)

EIC (The Ethanol Info Centre). (2010). Fuel ethanol and food supply. http://www.sentex.net/~crfa/ethafood.html. Accessed 21 January 2010.

Elert, G. (Editor). (2000). Density of Cooking Oil, The Physics Factbook (An educational website). http://hypertextbook.com/facts/2000/IngaDorfman.shtml. Accessed 15 November 2010.

Elward, M., McLaughlin, B., Alain, B. (2003). Livestock Feed Requirements Study 1999–2001. Catalogue No. 23-501-XIE, Statistics Canada, 84pp.

FAO (Food and Agriculture Organization of the United Nations). (2008). *Biofuels: Prospects, Risks and Opportunities. The State of Food and Agriculture*. Rome, Italy. ISSN 0081-4539, ISBN 978-92-5-105980-7. 138 pp.

Fritshe, U.R., Kampman, B. & Bergsma, G. (2009). Better use of biomass for energy. Position paper of IEA RETD and IEA Bioenergy, Oeko-Institut, December 2009.

GAO (United States Government Accountability Office). (2009). Biofuels - Potential Effects and Challenges of Required Increases in Production and Use. Report to Congressional Requesters. August 2009. GAO-09-446. 184 pp.

Halliday, L. (2003). Soybeans for Livestock. Farm Extension Services, Agriculture, Fisheries and Aquaculture Prince Edward Island. http://www.gov.pe.ca/af/agweb/index.php3?number=79367&lang=E

Hardin, B. (1996). USDA Researches Improved Ethanol Yield from Corn. USDA News and Events. October 25, 1996. http://www.ars.usda.gov/is/pr/1996/ethanol1096.htm. Accessed 18 November 2010.

IEA (International Energy Agency), (2004) Biofuels for Transport - An International Perspective. Paris. http://www.iea.org/textbase/publications/free_new_Desc.asp?PUBS_ID=1262 Accessed 18 September, 2008.

IRGC (International Risk Governance Council). (2008). Policy Brief - Risk Governance Guidelines for Bioenergy Policies, Geneva, 2008. ISBN 978-2-9700631-0-0. 68 pp.

Jaques, A.P. (1992). Canada's greenhouse gas emissions: estimates for 1990. Report ESP 5/AP/4. Environmental Protection, Conservation and Protection, Environment Canada, ISBN 0-662-20187-6. 78pp.

Karman, D., Rowlands, D. Patterson, N. & Smith, M. (2008). Technical and Policy Implications of Transportation Biofuel Regulatory Approaches - Final Report to: Natural Resources Canada. By: Carleton University, Ottawa, July 2008. 164pp.

Khanna, M., Hochman, G., Rajagopal, D., Sexton, S. & Ziberman, D. (2009). Sustainability of food, energy and environment with biofuels. *CAB Reviews: Perspectives in Agriculture, Veterinary Science, Nutrition and Natural Resources*, Vol.4, No.028, (April 2009), pp. 1-10, ISSN 1749-8848.

Klein, K.K. & LeRoy, D.G. (2007). The Biofuels Frenzy: What's in it for Canadian Agriculture? Green Paper Prepared for the Alberta Institute of Agrologists. Presented at the Annual Conference of Alberta Institute of Agrologists. Banf, Alberta, March 28, 2007. Department of Economics, University of Lethbridge. 46 pp.

Klein, K., Romain, R., Olar, M. & Bergeron, N. (2004). Ethanol Policies, Programs and Production in Canada. Presented at the Agriculture as a Producer and Consumer of Energy Conference, June 24-25, 2004, Sponsored by Farm Foundation and USDA's Office of Energy Policy and New Uses.

McKague, K. (2009). On-Farm Biodiesel Production. Ontario Ministry of Agriculture Food and Rural Affairs (OMAFRA). http://www.omafra.gov.on.ca/english/engineer/facts/biodiesel.htm. Creation Date: 04 March 2009.

Murphy, S. (2008). The multilateral trade and investment context for biofuels: Issues and challenges. International Insitute for Environment and Development (IIED). London, and Insitute for Agriculture and Trade Policy (IATP), Minneapolis. 32pp. www.iatp.org or www.tadeobservatory.com.

Neitzert, F., Olsen, K. & Collas, P. (1999). Canada's Greenhouse Gas Inventory–1997 Emissions and Removals with Trends. Air Pollution Prevention Directorate, Environment Canada. Ottawa, Canada. ISBN 0-622-27783-X. Cat No. En49-8/5~9E.

Otto, M. (Editor). (2009). Towards Sustainable Production and Use of Reources: Assesing Biofuels. ISBN: 978-92-807-3052-4. United Nations Environmental Programme (UNEP). 120 pp.

Peña, N. (2008). Biofuels for Transportation: A Climate Perspective. Solutions White Papers Series. PEW Center on Global Climate Change. June 2008. 32 pp. http://www.pewclimate.org/docUploads/BiofuelsFINAL.pdf. Accessed 18 November 2010.

Perry, R.H. & Green, D.W. (editors). (1987). Perry's Chemical Engineers' Handbook, 6th Ed, McGraw-Hill Publishers. ISBN 978-0-07-142294-9.

Pielke, R.A. Sr. (2005). Land use and climate change. *Atmospheric Science*, Vol.310, No.5754, pp.1625-1626.

Reijnders, L. (2008). Transport biofuels - a life-cycle assessment approach. *Perspectives in Agriculture, Veterinary Science, Nutrition and Natural Resources.* 3(071) 8pp. doi:10.1079/PAVSNNR20083071. http://www.cababstractsplus.org/cabreviews

Sawyer, D. (2008). Climate change, biofuels and eco-social impacts in the Brazilian Amazon and Cerrado. *Phil. Trans. R. Soc.* B (2008) Vol.363, pp.1747-1752 doi: 10.1098/rstb.2007.0030

Sawyer, K. (2007). Ontario's Growing Corn Acreage. BioEnergy Canada, August-September 2007 Issue. http://www.bioenergymagazine.ca/article.jsp?article_id=10&article_title=Ontario%27s+Growing+Corn+Acreage&q=&page=1. Accessed 3 November 2010.

Simpson, T. (2009). Biofuels: The Past, Present, and a New Vision for the Future. *BioScience* Vol. 59, No. 11, pp. 926-927. (Viewpoint December 2009) doi:10.1525/bio.2009.59.11.2. www.biosciencemag.org

Sinclair, T.R. & Sinclair, C.J. (2010). *Bread, Beer & the Seeds of Change- Agriculture's Imprint on World History*. CAB International. 193 pp.

Whyte, M. (2008). Ethanol craze raises concerns, thestar website. Sunday March 16, 2008. http://www.thestar.com/news/article/346536. Accessed 3 November 2010.

Vergé, X.P.C., Dyer, J.A., Desjardins, R.L. & Worth, D. (2007). Greenhouse gas emissions from the Canadian dairy industry during 2001. *Agricultural Systems*, Vol.94, No.3, pp.683-693.

Vergé, X.P.C., Dyer, J.A., Desjardins, R.L. & Worth, D. (2008). Greenhouse gas emissions from the Canadian beef industry. *Agricultural Systems*, Vol.98, No.2, pp.126–134.

Vergé, X.P.C., Dyer, J.A., Desjardins, R.L. & Worth, D. (2009a). Greenhouse gas emissions from the Canadian pork industry. *Livestock Science*, Vol.121, pp.92-101.

Vergé, X.P.C., Dyer, J.A., Desjardins, R.L. & Worth, D. (2009b). Long term trends in greenhouse gas emissions from the Canadian poultry industry. *Journal of Applied Poultry Research*, Vol.18, pp.210–222.

Vergé, X.P.C., Worth, D.E. Dyer, J.A. Desjardins, R.L. & McConkey, B.G. (2011). LCA of Animal Production. In: *Green Technologies in Food Production and Processing*. Chapter 4. Editors: Yves Arcand and Joyce Boye. Springer. New York, NY. (In press).

Yacentiuk, M. (2001). Full Fat Soybeans in Swine Rations. Manitoba Agriculture, Food and Rural Initiatives. http://www.gov.mb.ca/agriculture/livestock/pork/swine/bab02s57.html.

Zhang, Z. & Wetzstein, M. (2008). Biofuel economics from a US perspective: past and future. *Perspectives in Agriculture, Veterinary Science, Nutrition and Natural Resources*. 3(075) 15 pp. doi:10.1079/PAVSNNR20083075. http://www.cababstractsplus.org/cabreviews Accessed 10 November 2010

10

Uncertainty Analysis of the Life-Cycle Greenhouse Gas Emissions and Energy Renewability of Biofuels

João Malça[1,2] and Fausto Freire[1]
[1]*ADAI-LAETA, Dept of Mech. Engineering, University of Coimbra, Coimbra,*
[2]*Dept of Mech. Engineering, ISEC, Coimbra Polytechnic Institute, Coimbra*
Portugal

1. Introduction

Biofuels can contribute substantially to energy security and socio-economic development. However, significant disagreement and controversies exist regarding the actual energy and greenhouse gas (GHG) savings of biofuels displacing fossil fuels. A large number of publications that analyze the life-cycle of biofuel systems present varying and sometimes contradictory conclusions, even for the same biofuel type (Farrell et al., 2006; Malça and Freire, 2004, 2006, 2011; Gnansounou et al., 2009; van der Voet et al., 2010; Börjesson and Tufvesson, 2011). Several aspects have been found to affect the calculation of energy and GHG savings, namely land use change issues and modeling assumptions (Gnansounou et al., 2009; Malça and Freire, 2011). Growing concerns in recent years that the production of biofuels might not respect minimum sustainability requirements led to the publication of Directive 2009/28/EC in the European Union (EPC 2009) and the National Renewable Fuel Standard Program in the USA (EPA 2010), imposing for example the attainment of minimum GHG savings compared to fossil fuels displaced.

The calculation of life cycle GHG emission savings is subject to significant uncertainty, but current biofuel life-cycle studies do not usually consider uncertainty. Most often, life-cycle assessment (LCA) practitioners build deterministic models to approximate real systems and thus fail to capture the uncertainty inherent in LCA (Lloyd and Ries, 2007). This type of approach results in outcomes that may be erroneously interpreted, or worse, may promote decisions in the wrong direction (Lloyd and Ries, 2007; Plevin, 2010). It is, therefore, important for sound decision support that uncertainty is taken into account in the life-cycle modeling of biofuels. Under this context, this chapter has two main goals: i) to present a robust framework to incorporate uncertainty in the life-cycle modeling of biofuel systems; and ii) to describe the application of this framework to vegetable oil fuel in Europe. In addition, results are compared with conventional (fossil) fuels to evaluate potential savings achieved through displacement. Following this approach, both the overall uncertainty and the relative importance of the different types of uncertainty can be assessed. Moreover, the relevance of addressing uncertainty issues in biofuels life-cycle studies instead of using average deterministic approaches can be evaluated, namely through identification of important aspects that deserve further study to reduce the overall uncertainty of the system.

This chapter is organized in four sections, including this introduction. Section 2 presents the comprehensive framework developed to capture uncertainty in the life-cycle GHG emissions and energy renewability assessment of biofuels, addressing several sources of uncertainty (namely parameter and modeling choices). Section 3 describes and discusses the application of this framework to vegetable oil fuel in Europe. Section 4 draws the conclusions together.

2. Framework: Energy and GHG life-cycle modeling addressing uncertainty

This section presents the biofuel life-cycle modeling framework used in this chapter. The most relevant methodological issues and sources of uncertainty in the energy and GHG assessment of biofuels are also discussed.

2.1 Life-cycle assessment of biofuels

A Life-Cycle Assessment (LCA) study offers a comprehensive picture of the flows of energy and materials through a system and gives a holistic and objective basis for comparison. The LCA methodology is based on systems analysis, treating the product process chain as a sequence of sub-systems that exchange inputs and outputs. The results of an LCA quantify the potential environmental impacts of a product system over the life-cycle, help to identify opportunities for improvement and indicate more sustainable options where a comparison is made. The LCA methodology consists of four major steps (ISO 14044, 2006):

- The first component of an LCA is the definition of the goal and scope of the analysis. This includes the definition of a reference unit, to which all the inputs and outputs are related. This is called the functional unit, which provides a clear, full and definitive description of the product or service being investigated, enabling subsequent results to be interpreted correctly and compared with other results in a meaningful manner;
- The second component of an LCA is the inventory analysis, also Life-Cycle Inventory (LCI), which is based primarily on systems analysis treating the process chain as a sequence of sub-systems that exchange inputs and outputs. Hence, in LCI the product system (or product systems if there is more than one alternative) is defined, which includes setting the system boundaries (between economy and environment, and with other product systems), designing the flow diagrams with unit processes, collecting the data for each of these processes, leading with multifunctional processes and completing the final calculations. Its main result is an inventory table, in which the material and energy flows associated with the functional unit are compiled and quantified;
- The third component of an LCA is the Life-Cycle Impact Assessment (LCIA), in which the LCI input and output flows are translated into potential contributions to environmental impacts. Different methods and models are available to conduct this step, based on aggregating and reducing the large amount of LCI data into a limited number of impact categories;
- Finally, interpretation is the fourth component of an LCA. The results of the life-cycle study are analyzed, so that conclusions can be drawn and recommendations made, according to the scope and objectives of the study.

Life-cycle studies of biofuel systems can be classified into three groups (Liska and Cassman, 2008; Cherubini and Strømman, 2011):

- life-cycle energy analysis, focused on fossil fuel requirements, energy efficiency and/or characterizing biofuel renewability);

- life-cycle GHG assessment (calculating the GHG balance); and
- life-cycle assessment, in which a set of environmental impact categories are investigated.

Furthermore, concerning the particular purpose of the biofuel LCA studies, the following subdivision can be made (van der Voet et al., 2010):

- comparative LCA, in which biofuel systems are compared with their fossil fuel equivalents on a life-cycle basis (e.g. GHG calculators used by governments to support biofuel policies);
- biofuel LCA used to obtain insight into the main environmental impacts of a specific chain (e.g. for generation of data on new production processes); and
- biofuel LCA used to identify main hotspots in the chain, which are specially suited for biofuel production companies aiming at realizing improvements in their processes.

Important methodological challenges within the field of biofuel LCA can be identified, namely concerning the choice of functional unit and definition of system boundaries. The definition of a functional unit is an important step in a Life-Cycle Assessment (Cherubini, 2010): it is a quantified description of the identified functions (performance characteristics) of a product system and provides a reference to which all other data (inputs and outputs) in the assessment are related (ISO 14040, 2006). The definition of the functional unit in biofuel life-cycle studies is related to the scope and system boundaries of the study; therefore, there is no single or preferred functional unit for biofuel assessments. The most common functional units found in the literature are (van der Voet et al., 2010; Malça and Freire, 2011):

- Service-oriented, e.g. 1 km driven in a specific vehicle;
- Energy-oriented, e.g. 1 MJ of biofuel energy content;
- Mass-oriented, e.g. 1 kg of biofuel produced;
- Volume-oriented, e.g. 1 liter of biofuel produced; and
- Land area-oriented, e.g. 1 ha of land for energy crop production.

The option for mass- or volume-based functional units have been used in several studies (e.g. Shapouri et al., 1995; Kim and Dale, 2002; Shapouri et al., 2002). However, in most cases this is not an adequate basis for comparison of the function provided by different (bio)fuels.

The functional unit chosen for the application reported in this chapter is 1 MJ of the final (bio)fuel product, measured in terms of the lower heating value (LHV, heat of combustion excluding the latent heat in combustion products, i.e. the specific enthalpy of vaporization of water). This functional unit is consistent with the goal and scope, which is to calculate the life-cycle GHG intensity (g CO_2eq MJ^{-1}) and energy renewability efficiency of European rapeseed oil and compare these values with their fossil fuel equivalents. Therefore, the system has been modeled taking into account the energy and GHG emissions required to deliver the biofuel to the end user, namely biomass cultivation, processing, transportation and storage of raw materials, followed by biofuel production and distribution. Setting theses boundaries is appropriate, because the goal and scope is concerned with biofuel use as a generic energy carrier, without a particular transportation or energy conversion system being considered. This assessment enables life-cycle inventory results to be analyzed in a variety of different ways, including hotspot identification and optimization of the biofuel chain, as well as calculation of potential energy and GHG reductions over fossil fuels.

Calculation of energy and GHG savings of biofuel systems requires the establishment of an appropriate baseline. The definition of a reference system is particularly used by legislation, which sets minimum levels for GHG emission savings that biofuels must achieve (e.g. EPC, 2009; EPA, 2010). Most commonly, the reference system used is a fossil fuel pathway (gasoline or diesel). However, the EU directive 2009/28/EC (EPC, 2009) has adopted a generic reference value for fossil fuels used for transportation (83.8 g CO_2eq MJ^{-1}), not distinguishing between petrol and diesel. For bioliquids used for electricity production the reference value adopted is 91 g CO_2eq MJ^{-1}, for bioliquids used for heat production the value is 77 g CO_2eq MJ^{-1}, and for cogeneration is 85 g CO_2eq MJ^{-1}. A justification for adopting distinct values based on the type of final use and not on the fossil fuel displaced could not be found in directive 2009/28/EC. In this chapter, petroleum diesel is the reference system, and includes extraction, transport and refining of crude oil, and distribution of final fuel.

2.2 Energy analysis

Several reasons motivate the sometimes diverging results of life-cycle energy analyses of biofuel systems, namely (i) the quantification of energy fluxes either in terms of final energy or in terms of primary energy; and (ii) the use of different metrics for energy efficiency. These topics are explored in this section.

Energy resource depletion must be quantified in terms of primary energy - energy embodied in natural resources (e.g. coal, crude oil, uranium or biomass) that has not undergone any anthropogenic conversion or transformation. Primary energy is the sum of the final energy with all the transformation losses, with fuel primary energy values being greater than their final energy values. In fact, consumers buy final energy, but what is really consumed is primary energy, which represents the cumulative energy content of all resources (renewable and non-renewable) extracted from the environment. In the case of fuels, energy inputs required during the extraction, transportation and production processes measured in terms of primary energy ($E_{in,prim}$, MJ kg^{-1}), do not include the energy embodied in the final fuel, i.e. the fuel energy content (FEC, MJ kg^{-1}). Even though, the energy requirement of fossil fuels should also include the FEC, in which case the result is referred to as the gross energy requirement (GER, MJ kg^{-1}) (Mortimer et al., 2003):

$$GER = E_{in,non\text{-}renewable,prim} + FEC \quad (1)$$

In (bio)energy analysis studies it is essential to distinguish between non-renewable ($E_{in,non\text{-}renewable,prim}$) and renewable ($E_{in,renewable,prim}$) energy inputs, because we are concerned with the renewable nature of biofuels and the depletion of fossil fuels. Therefore, the essential comparison that needs to be made is between the non-renewable primary energy input to the biofuel life-cycle ($E_{in,non\text{-}renewable,prim}$) and the non-renewable primary energy requirements throughout the life-cycle of fossil fuels, including the fossil fuel energy content, i.e. the GER.

The life-cycle inventory results provide an opportunity to quantify the total energy demand and, therefore, the overall energy efficiency. Quantifying the overall energy efficiency of a biofuel is helpful to determine how much (non-renewable) energy must be expended to produce biomass and convert its energy to 1 MJ of available energy in the transportation fuel. The more non-renewable energy is required to make the biofuel, the less we can say the biofuel is "renewable". Thus, the renewable nature of a fuel can vary across the spectrum of "completely renewable" (i.e. zero non-renewable energy inputs) to non-renewable (i.e. non-renewable energy inputs as much or more than the energy output of the fuel) (Sheehan et al., 1998).

Within the energy analysis and LCA literature there is lack of consensus concerning the definition (and designation) of energy efficiency indicators to be used in a life-cycle perspective and, in particular, to characterize the energy requirements of renewable energy systems. In fact, various indicators have been used, often with the same meaning but different definition, or inversely, e.g. overall energy efficiency (Boustead and Hancock, 1979; Boustead, 2003); energy efficiency (ADEME, 2002); gross energy requirement and net energy requirement (Wilting 1996); energy requirement (Whitaker et al., 2010); overall energy balance (Armstrong et al., 2002); energy balance (Börjesson and Tufvesson, 2011); cumulative energy demand (Huijbregts et al., 2006); input/output energy balance, cumulative energy requirement, fossil energy requirement, and renewable energy requirement (Cherubini et al., 2009); net energy use, and energy substitution efficiency (Gnansounou et al., 2009); energy ratio (Liska and Cassman, 2008; Papong and Malakul, 2010); net energy yield (Liska and Cassman 2008); and energy return on investment[1] (Poldy, 2008). In particular, Sheehan et al. (1998) have used the life-cycle energy efficiency (LCEE), defined as the ratio between the biofuel energy content and the biofuel GER:

$$LCEE = \frac{\mathrm{FEC}}{(\mathrm{E}_{\mathrm{in,\ non-renewable,prim}} + \mathrm{FEC})} \tag{2}$$

The LCEE can be seen as a measure of the fraction of the GER (primary energy required throughout the biofuel life-cycle plus the biofuel energy content), which actually ends up in the fuel product. The same authors (and others, e.g. Lechón et al., 2009) have also adopted the fossil energy ratio (FER), defined as:

$$FER = \frac{\mathrm{FEC}}{E_{in,non-renewable,prim}} \tag{3}$$

According to this definition, if the fossil energy ratio is less than 1 the fuel is nonrenewable, as more energy is required to make the fuel than the energy available in the final fuel product. Biofuel with FER greater than 1 can be considered as (partially) renewable. In theory, a total renewable fuel would have no fossil energy requirement and, thus, its fossil energy ratio would be infinite. Other authors have also used the FER indicator, but under a different designation, for example "energy efficiency" (ADEME, 2002), whereas others have used the "energy requirement" (E_{req}), defined as the "primary energy input per delivered energy output" (Mortimer et al., 2003; Malça and Freire, 2004, 2006; Hoefnagels et al., 2010):

$$E_{req} = \frac{E_{in,non-renewable,prim}}{\mathrm{FEC}} \tag{4}$$

The energy requirement indicator is also used in Kim and Dale (2002) and Armstrong et al. (2002), but under the designation of "net energy" and "overall energy balance", respectively. It should be noted that E_{req} is the inverse of FER.

The "net energy value" (NEV), defined as the biofuel FEC minus the non-renewable energy required to produce the biofuel):

[1] To distinguish it from a financial measure, the energy return on investment (EROI) is sometimes called energy return on energy investment (EROEI) (Poldy, 2008).

$$NEV = FEC - E_{in,non-renewable,prim} \tag{5}$$

is used e.g. in Shapouri et al. (1995), Shapouri et al. (2002), Liska and Cassman (2008) and Papong and Malakul (2010)[2]. In this case, negative net energy values indicate that (bio)fuel is non-renewable, while positive values indicate the fuel is renewable to a certain extent.

According to Liska and Cassman (2008) and Cherubini et al. (2009), input-output ratios and primary energy requirements receive most attention when assessing the efficiency of bioenergy systems, because they provide a straightforward basis for comparison with conventional fossil fuel systems. Moreover, these metrics are usually thought as a surrogate for GHG emissions mitigation (Liska and Cassman, 2008). Nevertheless, intensity factors do not provide a measure of the "energy productivity" of a system on a land-area basis, which should be the chosen parameter when dedicated energy crops compete with food, feed or fiber under land-availability constraints (Liska and Cassman, 2008; Cherubini et al., 2009; Cherubini and Strømman, 2011). An example is the net energy yield NEY (GJ ha^{-1}) used by Liska and Cassman (2008), which combines energy efficiency and productivity into one single parameter.

Another metric, the Energy Renewability Efficiency, aiming at characterizing the renewability of (bio)fuel systems has been proposed by Malça and Freire (2004, 2006). The energy renewability efficiency (ERenEf) measures the fraction of final fuel energy obtained from renewable sources by subtracting from FEC all the inputs of non-renewable primary energy (Malça and Freire, 2006). It thus provides a more adequate means for quantifying the renewability degree (or its lack) of a particular energy system. ERenEf can be defined as:

$$ERenEF[\%] = \frac{\left(\mathrm{FEC} - \mathrm{E}_{\mathrm{non-renewable,prim}}\right)}{\mathrm{FEC}} \times 100 \tag{6}$$

A biofuel may be considered renewable if ERenEf assumes values between 0 and 100%. In case there were no inputs of non-renewable energy, the biofuel would be completely renewable with an ERenEf of 100%. If the ERenEf is lower than zero, then the biofuel should be characterized as non-renewable since the non-renewable energy required to grow and convert biomass into biofuel would be greater than the energy present in the biofuel final product. In this case, the biofuel is, indeed, not a fossil energy substitute and increasing its production does little to displace oil imports or increase the security of energy supply. By definition, non-renewable energy sources have negative values of ERenEf, with increasing negative values as life-cycle energy efficiency decreases. For example, fossil diesel (the fossil fuel displaced by rapeseed oil shows an average ERenEf value of -14.0%, meaning that the total primary energy required to produce fossil diesel is 14.0% greater than its final energy content.

2.3 GHG assessment

This section presents the methodology used for calculating the GHG balance of biofuel systems. Important issues in the GHG assessment of biofuels, such us carbon stock changes associated with land use change and soil emissions from land use, and how they are

[2] Papong and Malakul (2010) also use this net energy definition, although under the name "Net Energy Gain".

addressed in the practical modeling of the life-cycle are discussed. Generic assumptions concerning GHG accounting are also formulated.

The life-cycle GHG balance of biofuel systems can be calculated by summing up the GHG emissions of the several process steps, namely land use change, cultivation of raw materials (soil preparation, fertilization, sowing, weed control, and harvesting) and biofuel production (transport, storage and drying of feedstock, processing of feedstock into biofuel, and biofuel transport to the final user). Biofuel use (combustion in engines or boilers) is not explicitly modeled, but is assumed that tailpipe CO_2 emissions from biofuel combustion are neutral, being balanced by the CO_2 sequestered during crop growth, which does not occur for fossil fuels. An alternative approach would be to distinguish between fossil and biogenic CO_2 emissions throughout the life-cycle (see e.g. Rabl et al. 2007; Guinée et al. 2009; Luo et al. 2009).

One emerging but highly controversial issue in the GHG balance of biofuels is indirect land use change (iLUC) (Anex and Lifset, 2009; Liska and Perrin, 2009). In the approach proposed in this chapter iLUC is not considered, but a brief discussion is presented at the end of this section. The greenhouse gases considered are carbon dioxide (CO_2), methane (CH_4) and nitrous oxide (N_2O), with average global warming potentials (100 year time horizon) of GWP_{CH4}=25 and GWP_{N2O}=298. Other GHG emissions from biofuel systems were found to be negligible and were not pursued. Global Warming Potentials used by the IPCC provide "CO_2 equivalence" factors for greenhouse gases other than CO_2, which allows aggregation of emissions of different gases into a single metric (IPCC, 2007). In terms of global warming, GWP_{CH4}=25 means that 1 g of methane released to the atmosphere is equivalent to the release of 25 g of carbon dioxide. In practical terms, GHG emissions in each step are multiplied by the respective equivalence factors and summed up yielding a single figure in CO_2 equivalents. Finally, the GHG emissions of the overall biofuel chain can be calculated.

GHG emissions for feedstock and energy inputs are calculated by using suitable emission factors (Mortimer and Elsayed, 2006; Malça and Freire, 2010).

For comparative and decision purposes, GHG emission savings can be calculated by comparing the life-cycle GHG emissions of biofuels with the equivalent emissions of fossil fuels, following the methodology used e.g. in EPC (2009):

$$GHG_{emission\ savings}\ [\%] = \frac{(\text{Fossil Fuel}_{\text{emissions}} - \text{Biofuel}_{\text{emissions}})}{\text{Fossil Fuel}_{\text{emissions}}} \times 100 \qquad (7)$$

DIRECT LAND USE CHANGE AND LAND USE. Soil carbon stock change is an emergent topic in the literature and can contribute significantly to biofuel GHG intensity (EC, 2010a). However, it is site specific and highly dependent on former and current agricultural practices, climate and soil characteristics and, thus, previous biofuel LCA studies have neglected this issue (Larson, 2006; Malça and Freire, 2011). A change in land use (for example, set-aside land to cropland) or in agronomic practices (change to low tilling, for example) can liberate carbon that had previously been sequestered over a long period of time or, conversely, lead to a carbon build-up in the soil (Cherubini and Strømman, 2011). Moreover, soil organic carbon (SOC) stock exchange is a relatively slow process and thus difficult to measure (Heller et al., 2003). IPCC (2006) guidelines indicate a default time period for transition between equilibrium SOC values (i.e. soil carbon levels from which there is no further net accumulation or degradation) of 20 years.

Annualized soil carbon stock variations due to land use change and practices ΔC_{LUC-a} (tonnes per hectare per year, t C ha^{-1} yr^{-1}) are given by (EPC, 2009)

$$\Delta C_{LUC-a} = \frac{CS_R - CS_A}{T_{LUC}} \tag{8}$$

in which CS_R (tC ha^{-1}) is the carbon stock (CS) per unit area of the reference land use (cropland, set-aside land or grassland), CS_A (tC ha^{-1}) is the carbon stock per unit area associated with the arable use of soils, and T_{LUC} (yr) is the time period for transition between equilibrium carbon stocks. Actually, set-aside lands and grasslands placed in cultivation lose soil carbon at an exponential rate (JEC, 2007): most of the carbon loss occurs within the first few years following initial cultivation. A discussion of the temporal dynamics of GHG emissions caused by land use change is, however, beyond the scope of this chapter.

Carbon stocks per unit area CS_R and CS_A include both soil and vegetation and can be calculated according to EC (2010) rules. The soil organic carbon (SOC) content is given by $SOC = SOC_{ST}.F_{LU}.F_{MG}.F_I$, in which SOC_{ST} is the standard soil organic carbon in the 0-30 cm topsoil layer, F_{LU} is a factor reflecting the type of land use, F_{MG} reflects the adopted soil management practices and F_I quantifies the level of carbon input to soil. Carbon stock values concerning above and below ground vegetation as provided in EC (2010) guidelines are also included in the calculation of the overall land carbon stock.

Several authors call the amount of CO_2 emissions from land use change the "carbon debt" of land conversion (Fargione et al., 2008). Over time, this carbon debt can be gradually compensated if GHG emission savings of growing biofuels while displacing fossil fuels are realized. The period of time that biofuel production takes to repay the carbon debt is called the carbon payback time; it is calculated by dividing the net carbon loss from LUC per hectare by the amount of carbon saved per hectare and per year by the use of biofuels, excluding LUC emissions (Wicke et al., 2008).

The calculation of GHG emissions also includes emissions of nitrous oxide (N_2O) from soil. The assessment of N_2O emissions from soil has recently proven to be an important issue in the GHG balance of biofuels (Crutzen et al., 2008; Reijnders and Huijbregts, 2008). Agricultural practices, and particularly the use of fertilizers containing nitrogen, are important issues affecting the emission of N_2O from soils (Kaiser et al., 1998; Reijnders and Huijbregts, 2008). Generally, a small amount of the nitrogen in the fertilizer ends up being released to the atmosphere as N_2O, both *i)* directly, from nitrification of nitrogen in the fertilizer and from crop residues; and *ii)* indirectly, following volatilization of NH_3 and NO_x and after leaching and runoff of N from managed soils (IPCC, 2006). Because N_2O has a high impact on global warming, its emissions from agricultural soils cannot be neglected. The contribution to net emissions of N_2O from nitrogen fertilizer application is one of the most uncertain variables due to the number of parameters that can affect its value (Larson, 2006). Actual emissions from fields vary depending on soil type, climate, tillage method, fertilizer application rates and crop type (Larson, 2006; Reijnders and Huijbregts, 2008; Stephenson et al., 2008; Crutzen et al., 2008).

INDIRECT LAND USE CHANGE. An aspect that requires a consequential approach in life-cycle studies is the assessment of indirect land use change associated with biofuels. Increased biofuel demand may lead to an expansion of cropped area at the expenses of other land uses. The displacement of prior crop production to other areas (indirect LUC) may contribute to important environmental impacts, namely GHG emissions (Fargione et al.,

2008; Searchinger et al., 2008; Wicke et al., 2008), which has recently been the subject of important controversy among the scientific community. This builds on the fact that market mechanisms should be taken into account when modeling all the consequences of increased consumption of biofuels, which requires subjective assumptions and leads to potentially higher complexity and uncertainty.

A report by Croezen et al. (2010) discussed the use of different agro-economic models - simulating global agricultural markets, trade, intensification, possible crop replacements - to estimate iLUC implications and showed that overall emissions from iLUC are within 10 to 80 g CO_2 MJ^{-1} of biofuel produced. Other attempts for addressing indirect land use change and its influence on life-cycle results, namely through the use of single CO_2 emission factors -the iLUC factor approach-, have also been conducted (e.g. Bowyer, 2010; Fritsche et al., 2010). Nevertheless, these models likely estimate GHG emissions from iLUC with significant inaccuracy (Cherubini and Strømman, 2011). Further work is still required to address the practical modeling of indirect LUC associated with biofuels, as stated e.g. by Anex and Lifset (2009), Liska and Perrin (2009) and Kløverpris et al. (2008), so that a harmonized methodology can be established. Also, the EU recognizes in a report published on December 2010 (EC, 2010b) that a number of uncertainties associated with iLUC modeling remain to be addressed, which could significantly impact the results. Therefore, indirect LUC is beyond the scope of this chapter.

2.4 Uncertainty analysis

Uncertainty analysis is a systematic procedure to determine how uncertainties in data and assumptions propagate throughout a life-cycle model and how they affect the reliability of the life-cycle study outcomes. Uncertainties may occur in the several phases of an LCA, namely in the goal and scope definition, inventory analysis and impact assessment. Examples are provided e.g. in Björklund (2002), Huijbregts (1998), Heijungs and Huijbregts (2004), and Geisler et al. (2005).

In general, results of a life-cycle study can be uncertain for a variety of reasons (Morgan and Henrion, 1990; Huijbregts, 1998; Björklund, 2002; Huijbregts et al., 2003; Heijungs and Huijbregts, 2004; Lloyd and Ries, 2007), and different typologies can be used to describe the uncertainties considered. According to Huijbregts (1998), the following sources of uncertainty in LCA can be distinguished:

- parameter uncertainty, which arises from lack of data, empirical inaccuracy (imprecise measurements), and unrepresentativity of data (incomplete or outdated measurements);
- uncertainty due to choices (or scenario uncertainty), which reflects the inherent dependence of outcomes on normative choices in the modeling procedure (e.g. choice of functional unit, definition of system boundaries, or selection of allocation methods); and
- model uncertainty, due to the use of mathematical relationships between model inputs and outputs that simplify real-world systems.

In general, parameter and model uncertainty are characterized by means of probability distributions, whereas uncertainty due to choices is addressed through the development of unique scenarios (Lloyd and Ries, 2007; Malça and Freire, 2010).

PARAMETER UNCERTAINTY. Every type of modeling is associated with uncertainties in its parameters (Schade and Wiesenthal, 2011). In this article, a robust approach is used to address and incorporate parameter uncertainty in the life-cycle modeling of rapeseed oil. The main steps of this approach can be summarized as follows:

- firstly, a preliminary sensitivity analysis is conducted, in which single parameter variations are tested to see how the results are affected. The merit of this step is to identify the parameters with the highest impact on the model outputs, and thus the parameters that require particular attention in the next steps;
- secondly, a literature review is conducted to identify variation ranges and assign appropriate probability density functions for the most influential parameters;
- thirdly, an uncertainty propagation method is used (with Monte-Carlo simulation) for calculating probability distributions of output variables based on the uncertainty within selected input parameters;
- finally, an uncertainty importance analysis is conducted in order to identify the parameters that contribute most to the overall output variance.

Although widely used, single sensitivity analysis generally underestimates the uncertainty in a model (Plevin, 2010), as e.g. with non-linear models, where the sensitivity to a specific parameter depends on the nominal values assigned to other variables (Saltelli et al., 2006). This case requires that sensitivity is assessed with parameters varying simultaneously, i.e. using global sensitivity analysis. A common technique for global sensitivity analysis is Monte-Carlo simulation. Monte-Carlo simulation is based on the repetition of many individual model iterations (typically from hundreds to thousands), with each iteration using a randomly constructed set of values selected from each parameter probability distribution. The set of model outputs computed by the simulation is then aggregated into a probability distribution. The Oracle Crystal Ball software package was used to perform Monte-Carlo simulation (Oracle, 2010).

To compare the relative importance of the uncertainty in input parameters to the model output uncertainty, an uncertainty importance analysis is performed. Generally, a limited number of parameters account for the majority of uncertainty in the model outputs (Morgan and Henrion, 1990). The merit of estimating uncertainty importance is to identify these parameters, and thus guide further research to reduce their uncertainty. Moreover, the remaining parameters (typically a much larger set), which contribute negligibly to the overall variance, can be treated as uncertain, simplifying the model and saving computation time.

UNCERTAINTY OF GLOBAL WARMING POTENTIALS. Several time horizons can be adopted for the estimation of GHG emissions. Taking into account the short- to mid-term implications of first generation biofuels in terms of global warming effect, the most commonly used time horizon of 100-years has been chosen for GWP estimation in the application presented in section 3. Nonetheless, other time horizons can be adopted. Results with GHG emissions for various time horizons (20, 100 and 500-year) have been calculated by the authors of this chapter and it has been concluded that 500-yr GHG emissions are lower due to a significantly lower GWP of nitrous oxide (153 vs. 298 kg CO_2eq for 500- and 100-yrs, respectively). Moreover, uncertainty ranges for a 500-yr timeframe are narrower than corresponding 100-year values, because of the lower uncertainty in the estimation of GWP_{N2O}. On the other hand, calculated RO GHG emissions for 20- and 100-yr time horizons are similar, because 20- and 100-yr GWPs of N_2O are also very similar. Since methane (CH_4) hardly contributes to the life-cycle GHG emissions of RO (Malça and Freire, 2009), the implications of GWP_{CH4} variation between different time horizons are not significant. An uncertainty of ±35% for the 90% confidence range has been considered for GWP_{CH4} and GWP_{N2O}, according to IPCC (2007).

MULTIFUNCTIONALITY (Scenario Uncertainty). Most industrial and agricultural processes are multifunctional. In particular, many of the feedstocks for biofuels are either co-produced with other products or are from by-products from other production processes. Biofuel production systems generate large quantities of co(by)-products and thus LCA practitioners are faced with the problem that the product system under study provides more functions than that which is investigated in the functional unit of interest. This leads to the following central question: how should the resource consumption and energy used be distributed over the various co(by)-products? An appropriate procedure is required to partition the relevant inputs and outputs to the functional unit under study.

The international standards on LCA include several options for dealing with co-production (ISO 14044, 2006): i) sub-dividing the process into two or more sub-processes; ii) expanding the product system to take into account potential effects of providing a new use for the co-products on systems currently using the co-products - known as system boundary expansion - and iii) allocating inputs and outputs between product streams based on causal relationships.

Although allocation methods are straightforward to implement, they "arbitrarily" allocate inputs and outputs on the basis of specific relationships between co-products (Weidema, 2003). For this reason, ISO standards on LCA indicate that allocation[3] should be avoided, wherever possible, in favor of subdividing the system in sub-processes (often not possible) or by expanding the system (system boundary expansion). As explained by Guinée et al. (2009), system expansion (also called system extension) means extending the product system to include additional functions related to the co-products. As a result, the system includes more than one functional unit. Sometimes the expression "system extension" refers to what actually is the "substitution method" (also called "replacement method", "displacement method" or "avoided-burdens" approach). Substitution refers to expanding the product system with "avoided" processes to remove additional functions related to the functional flows of the system. In this case, energy and emission credits can be assumed equal to those required to produce a substitute for the co-products.

Allocation can be based on physical properties of the products, such as mass, volume, energy, carbon content, because data on the properties are generally available and easily interpreted. Where such physical causal relationships cannot be used as the basis for allocation, the allocation should reflect other relationships between the environmental burdens and the functions. Many biofuel life-cycle studies use the mass of co-products as the basis for partitioning the system (e.g. ADEME, 2002; Neupane et al., 2011). Other studies use the energy content (e.g. Janulis, 2004; Wagner et al., 2006). However, the main reason for using mass seems to arise because both main and co-products can be weighted, and the use of energy content would only be relevant if both main and co-products were actually burned as fuels. Nonetheless, mass and energy allocation factors do not change over time, like economic factors or substituted product types do (Hoefnagels et al., 2010). At the European policy level, energy allocation has been selected as the method for the regulation of individual economic operators, because it is easy to apply, is predictable over time and

[3] The meaning of allocation in LCA is often used misleading. According to ISO 14044:2006, sub-division and system boundary expansion are not formally part of the allocation procedure.

minimizes counter-productive incentives (EPC, 2009). Allocation can also be based on the exergy (e.g. Frischknecht, 2000; Dewulf et al., 2005) or carbon (e.g. Gnansounou et al., 2009) content of the co-products. Allocation based on the relative economic value (market price) of main and co-products is used e.g. by Guinée et al. (2004), Zah et al. (2007), Reijnders and Huijbregts (2008), and Menichetti and Otto (2008). The rationale for economic allocation is that demand is the driving force of production systems and thus their environmental burdens should be allocated according to market principles (Gnansounou et al., 2009). Compared to physical allocation, economic allocation produces results that are more rational when large quantities of by-products with low economic value are produced (Börjesson and Tufvesson, 2011). Nevertheless, the volatility of market prices, subsidies and market interferences are pointed out as the main drawbacks of this method, as they may strongly influence the calculation of allocation parameters and thus the results of the life-cycle study (Gnansounou et al., 2009). Finally, some authors (e.g. Huo et al., 2009) use a mix of allocation and/or substitution methods to address co-product credits in biofuel chains, i.e. they use a hybrid approach.

The issue of the most suitable allocation method is still open (Cherubini, 2010). In most studies no discussion is provided regarding the selection of the allocation procedure and, in general, no complete justification can be found concerning the reason to choose one and not a different allocation procedure. In fact, it is important to recognize that there is no single allocation procedure deemed appropriate for all biofuel processes (Mortimer et al., 2003). Therefore, whenever several alternative allocation procedures seem applicable, a sensitivity analysis should be conducted (ISO 14044:2006).

Several authors demonstrate that the choice and justification of allocation procedures are major issues in biofuel life-cycle studies, as they can have a significant influence on the results (Malça and Freire, 2004, 2006, 2010; Cherubini et al., 2009; Gnansounou et al., 2009; van der Voet et al., 2010). Moreover, the large influence of methodological choices (including allocation methods) may override many other types of uncertainty, as pointed out by Björklund (2002). This opinion is shared by Morgan and Henrion (1990) and Krupnick et al. (2006), who state that in some models the differences between scenarios may overcome parameter uncertainty and variability. Nevertheless, uncertainty due to choices cannot be eliminated, but can be rather easily illustrated by identifying the relevant alternatives and performing sensitivity analysis.

Section 3 presents an application of the approach presented and discussed in this section. Energy renewability efficiency and GHG intensity of rapeseed oil have been calculated capturing parameter uncertainty and alternative co-product treatment approaches.

3. An application to vegetable oil fuel in Europe

3.1 Vegetable oil use

Pure vegetable oil, also known as pure plant oil or straight vegetable oil, is an alternative fuel for diesel engines in transportation and also stationary applications, namely for heating purposes and/or electricity generation. The use of vegetable oils in internal combustion engines dates back to the beginning of the XX century, when a compression ignition engine, first developed by Rudolf Diesel, worked on peanut oil at the 1900's World Exhibition in Paris (Knothe, 2001). Vegetable oils were used in diesel engines for only a few years, however, until manufacturers optimized the engine design for low-grade fractions of petroleum in the 1920´s (Luque et al., 2008). Oil shortages in the 1930´s and 1970´s promoted once more research into the use of vegetable oil for energy purposes, as well as during World War II when vegetable oils were used as emergency fuels. An interesting aspect in the historical development and promotion of vegetable oils is that environmental issues were set aside and no emission studies were conducted (Knothe, 2001).

Main applications of vegetable oils include motor vehicles, e.g. passenger cars and agricultural machinery equipped with compression ignition engines, and stationary applications, like power generation (with diesel engine or gas turbine generators) and boiler heating systems (Cocco, 2009; Chiaramonti and Tondi, 2003). Concerning vehicle applications, vegetable oils represented almost 10% of the European biofuel consumption for road transport in 2007. Germany was the leading country in using motor fuels based on pure vegetable oils, with a consumption of approximately 1 Mtonnes in 2006 and 750 ktonnes in 2007 (EurObserv'ER, 2008). Other European countries using pure vegetable oils for automotive purposes include the Netherlands and Ireland, with several projects benefiting from excise duty relief and promoting the use of straight vegetable oil among transport operators, particularly vehicle fleets (DCENR, 2007; DMFA, 2007).

Direct utilization of vegetable oil does not pose significant technical challenges, if the energy system is conveniently adapted to the characteristics of the fuel (Chiaramonti and Tondi, 2003). Several properties of vegetable oils differ significantly from petroleum diesel, namely kinematic viscosity, thus requiring adjustments on fossil fuel-based systems for reliable operation. The kinematic viscosity of vegetable oils is about one order of magnitude greater, which thickens the lubricating oil and causes poor atomization in the combustion chamber, leading to incomplete combustion and carbon deposits (Chiaramonti and Tondi, 2003). The high viscosity of vegetable oils also limits their use during wintertime. To overcome the difficulties associated with straight vegetable oil (SVO) use, the viscosity must be reduced, which is usually achieved by preheating the vegetable oil. Typical conversion kits include a heat exchanger in which waste heat from the engine is supplied to the oil. This heat source can be supplemented by an electric booster (Mondal et al., 2008). An additional (small) fuel tank for fossil diesel and a three-way valve to switch between the main SVO tank and the small fossil diesel tank complete the package.

The need for modified diesel engines and the lack of a fuel distribution system are the main barriers for the dissemination of vegetable oil-powered vehicles. Moreover, there is no consensus on the suitability of SVO use in diesel vehicles equipped with direct injection engines or electronic injection pumps (Sidibé et al., 2010; Misra and Murthy, 2010). Large stationary applications, on the other hand, seem more feasible for the use of vegetable oil as diesel fuel. Actually, diesel engines and boilers that are able to burn low grade (and high viscosity) fuel oil can be easily switched to vegetable oil, owing to fuel and plant operation similarities. Furthermore, vegetable oils may not require complex or costly upgrading processes when used in heating systems. The main barrier for SVO penetration in large scale applications is rather economic, as a result of the low market prices of the fuel oils traditionally used (Chiaramonti and Tondi, 2003).

Several factors determine which vegetable oils are of most interest for energy purposes, namely geographic region, climate and economics. The main feedstock in cold, temperate European regions is rapeseed (canola); in the United States and some countries in Latin America soybean is the most used raw material; and in tropical countries, palm is the preferred feedstock. This article focuses on the direct use of rapeseed oil (pure or blended) as an energy carrier. Table 1 lists European data on rapeseed and rapeseed oil production, including the top 4 producing countries.

	2009 rapeseed production (10³ tonne)	World ranking [(a)]	2009 area harvested (10³ ha)	2009 crop yield (tonne/ha)	2005-2009 avg. crop yield [(b)] (tonne/ha)	2009 rapeseed oil production (10³ tonne)
Germany	6306.7	4th	1471.2	4.29	3.80 (3.44–4.29)	3345.3
France	5584.1	5th	1480.8	3.77	3.32 (2.90–3.77)	1742.6
Poland	2496.8	7th	810.0	3.08	2.75 (2.64–3.08)	870.8
UK	1951.0	8th	580.6	3.36	3.25 (3.10–3.36)	779.0
Total (EU-27)	**21417.6**	**-**	**6015.9**	**2.92**	**-**	**8466.7**
Top-4 share (%)	**76.3**	**-**	**72.2**	**-**	**-**	**79.6**

[(a)] World rankings for 2008; [(b)] minimum and maximum rapeseed yields in brackets.

Table 1. Rapeseed and rapeseed oil production in the EU-27, including major producers. (FAOSTAT, 2011; EUROSTAT, 2011)

Vegetable oils are currently used as diesel fuel for automotive purposes, and in thermal and power plants for heat and electricity production. Even though technological challenges for the use of straight vegetable oils have been overcome, there are still several non-technical barriers, namely the need for systems adaptation to run on SVO and the lack of a fuel distribution network, which do not occur with fossil fuels. Moreover, higher vegetable oil costs in comparison to fossil fuels also halt the market penetration of stationary SVO applications, as shown by the lower prices of heavy fuel oil for industry (Tables 2 and 3). For automotive applications, however, fuel costs work as an incentive for the promotion of SVO, with SVO prices lower than automotive fossil diesel prices.

On the other hand, the use of SVO seems very promising in developing countries, where self energy production at minimal costs is of greatest importance. The development of vegetable oil production chains, combining simpler production technology with lower production costs - e.g. mechanical oilseed presses, hand- or fuel-driven - is an approach that may greatly contribute for the socio-economic welfare of populations in these countries.

Year	Rapeseed oil
2005	669.4
2006	793.6
2007	970.0
2008	1329.2
2009	858.7
2010	951.1

[(a)] Prices paid at the farm gate.

Table 2. Annual average prices (US$/tonne) of rapeseed oil [(a)] (FAOSTAT, 2011)

Country	Heavy fuel oil for industry	Automotive diesel fuel [(a)]
Germany	515.2	1594.1
France	542.0	1483.5
Poland	590.0	1332.9
United Kingdom	n/a	1785.9

n/a: data not available; [(a)] A density of 0.85 kg/liter for diesel fuel has been used.

Table 3. Retail prices of selected fuels (US$/tonne) for the 1st quarter of 2010 (IEA, 2010)

3.2 Life-cycle modeling and inventory incorporating uncertainty

3.2.1 RO life-cycle chain

The life-cycle stages of the RO chain include rapeseed cultivation, harvesting, transport and drying of the seeds, crushing and extraction of the oil, oil degumming and refining. These steps are illustrated in the flowchart of Fig. 1. A detailed description of the RO production system can be found, for example, in Mortimer and Elsayed (2006), Stephenson et al. (2008) and Malça and Freire (2009, 2010).

Rape (Brassica napus L.), also known as Rapeseed, Oilseed Rape or Canola, is a yellow-flowered member of the family Brassicaceae widely cultivated throughout the world for the production of vegetable oil for human food consumption, but increasingly used for energy. Different cultivation methods may be used, namely in terms of soil management and soil inputs, depending on the climate region, soil type, and established agricultural practices. The cultivation step includes soil preparation, fertilization, sowing, weed control, and harvesting. Seeds are separated from the rest of the plant during harvesting. The straw, consisting of stalks, pods and leaves, is usually ploughed back into the field (SenterNovem, 2005; JEC, 2007; UFOP, 2008; Börjesson and Tufvesson, 2010). Several studies point out the incorporation of straw in the soil as a farm management activity with several benefits, namely the return and cycling of nutrients, the building of soil organic matter and the prevention of soil erosion.

Following harvesting, oilseeds are cleaned and dried. The typical moisture content of oilseeds is reduced, as required by oil extraction facilities and to ensure stability in storage. Moreover, large scale oil extraction is usually preceded by grinding and cooking of the seeds, to facilitate the oil extraction process. Vegetable oil may be extracted from the seeds by physical and/or chemical extraction. Different types of mechanical extraction devices can be used, namely the screw press and the ram press (Tickell et al., 2003). The first uses a screw inside a metal housing; as the screw turns, the oil is squeezed out of the seeds. The ram press uses a piston-cylinder set to crush the oilseeds. After mechanical pressing, protein-rich cake is also produced and can be used in animal feed. The press cake has, however, high oil content and a further (chemical) extraction step is usually conducted to extract the remaining oil, in order to increase the overall vegetable oil yield. Chemical extraction uses a petroleum-derived solvent, usually hexane; this is the extraction method considered in this chapter.

When solvent extraction is used, the oil goes through a distillation process to recover the hexane, which is recycled back to the oil extraction process. The final step in the production of vegetable oils is oil refining, which includes degumming, neutralization and drying. Gums are precipitated by the addition of hot water and phosphoric (or equivalent) acid and separated out by centrifugal separation. Free fatty acids in the oil are converted to soap using an alkali solution of sodium hydroxide, which is subsequently removed by continuous centrifugation. Finally, the oil is vacuum dried to remove any traces of water.

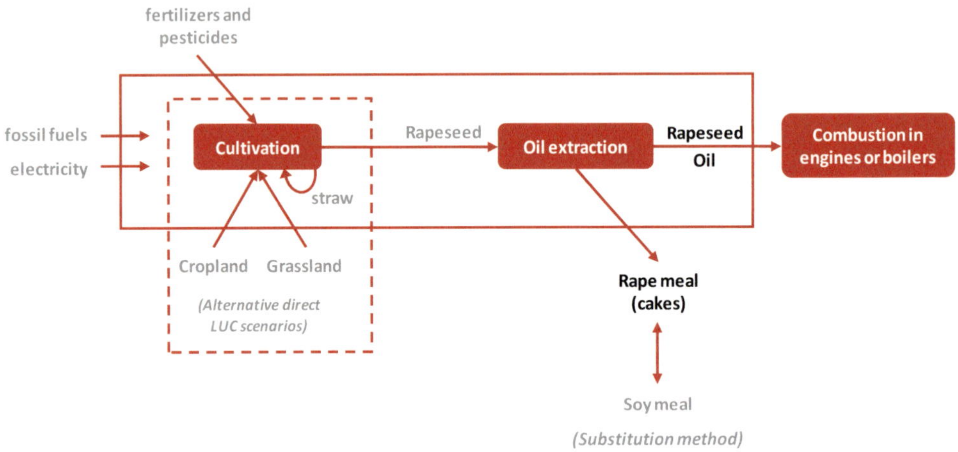

Fig. 1. Flow chart illustrating the life-cycle chain (well-to-tank) of Rapeseed Oil

The multifunctionality of biofuel systems is considered a critical issue in biofuel life-cycle studies, as discussed in section 2. For the RO production system, in particular, one valuable co-product is obtained: rape meal. Different approaches are addressed here for dealing with this co-production): i) the substitution method, in which the system is expanded with the avoided process - (soy meal production); ii) allocation, i.e. splitting up the process into two single-functional processes (RO production + rape meal production) on the basis of underlying relationships (physical: mass, energy; and economic); and iii) the no allocation, in which rape meal is ignored, i.e. all burdens (energy and material inputs, and related emissions) are fully allocated to RO.

Concerning the application of the substitution method, it is considered that the RO co-product rapeseed meal replaces imported soybean meal in animal feed. The technical feasibility of replacing soybean meal with rapeseed meal for feeding pigs and piglets has already been demonstrated (e.g. Kracht et al., 2004). Research recently conducted in France has also concluded that replacing soybean meal with rapeseed meal in the feed rations for dairy cows and for fattening beef cattle is technically feasible (GAIN, 2005). Actually, rape meal from oilseed crushing is replacing soybean meal imports as a high-protein animal feed (GAIN, 2007; Ceddia and Cerezo, 2008). This substitution approach is also considered in other works (e.g. Bernesson et al., 2004; JEC, 2007; Lechón et al., 2009; Soimakallio et al., 2009).

3.2.2 Key issues affecting soil carbon exchange

Several issues influence soil carbon exchange, namely land use change scenarios, agricultural practices and geographic region. Concerning land use change, two reference land uses have been considered in this article: (i) grassland; and (ii) long-term cultivated cropland. Appropriate land use factors F_{LU}, which reflect the difference in soil organic carbon associated with the type of land use compared to a standard soil organic carbon SOC_{ST}, have been taken from EC (2010), IPCC (2006).

EC (2010) differentiates three alternative management practices for cropland - full-tillage; reduced or low-tillage; and no-till – based on the level of soil disturbance during cultivation, respectively substantial, reduced or minimal. Full- and reduced-tillage have been considered for the reference land use, whereas low-tillage has been assumed for the actual land use (rapeseed cultivation). Concerning grassland, the management scenario that most contributes to carbon sequestration in the soil is improved grassland (according to EC 2010), which has been used in our assessment. The alternatives in soil management practices have been quantified through F_{MG}, a factor that reflects the difference between the soil organic carbon associated with the main management practice and the standard soil organic carbon SOC_{ST} (EC, 2010a).

The level of carbon input to the soil may also differ depending on the return of crop residues to the field and the adoption of other agricultural practices (EC, 2010a). To quantify extreme scenarios in terms of soil carbon content in the reference land use, high and low carbon inputs have been considered, respectively for grassland and cropland, whereas in the actual land use the option for medium inputs to rapeseed cultivation has been selected. The input factor F_I, which reflects the difference in soil organic carbon associated with different levels of carbon input to soil compared to the standard soil organic carbon SOC_{ST}, has been used (EC, 2010a; IPCC, 2006).

The geographic region is another key aspect for assessing the GHG emissions of a specific crop, since climate and soil type are two important factors affecting the calculation of land carbon stocks. Main rapeseed oil producers in Europe are France and Germany (see Table 1). A cool temperate moist climate has been selected as representative of main rapeseed production in Europe, according to the classification made in EC (2010). Concerning soil type, EC (2010) shows that high activity clay soil is the most representative soil type for countries involved in rapeseed cultivation. Active soils are also indicated in JEC (2007) as the most likely soil type to be converted to arable cropping.

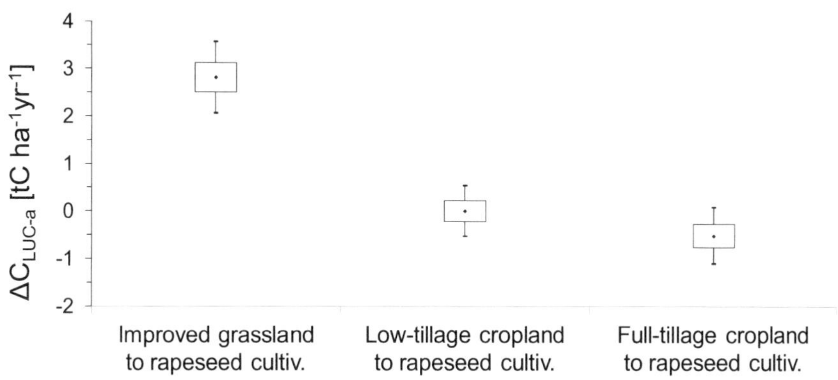

Fig. 2. Soil carbon exchange associated with LUC scenarios for Rapeseed Oil. The boxes show the interquartile range, the mark is the median and the ends of the whiskers are the 5th and 95th percentiles. Same notation is used in figs. 3 and 5

As shown in the above discussion, a large degree of variability exists concerning the management practices and input levels associated with rapeseed cultivation. The guidance provided in EC (2010) concerning the selection of the appropriate coefficients F_{LU}, F_{MG} and F_I for land use and management has been followed in this article. Moreover, appropriate probability distributions have been assigned to ΔC_{LUC-a}, based on the error ranges provided in IPCC (2006) for each LUC scenario (Fig. 2).

3.3 Results and discussion

Rapeseed oil life-cycle energy renewability efficiency and GHG intensity incorporating uncertainty are presented in section 3.3. GHG emission savings of displacing petroleum diesel are also evaluated. As discussed in section 2, a "well-to-tank" approach has been used, in which energy and GHG emissions are assessed from the very first production stage until the final fuel distribution depot. The functional unit chosen is 1 MJ of fuel energy content (FEC), measured in terms of the lower heating value (LHV).

3.3.1 Energy Renewability Efficiency

The life-cycle energy renewability efficiency ERenEf of rapeseed oil is displayed in the box plot of Fig. 3. The output distributions are divided in the 5th, 25th, 50th, 75th, and 95th percentiles. Scenario uncertainty has been considered regarding the modeling choice of how co-product credits are accounted for, namely using mass, energy and market value allocation approaches and the substitution method. A comparison with fossil diesel shows that rapeseed oil clearly contributes to non-renewable primary energy savings as opposed to its fossil reference. RO ERenEf is clearly positive, which indicates that an important fraction of the biofuel energy content (from 60% to 85%, depending on the approach for dealing with co-products, Fig. 3) comes from renewable energy sources.

Comparing the three allocation methods used, Fig. 3 shows that mass allocation results have the lowest uncertainty range, whereas economic allocation results are more uncertain because they depend on the variability of market prices. System expansion shows the highest degree of uncertainty due to differences in credits for soy meal substitution by rape meal.

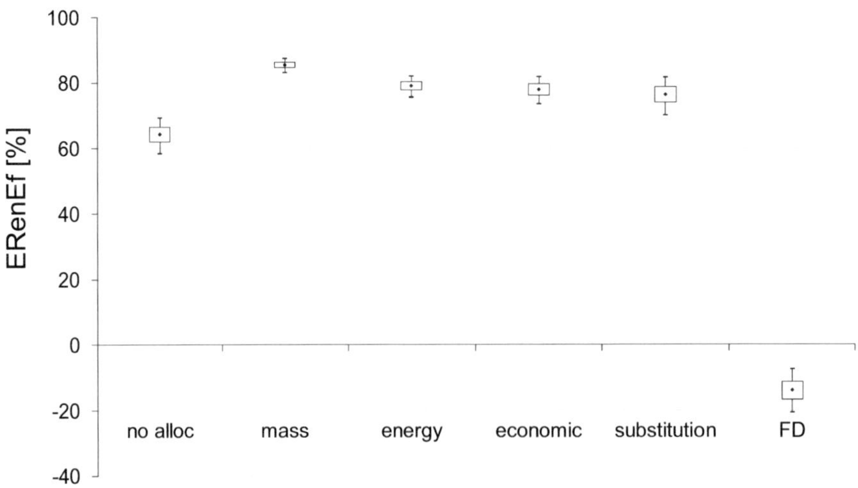

Fig. 3. RO life-cycle ERenEf results: scenario and parameter uncertainty

Moreover, mass allocation shows the highest results, which is explained by the relatively high mass share of rape meal in the oil extraction stage (approximately 1.5 kg of rape meal per kg of RO produced). Although it is a straightforward method, mass allocation is very often a meaningless approach, namely when energy systems or market principles come into play. Allocations based on energy and economic value show lower ERenEf values, due to the higher heating value and market price of RO in comparison to rape meal.

Figure 4 shows which parameters are most significant in the overall uncertainty of RO ERenEf. The uncertainty importance analysis that has been conducted shows that several parameters have important contributions in the uncertainty, namely diesel fuel use in agricultural machinery, N fertilizer application rate and energy use in N fertilizer production. In particular, Fig. 4(b) for economic allocation shows that market prices (and their inherent volatility) also affect the variance of ERenEf.

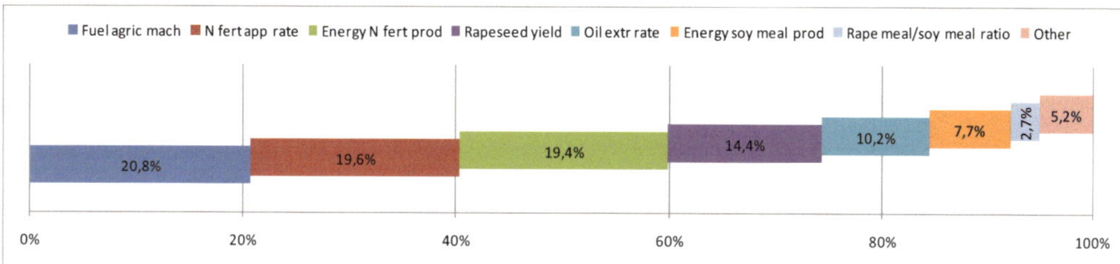

(a)

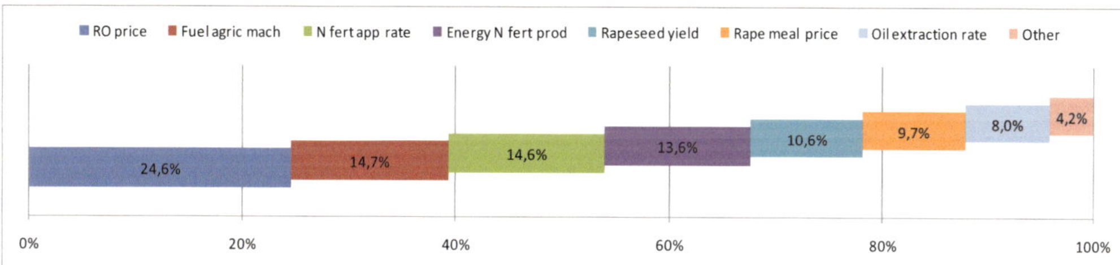

(b)

Fig. 4. Contribution of input data to the variance of RO life-cycle ERenEf: (a) substitution method; (b) economic allocation

3.3.2 GHG savings

Life-cycle GHG emission savings of RO displacing petroleum diesel are shown in Fig. 5. The uncertainty associated with the life-cycle GHG emissions of petroleum diesel has been considered using a normal probability distribution (μ=82 g CO_2eq MJ^{-1}; σ=3 g CO_2eq MJ^{-1}). An important conclusion from Fig. 5 is that parameter uncertainty is significantly higher in the case of RO GHG emissions when compared to ERenEf values of Fig. 3. An uncertainty importance analysis will put into evidence the parameters that most contribute to this higher magnitude of uncertainty.

Figure 5 shows that RO GHG emissions are considerably higher than fossil diesel (FD) GHG emissions if the most severe land use change scenario (improved grassland to rapeseed cultivation) is considered, i.e. FD substitution by RO results in negative GHG savings. This outcome contrasts with the other two LUC scenarios (conversion from full-tillage or low-tillage croplands) in which rapeseed oil GHG savings are positive. Moreover, these savings are above the 35% GHG saving target of the European renewable energy directive (EPC, 2009), regardless of the co-product method used.

Fig. 5 also shows that in the "low-tillage cropland to rapeseed cultivation" LUC scenario, the parameter uncertainty range overcomes the differences between calculated median values for the various scenarios of co-product treatment. Soil carbon sequestration associated with conversion of "full-tillage cropland to rapeseed cultivation" results in very low RO life-cycle GHG emissions, complying with the 2018 target of 60% GHG savings over fossil diesel of EPC (2009). In this scenario, differences between co-product approaches become negligible.

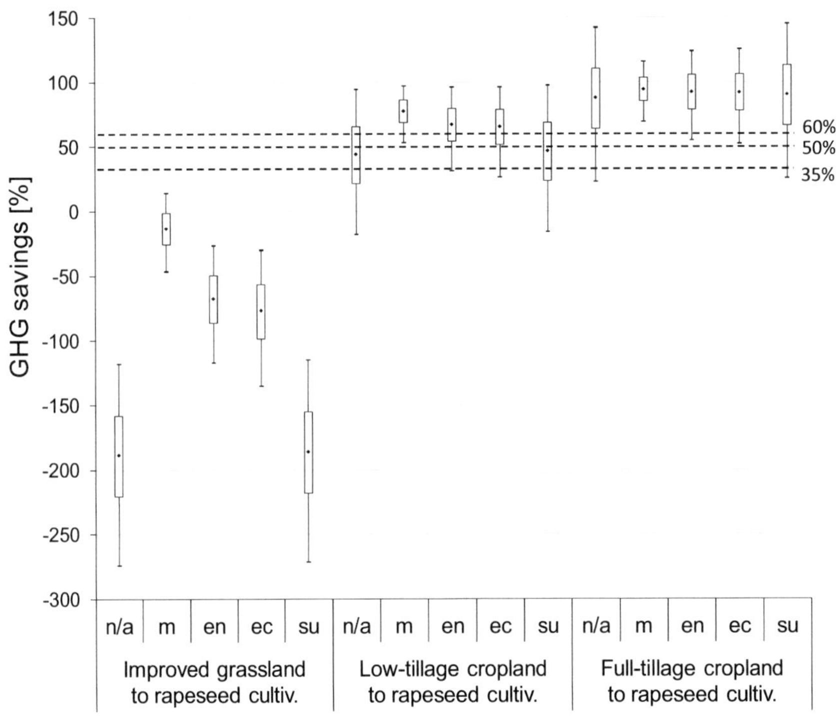

Fig. 5. RO life-cycle GHG emission savings: LUC scenarios and co-product approaches (n/a: no allocation; m: mass; en: energy; ec: economic; su: substitution). Dashed lines indicate minimum levels of GHG savings (EPC, 2009)

Figure 6 shows which parameters are most significant in the overall uncertainty of RO GHG emissions for the three LUC scenarios considered. The highest sources of uncertainty arise in the cultivation stage. Soil carbon emissions from land use change are the main contributor to the uncertainty of RO GHG intensity, with nitrous oxide emissions from cultivated soil as the second most important aspect. Agricultural yield and oil extraction efficiency (amount of rapeseed oil that can be extracted per kg of processed seed) are also important in the "grassland to rapeseed" LUC scenario. The remaining parameters hardly contribute to the variance of GHG emissions. Further research work must focus on the most important sources of uncertainty, in order to reduce the overall uncertainty of the rapeseed oil chain and improve the reliability of RO life-cycle studies outcomes.

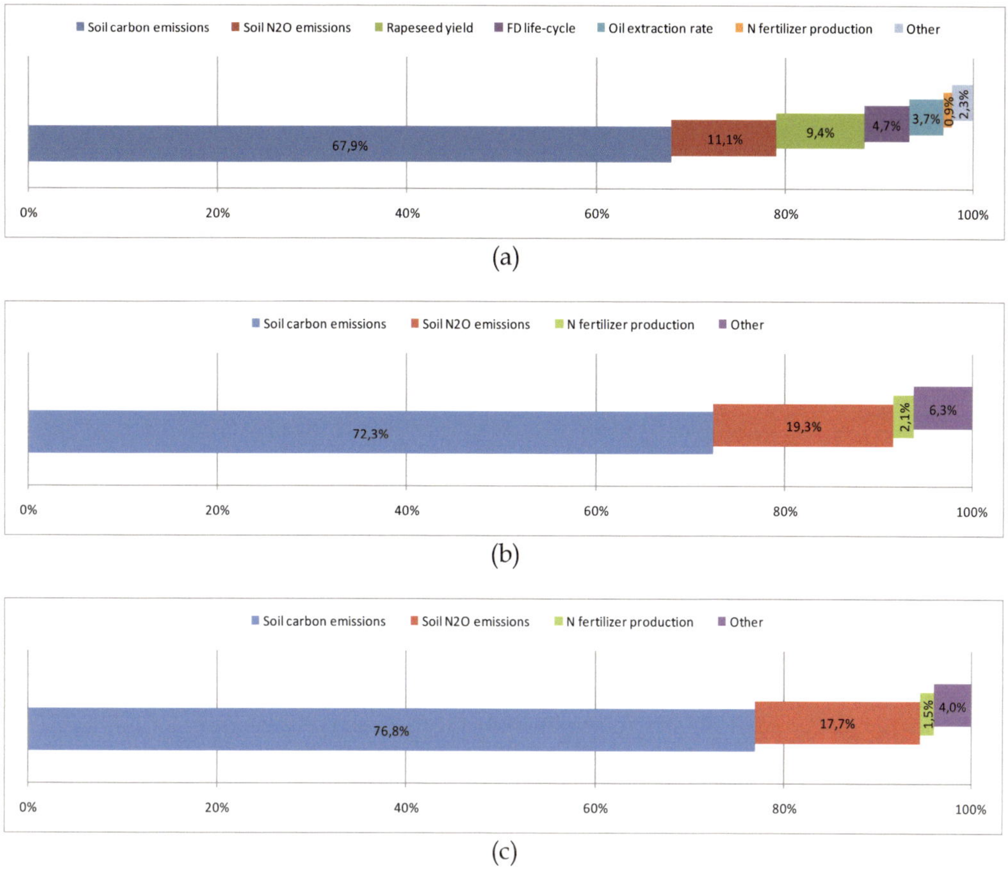

Fig. 6. Contribution of input data to the variance of RO life-cycle GHG emission savings (substitution method). Land use change scenarios: (a) improved grassland to rapeseed cultivation; (b) low-tillage cropland to rapeseed cultivation; (c) full-tillage cropland to rapeseed cultivation

4. Conclusions

This chapter has two main goals: i) to present a robust framework to incorporate uncertainty in the life-cycle modeling of biofuel systems; and ii) to describe the application of the framework to vegetable oil fuel in Europe. The chapter also compares rapeseed oil life-cycle results (energy renewability efficiency and GHG emissions) with its fossil fuel equivalent (diesel), in order to evaluate potential savings achieved through displacement.

A comprehensive assessment of uncertainty in the life-cycle of rapeseed oil has been conducted. Several sources of uncertainty have been investigated, namely related to parameters, global warming potentials and concerning how co-product credits are accounted for. It has been shown that depending on whether or not uncertainty in parameters is taken into account, and what modeling choices are made, results and conclusions from the life-cycle study may vary quite widely. In particular, it has been reported that the net GHG balance is strongly influenced by soil carbon stock variations due to land use change and by the magnitude of nitrous oxide emissions from cultivated soil.

Depending on prior land use, GHG emissions may comply with the European directive target of 35% GHG emission savings or, conversely, may completely offset carbon gains attributed to rapeseed oil production. These results contrast with the energy balance of rapeseed oil, which shows a high degree of energy renewability efficiency, regardless of parameter uncertainty and modeling choices made. Moreover, non-renewable primary energy savings are always achieved with rapeseed oil use, as opposed to fossil diesel use.
The benefits of using rapeseed oil to displace fossil diesel have been demonstrated, but special attention is needed to reduce emissions from carbon stock changes and nitrogen fertilizer application, in order to ensure that rapeseed oil use avoids GHG emissions. Only through a comprehensive evaluation of the life-cycle of biofuels, capturing uncertainty issues, it is possible to ensure reliable outcomes and guarantee the environmental sustainability of biofuel production systems.

5. Acknowledgements

The research presented in this article has been supported by the Portuguese Science and Technology Foundation (FCT) projects PTDC/TRA/72996/2006 *"Biofuel systems for transportation in Portugal: a well-to-wheels integrated multi-objective assessment"*, MIT/SET/0014/2009 *"Biofuel capturing uncertainty in biofuels for transportation: resolving environmental performance and enabling improved use"*, and MIT/MCA/0066/2009 *"Economic and Environmental Sustainability of Electric Vehicle Systems"*.

6. References

ADEME (Agence de l'Environnement et de la Maitrise de l'Energie). Energy and greenhouse gas balances of biofuels' production chains in France, executive summary, Paris; December 2002.

Anex R, Lifset R. 2009. Assessing Corn Ethanol: Relevance and Responsibility. *Journal of Industrial Ecology* 13(4):479-482.

Armstrong A, Baro J, Dartoy J, Groves A, Nikkonen J, Rickeard D, Thompson D, & Larivé J. Energy and greenhouse gas balance of biofuels for Europe - an update, report no. 2/02. Brussels: CONCAWE, 2002.

Bernesson S, Nilsson D, & Hansson PA. 2004. A limited LCA comparing large- and small-scale production of rape methyl ester (RME) under Swedish conditions. *Biomass & Bioenergy* 26(6):545–559.

Björklund A. 2002. Survey of Approaches to Improve Reliability in LCA. *Int. Journal of Life Cycle Assessment* 7(2):64-72.

Börjesson P, & Tufvesson L. 2011. Agricultural crop-based biofuels – resource efficiency and environmental performance including direct land use changes. *Journal of Cleaner Production* 19:108-120.

Boustead I, & Hancock G. Handbook of Industrial Energy Analysis. Ellis Horwood ltd, John Wiley and Sons, 1979.

Boustead I. Eco-Profiles of the European plastics industry. Methodology. Report. Brussels: Association of Plastics Manufacturers in Europe, 2003.

Bowyer C. 2010. Anticipated Indirect Land Use Change Associated with Expanded Use of Biofuels and Bioliquids in the EU – An Analysis of the National Renewable Energy Action Plans. Institute for European Environmental Policy, London, UK.

Ceddia M, & Cerezo E. A Descriptive Analysis of Conventional Organic and GM crop and Certified Seed Production in the EU. Luxembourg: Joint Research Centre of the European Commission; 2008.

Cherubini F, Birda N, Cowie A, Jungmeier G, Schlamadinger B, & Woess-Gallasch S. 2009. Energy- and greenhouse gas- based LCA of biofuel and bioenergy systems: Key issues, ranges and recommendations. *Resources Conservation & Recycling* 53(8):434-447.

Cherubini F. 2010. GHG balances of bioenergy systems - Overview of key steps in the production chain and methodological concerns. Renewable Energy 35:1565–1573.

Cherubini F, & Strømman AH. 2011. Life cycle assessment of bioenergy systems: State of the art and future challenges. *Bioresource Technology* 102:437-451.

Chiaramonti D, & Tondi G. Stationary Applications of Liquid Biofuels, Final Report, ETA Renewable Energies, December, Firenze, 2003.

Ciroth A, Fleischer G, & Steinbach J. 2004. Uncertainty Calculation in Life Cycle Assessments: A Combined Model of Simulation and Approximation. *International Journal of Life Cycle Assessment* 9(4): 216-226.

Cocco D. 2009. Predicted performance of integrated power plants based on diesel engines and steam cycles fuelled with a rapeseed oil chain. Proceedings of the Institution of Mechanical Engineers Part A - *Journal of Power and Energy* 223(A5):477–485.

Croezen H, Bergsma G., Otten M., & van Valkengoed M. Biofuels: indirect land use change and climate impact. CE Delft, Delft, the Netherlands, June 2010.

Crutzen PJ, Mosier AR, Smith KA, & Winiwarter W. 2008. N2O release from agro-biofuel production negates global warming reduction by replacing fossil fuels. *Atmospheric Chemistry and Physics* 8(2):389–395.

DCENR (Department of Communications, Energy and Natural Resources). Report on measures taken to promote the use of biofuels or other renewable fuels to replace diesel or petrol: compliance with Directive 2003/30/EC. July, Ireland, 2007.

Dewulf A, van Langenhove H, & van de Velde B. 2005. Exergy-Based Efficiency and Renewability Assessment of Biofuel Production. *Environmental Science & Technology* 39:3878-3882.

DMFA (Dutch Ministry of Foreign Affairs). Report from the Netherlands for 2006 pursuant to Article 4(1) of Directive 2003/30/EC on the promotion of the use of biofuels or other renewable fuels for transport. July, The Netherlands, 2007.

EC (European Commission). 2010a. Commission decision 2010/335/EU of 10 June 2010 on guidelines for the calculation of land carbon stocks for the purpose of Annex V to Directive 2009/28/EC.

EC 2010b. Report from the Commission on indirect land-use change related to biofuels and bioliquids. COM(2010) 811 final, December 22.

EPA (Environmental Protection Agency). Regulation of Fuels and Fuel Additives: Modifications to Renewable Fuel Standard Program. Federal Register, Vol. 75, No. 244, December 21, 2010.

EPC. Directive 2009/28/EC of the European Parliament and of the Council of 23 April 2009 on the promotion of the use of energy from renewable sources and amending and subsequently repealing Directives 2001/77/EC and 2003/30/EC; 2009.

EurObserv'ER. Biofuels Barometer. Systèmes Solaires - Le journal des énergies renouvelables 185; June 2008, p. 49-66.

Eurostat (Statistical Office of the European Communities). http://epp.eurostat.ec.europa.eu (accessed March 2011).

FAOSTAT (FAO statistical database). http://faostat.fao.org (accessed January 2011).

Fargione J, Hill J, Tilman D, Polasky S, & Hawthorne P. 2008. Land Clearing and the Biofuel Carbon Debt. *Science* 319(5867):1235-1238.

Farrell AE, Plevin RJ, Turner BT, Jones AD, O'Hare M, & Kammen DM. 2006. Ethanol Can Contribute to Energy and Environmental Goals. *Science* 311, pp. 506-508, Jan 27.

Frischknecht R. 2000. Allocation in Life Cycle Inventory Analysis for Joint Production. *Int. Journal of Life Cycle Assessment* 5 (2):85-95.

Fritsche U, Sims R, & Monti A. 2010. Direct and indirect land-use competition issues for energy crops and their sustainable production - an overview. *Biofuels, Bioproducts & Biorefining* 4:692–704.

GAIN (Global Agriculture Information Network). 2005. France explores substituting soybean meal with rapeseed meal. USDA Foreign Agricultural Service. Paris.

GAIN (Global Agriculture Information Network). Impacts on Oilseed Industry following Biofuel Boom. Paris: USDA Foreign Agricultural Service; 2007.

Geisler G, Hellweg S, & Hungerbuhler K. 2005. Uncertainty Analysis in Life Cycle Assessment (LCA): Case Study on Plant-Protection Products and Implications for Decision Making. *International Journal of Life Cycle Assessment* 10(3): 184-192.

Gnansounou E, Dauriat A, Villegas J, & Panichelli L. 2009. Life cycle assessment of biofuels: Energy and greenhouse gas balances. *Bioresource Technology* 100(21):4919-4930.

Guinée JB, Heijungs R, & Huppes G. 2004. Economic Allocation: Examples and Derived Decision Tree. *Int Journal of Life Cycle Assessment* 9(1):23-33.

Guinée JB, Heijungs R, & van der Voet E. 2009. A greenhouse gas indicator for bioenergy: some theoretical issues with practical implications. *Int Journal of Life Cycle Assessment* 14(4):328-339.

Heijungs R, & Huijbregts M. 2004. A Review of Approaches to Treat Uncertainty in LCA. In: C. Pahl-Wostl, S. Schmidt, A.E. Rizzoli, and A.J. Jakeman (Eds). Complexity and Integrated Resources Management. Transactions of the 2nd Biennial Meeting of the International Environmental Modelling and Software Society, Vol 1, Osnabrück.

Heijungs, R. 1996. Identification of key issues for further investigation in improving the reliability of life-cycle assessments. *Journal of Cleaner Production* 4(3–4):159–166.

Hekkert M, Hendriks F, Faaij A, & Neelis M. 2005. Natural gas as an alternative to crude oil in automotive fuel chains well-to-wheel analysis and transition strategy development. *Energy Policy* 33:579–594.

Heller M, Keoleian G, & Volk T. 2003. Life cycle assessment of a willow bioenergy cropping system. *Biomass and Bioenergy* 25:147–165.

Hoefnagels R, Smeets E, & Faaij A. 2010. Greenhouse gas footprints of different biofuel production systems. *Renewable & Sustainable Energy Reviews* 14:1661-1694.

Huijbregts, M. 1998. Application of uncertainty and variability in LCA. Part I: A General Framework for the Analysis of Uncertainty and Variability in Life Cycle Assessment. *International Journal of Life Cycle Assessment* 3(5): 273–280.

Huijbregts M, Gilijamse W, Ragas A, & Reijnders L. 2003. Evaluating Uncertainty in Environmental Life-Cycle Assessment. A Case Study Comparing Two Insulation Options for a Dutch One-Family Dwelling. *Environmental Science & Technology* 37:2600-2608.

Huijbregts M, Rombouts L, Hellweg S, Frischknecht R, Hendriks A, van de Meent D, Ragas A, Reijnders L, & Struijs J. 2006. Is Cumulative Fossil Energy Demand a Useful Indicator for the Environmental Performance of Products? *Environmental Science & Technology* 40(3):641-648.

Huo H, Wang M, Bloyd C, & Putsche V. 2009. Life-Cycle Assessment of Energy Use and Greenhouse Gas Emissions of Soybean-Derived Biodiesel and Renewable Fuels. *Environmental Science & Technology* 43:750-756.

IEA (International Energy Agency). 2010 key world energy statistics, Paris, 2010.

IPCC (Intergovernmental Panel on Climate Change). Climate Change 2007 - The Physical Science Basis. Fourth Assessment Report. New York: Cambridge University Press; 2007.

IPCC. 2006 IPCC Guidelines for National Greenhouse Gas Inventories. Vol.4, Ch. 11: N2O Emissions from Managed Soils and CO_2 Emissions from Lime and Urea Application. Prepared by the National Greenhouse Gas Inventories Programme, Eggleston HS, Buendia L, Miwa K, Ngara T, Tanabe K, editors. Japan: Institute for Global Environmental Strategies; 2006.

ISO (International Organization for Standardization). ISO 14040: Environmental management - Life cycle assessment - Principles and framework. Genève, Switzerland; 2006.

ISO. ISO 14044: Environmental management - Life cycle assessment - Requirements and guidelines. Genève, Switzerland; 2006.

Janulis P. 2004. Reduction of energy consumption in biodiesel fuel life cycle. *Renewable Energy* 29:861-871.

JEC (JRC/EUCAR/CONCAWE Consortium; CONCAWE: The oil companies' European association for environment, health and safety in refining and distribution; EUCAR: European Council for Automotive R&D; JRC: Joint Research Centre of the European Commission). Well-to-wheels analysis of future automotive fuels and powertrains in the European context. Well-to-tank and Well-to-wheels report, Version 2c, Brussels; March 2007.

Kaiser E, Kohrs K, Kucke M, Schnug E, Heinemeyer O, & Munch J. 1998. Nitrous oxide release from arable soil: importance of N-fertilization, crops and temporal variation. *Soil Biology & Biochemistry* 30(12):1553-1563.

Kim S, & Dale B. 2002. Allocation Procedure in Ethanol Production System from Corn Grain. I-System Expansion. *Int. Journal of Life Cycle Assessment* 7(4):237-243.

Kløverpris J, Wenzel H, Banse M, Milà i Canals L, & Reenberg A. 2008. Conference and Workshop on Modelling Global Land Use Implications in the Environmental Assessment of Biofuels. *Int Journal of Life Cycle Assessment* 13(3):178-183.

Knothe G. Historical perspectives on vegetable oil-based diesel fuels. *Industrial Oils* 12: 1103-1107, 2001.

Kracht W, Nicke S, Kluge H, Keller K, Matzke W, Hennig U & Schumann W. 2004. Effect of Dehulling of Rapeseed on Feed Value and Nutrient Digestibility of Rape Products in Pigs. *Archives of Animal Nutrition* 58(5):389-404.

Krupnick A, Morgenstern R, Batz M, Nelson P, Burtraw D, Shih J, & McWilliams M. 2006. Not a sure thing: Making regulatory choices under uncertainty. Technical report, Resources for the Future, Washington DC.

Larson E. 2006. A review of life-cycle analysis studies on liquid biofuel systems for the transport sector. *Energy & Sustainable Development* 10(2):109–126.

Lechón Y, Cabal H, de la Rúa C, Caldés N, Santamaría M, & Sáez R. 2009. Energy and greenhouse gas emission savings of biofuels in Spain's transport fuel. The adoption of the EU policy on biofuels. *Biomass & Bioenergy* 33(6-7):920-932.

Liska AJ, & Cassman KG. 2008. Towards Standardization of Life-Cycle Metrics for Biofuels: Greenhouse Gas Emissions Mitigation and Net Energy Yield. *Journal of Biobased Materials and Bioenergy* 2:187–203.

Liska AJ, & Perrin RK. 2009. Indirect land use emissions in the life cycle of biofuels: regulations vs science. *Biofuels, Bioproducts & Biorefining* 3:318–328.

Lloyd S, & Ries R. 2007. Characterizing, Propagating, and Analyzing Uncertainty in Life-Cycle Assessment: A Survey of Quantitative Approaches, *Journal of Industrial Ecology* 11(1): 161–179.

Luque R, Davila L, Campelo JM, Clark JH, Hidalgo JM, Luna D et al. 2008. Biofuels: a technological perspective. *Energy and Environmental Science* 1(5):542-564.

Luo L, van der Voet E, Huppes G, & Udo de Haes H. 2009 Allocation issues in LCA methodology: a case study of corn stover-based fuel ethanol. *Intl Journal of Life Cycle Assessment* 14:529-539.

Malça J, & Freire F. 2004. Life cycle energy analysis for bioethanol: allocation methods and implications for energy efficiency and renewability. 17th International Conference on Efficiency, Costs, Optimization, Simulation and Environmental Impact of Energy and Process Systems (ECOS 2004), July 07-09, Guanajuato Mexico.

Malça J, & Freire F. 2006. Renewability and life-cycle energy efficiency of bioethanol and bioethyl tertiary butyl ether (bioETBE): Assessing the implications of allocation. *Energy* 31(15):3362-3380.

Malça J, & Freire F. 2009. Energy and environmental benefits of rapeseed oil replacing diesel. *International Journal of Green Energy* 6(3):287-301.

Malça J, & Freire F. 2010. Uncertainty Analysis in Biofuel Systems: An Application to the Life Cycle of Rapeseed Oil. *Journal of Industrial Ecology* 14(2):322-334.

Malça J, & Freire F. 2011. Life-cycle studies of biodiesel in Europe: A review addressing the variability of results and modeling issues. *Renewable & Sustainable Energy Reviews* 15(1):338-351.

Menichetti E, & Otto M. 2008. Energy Balance & Greenhouse Gas Emissions of Biofuels from a Life Cycle Perspective. In: R.W. Howarth and S. Bringezu (eds) Biofuels: Environmental Consequences and Interactions with Changing Land Use. Proceedings of the Scientific Committee on Problems of the Environment (SCOPE) International Biofuels Project Rapid Assessment, September 22-25 2008, Gummersbach, Germany.

Misra R, & Murthy M. 2010. Straight vegetable oils usage in a compression ignition engine—A review. *Renewable and Sustainable Energy Reviews* 14:3005–3013.

Mondal P, Basu M, & Balasubramanian N. 2008. Direct use of vegetable oil and animal fat as alternative fuel in internal combustion engine. *Biofuels, Bioproducts & Biorefining* 2:155–174.

Morgan MG & Henrion M. 1990. A Guide to Dealing with Uncertainty in Quantitative Risk and Policy Analysis. New York: Cambridge University Press.

Mortimer N, Cormack P, Elsayed M, & Horne R. Evaluation of the Comparative Energy, Global Warming and Social Costs and Benefits of Biodiesel, report. UK: Resource Research Unit, Sheffield Hallam University, 2003.

Mortimer ND, Elsayed MA. North East Biofuel Supply Chain Carbon Intensity Assessment. Sheffield, UK: North Energy Associates Ltd; 2006.

Neupane B, Halog A, & Dhungel S. 2011. Attributional life cycle assessment of woodchips for bioethanol production. *Journal of Cleaner Production* 19:733-741.

Oracle. 2010. Oracle Crystal Ball software v.11.1.

Papong S, & Malakul P. 2010. Life-cycle energy and environmental analysis of bioethanol production from cassava in Thailand. *Bioresource Technology* 101:S112–S118.

Plevin R. 2010. Life Cycle Regulation of Transportation Fuels: Uncertainty and its Policy Implications. PhD thesis. University of California, Berkeley, USA.

Poldy F. 2008. Net energy and strategic decision-making. *Biofuels, Bioproducts & Biorefining* 2:389–392.

Rabl A, Benoist A, Dron D, Peuportier B, Spadaro J, & Zoughaib A. 2007. How to Account for CO_2 Emissions from Biomass in an LCA. *Intl Journal of Life Cycle Assessment* 12(5):281.

Reijnders L, & Huijbregts M. 2008. Biogenic greenhouse gas emissions linked to the life cycles of biodiesel derived from European rapeseed and Brazilian soybeans. *Journal of Cleaner Production* 16(18):1943–1948.

Reijnders L. 2009. Transport biofuels: Can they help limiting climate change without an upward impact on food prices? *Journal of Consumer Protection and Food Safety* 4:75-78.

Saltelli A, Ratto M, Tarantola S, & Campolongo F. 2006. Sensitivity analysis practices: Strategies for model-based inference. *Reliability Engineering & System Safety* 91(10-11):1109–1125.

Schade B, & Wiesenthal T. 2011. Biofuels: A model based assessment under uncertainty applying the Monte Carlo method. *Journal of Policy Modeling* 33:92–126.

Searchinger T, Heimlich R, Houghton RA, Dong F, Elobeid A, Fabiosa J, et al. 2008. Use of U.S. Croplands for Biofuels Increases Greenhouse Gases through Emissions from Land-Use Change. *Science* 319(5867):1238-1240.

SenterNovem (Agency of the Dutch Ministry of Economic Affairs for Innovation and Sustainable Development). The road to pure plant oil? The technical, environment-hygienic and cost-related aspects of pure plant oil as a transport fuel. Report 2GAVE-05.05. The Netherlands, 2005.

Shapouri H, Duffield J, & Graboski M. Estimating the Net Energy Balance of Corn Ethanol, report no. 721. US Dept. of Agriculture, 1995.

Shapouri H, Duffield J, & Wang M. The Net Energy Balance of Corn Ethanol: an Update, report no. 813. US Dept. of Agriculture, 2002.

Sheehan J, Camobreco V, Duffield J, Graboski M, & Shapouri H. Life Cycle Inventory of Biodiesel and Petroleum Diesel for Use in an Urban Bus, Final Report. Golden, CO: National Renewable Energy Laboratory, 1998.

Sidibé S, Blin J, Vaitilingom G, & Azoumah Y. 2010. Use of crude filtered vegetable oil as a fuel in diesel engines state of the art: Literature review. *Renewable and Sustainable Energy Reviews* 14:2748–2759.

Soimakallio S, Makinen T, Ekholma T, Pahkala K, Mikkola H, & Paappanen T. 2009. Greenhouse gas balances of transportation biofuels, electricity and heat generation in Finland: Dealing with the uncertainties. *Energy Policy* 37:80-90.

Stephenson AL, Dennis JS, & Scott SA. 2008. Improving the sustainability of the production of biodiesel from oilseed rape in the UK. *Process Safety and Environmental Protection* 86:427-440.

Tickell J. 2003. From the fryer to the fuel tank - The complete guide to using vegetable oil as an alternative fuel, 3rd ed., Joshua Tickell Publications, New Orleans, Louisiana.

UFOP (Union zur Förderung von Oel-und Proteinpflanzen e.V.). *Rapeseed Magazine*, Rape Blossom, 2008.

van der Voet E, Lifset RJ, & Luo L. 2010. Life-cycle assessment of biofuels, convergence and divergence. *Biofuels* 1(3):435-449.

Wagner U, Eckl R, & Tzscheutschler P. 2006. Energetic life cycle assessment of fuel cell powertrain systems and alternative fuels in Germany. *Energy* 31(14):3062-3075.

Weidema B, Fress N, Petersen E, & Ølgaard H. Reducing Uncertainty in LCI: Developing a Data Collection Strategy, Environmental Project No. 862. Denmark, 2003.

Whitaker J, Ludley KE, Rowe R, Taylor G, & Howard DC. 2010. Sources of variability in greenhouse gas and energy balances for biofuel production: a systematic review. *Global Change Biology Bioenergy* 2, pp. 99–112.

Wicke B, Dornburg V, Junginger M, & Faaij A. 2008. Different palm oil production systems for energy purposes and their greenhouse gas implications. *Biomass & Bioenergy* 32(12):1322–1337.

Wilting H. An energy perspective on economic activities. PhD thesis. Groningen, 1996.

Zah R, H Böni, M Gauch, R Hischier, M Lehmann, & P Wäger. Ökobilanz von Energieprodukten: Ökologische Bewertung von Biotreibstoffen (Life Cycle Assessment of Energy Products: Environmental Impact Assessment of Biofuels). EMPA, St. Gallen, Switzerland, 2007.

11

Biofuel Programs in East Asia: Developments, Perspectives, and Sustainability

Tatsuji Koizumi
Policy Research Institute, Ministry of Agriculture, Forestry and Fisheries
Japan

1. Introduction

The governments of East Asian countries and the region are promoting biofuel programs to address energy security and environmental problems as well as to increase farm income. This chapter covers East Asian biofuel programs, including China (People's Republic of China), Japan, Korea (Republic of Korea), and Taiwan. China has 205 thousand kℓ of fuel bioethanol. It is the third-largest biofuel producing country after the U.S. and Brazil (F.O.Licht, 2010). Verification tests and large-scale projects for biofuel production are currently underway in China. With Chinese oil imports rising rapidly as a result of motorization, the Chinese government is expected to expand its bioethanol program in the future. This expansion is expected to mitigate the country's dependence on oil imports and reduce air pollution problems. Although corn is the main feedstock for bioethanol production, the Chinese government aims to diversify bioethanol production, especially from cassava, instead of relying on expanded grain-based bioethanol production.

Japan has a long history of producing bioethanol. However, the technologies it once used were forgotten and remained unused for more than half a century. The enforcement of the Kyoto Protocol required Japan to start a biofuel program and influenced the start of biofuel programs in Korea and Taiwan. Japan promotes biofuel production from rice straw, wooden biomass, and algae. The R&D of second-generation biofuel that is developing in Japan includes improving varieties of energy resource crops, developing technologies for manufacturing biofuel, and developing cultivation methods.

The governments of East Asian countries and the region are promoting biofuel programs that rely on various feedstocks (Table 1), but this reliance and the escalating consumption of biofuel is competing with food and feed in these countries and the region. Consequently, the governments of East Asian countries and the region are developing biofuel programs that will not compete with their food availability.

Several studies have addressed East Asian biofuel production and programs. Koizumi and Ohga (2007) and Koizumi (2008) examined an economic analysis of the available supplies of domestically produced biofuel in Asian countries. Wang et al., (2009) examined the distribution and development of biofuel crops and the bioenergy industry in China. Chaves et al., (2010) reviewed technical and policy development of Chinese biofuel, while more recently Wang (2011) reviewed non-food biofuel commercialization in China.

Matsumoto et al., (2009) reviewed biofuel initiatives, strategies, policies, and the future potential of biofuel in Japan. Koizumi (2009) used econometric models to examine how Chinese bioethanol imports would impact the Brazilian and world sugar markets.

However, these studies for Japan and Asian countries need to update R&D for second-generation biofuel production. In addition, none of these studies has covered sustainability criteria for biofuel production. This chapter reviews not only East Asian biofuel production and programs, but also R&D for second-generation biofuel production and sustainability criteria for biofuel production in East Asian countries and the region. It also examines the impacts Chinese and Japanese biofuel import expansion would have on world sugar markets by applying developed econometric models. The next section covers biofuel production and policies in East Asian countries and the region. The third section discusses the impact of biofuel programs on agricultural markets. The fourth section discusses securing biofuel production, R&D for second-generation biofuel, and the sustainability of biofuel production. The last section summarizes the conclusion.

	Fuel Bioethanol		biodiesel	
	Annual Production (1,000kℓ)	**Current Main Feedstock**	**Annual Production (1,000kℓ)**	**Current Main Feedstock**
China	2,050	Corn, Wheat and Cassava	191	Used cooking oil
Japan	0.2	Sugarcane molasses, wheat unsuitable for food consumption, and others	10	Used cooking oil
Korea	—	—	300	Soybean oil, palm oil and used cooking oil
Taiwan	—	—	36	Used cooking oil

Table 1. Fuel biofuel production and feedstock in East Asia

Source: Chinese and Taiwan's biofuel production data were derived from F.O.Licht (2010), Japanese biofuel production data were derived from Ministry of Agriculture, Forestry and Fisheries (2010), and Korean biofuel data were derived from USDA-FAS (2010).

Note:

1. Chinese bioethanol production was 7.3 million kℓ, Japanese bioethanol production was 100 thousand kℓ, Korean bioethanol production was 169 thousand kℓ and Taiwan's bioethanol production was 10 thousand kℓ in 2009 (F.O.Licht, 2010). However, these data nclude industrial, fuel, and other uses.
2. "- " means unknown.

2. Biofuel production and policies in East Asia

2.1 China

2.1.1 Chinese biofuel program

In China, petroleum consumption is increasing rapidly and imports of crude oil are rising. The increase in petroleum consumption is causing a serious air pollution problem. In addition, excessive stocks of grain, especially corn, were crucial problems from 1996 to 2000. To deal with energy security, air pollution, and excessive grain stocks, the Chinese government strongly promoted the national bioethanol program.

As a result of high economic growth in China, the number of cars there is increasing rapidly. From 1990 to 2008, the market for passenger cars grew from 0.51 to 9.38 million. The Chinese car market has overtaken that of Japan to become the second-largest car market in the world, with sales of 7.28 million vehicles in 2006 (Wang, 2011). Chinese petroleum consumption increased from 164 million tons in 1990 to 553 million tons in 2008; and crude oil imports rose from 2.9 million tons in 1990 to 178.9 million tons in 2008 (National Bureau of Statistics of China, 2009). After the USA, China is the second-largest petroleum consumer in the world (International Energy Agency (IEA), 2008). Increasing oil consumption led China to become a net oil importer from 1994. The IEA has projected that Chinese oil consumption for transportation use would increase by 5.3% per annum from 2006 to 2030 (IEA, 2008). It is assumed that Chinese oil consumption will expand in the future. However, a shortage of energy, including petroleum, has been a serious problem since the 1990s. Proved oil reserves in China amounted to only 1.2% of the total world proved oil reserves at the end of 2008 (BP, 2009). In addition, rising crude oil prices since 2003 have had a negative impact on Chinese energy markets, as well as other regions.

The increase in petroleum consumption has caused air pollution problems. Next to the USA, China is the largest CO_2 emission country in the world (IEA, 2008). The Chinese Environmental Protection Agency estimated that 79 percent of air pollution originated from vehicle exhausts (Institute of Chinese Affairs, 2010). The Chinese government wants to improve the air pollution situation. From 1996 to 2000, it is estimated China had excessive ending stocks for grain, especially for corn. China is now estimated to have 123.8 million tons of corn ending stock, which is equivalent to 92.6% of the production level in 1999/2000 (USDA-FAS, 2011). Dealing with excessive ending stocks was one of the crucial problems for the Chinese government at that time.

In China, the concept of alternative energy was expressly stated in the Five-Year Plan of 1982. In 2001, the promotion of biomass energy was expressly stated in the Five-Year Plan for the period 2001-2005. In June 2002, the Chinese government started to mandate the use of bioethanol blend gasoline in five cities of Heilongjiang and Hernan. In October 2004, the government introduced the compulsory use of a 10 percent blend of bioethanol to gasoline (E10) in all areas of Heilongjiang, Jilin, Liaoning, Hernan, and Anhui. The government expanded the E10 program in 27 cities of Shandong, Jiangsu, Hebei, and Hubei from 2006.

2.1.2 Biofuel production

In the Chinese government, the Energy Bureau of the National Development and Reform Commission (NDRC) leads this whole program; the Ministry of Science and Technology takes part in technical affairs; the State Grain Administration takes part in the supply of agricultural feedstock; and the Ministry of Agriculture participates in the rural energy policy. In China, corn and wheat comprise a major part of the feedstock for bioethanol. Bioethanol is produced from corn in Heilongjiang, Jilin, and Anhui. It is also produced from wheat in Hernan. In addition, bioethanol is produced from cassava in Guangxi. Currently, five bioethanol production plants in China (Table 1) have operating licenses from the government.

China also produces biodiesel for fuel use. There are four major plants in Fujiang, Jiangsu, Hebei, and Beijing. Although China's production capacity has been estimated at 954.2 thousand kℓ (USDA-FAS, 2009a), it produced only 191 thousand kℓ in 2009, because of a lack of feedstock availability. The main feedstock for biodiesel is used cooking oil. Although Chinese mills prefer to produce biodiesel from vegetable oil, securing vegetable oil for biodiesel use can be difficult because China is a net importer of oilseed and vegetable oil. Securing feedstock is a crucial problem for expanding biodiesel production in China. Biofuel is sold only to two state-owned companies, China Petroleum and Chemical Corporation

(Sinopec) and China National Petroleum Corporation (CNPC) for blending with gasoline (Zhou and Thomson, 2009).

Location	Company	Main Feedstock	2008 Production (Estimated: tons)	2009 Production Capacity (tons)	Supply Location
Heilongjiang, Zhaodong	China Resources Alcohol Co.	Corn	163,296	163,296	Heilongjiang
Jilin, Jilin	Jilin Fuel Ethanol Co.	Corn	426,384	453,600	Jilin and Liaoning
Henan, Nanyang	Henan Tian Guan Fuel-Ethanol Co.	Wheat	371,952	408,240	Henan, Hubei (9 Cities) and Hebei (4Cities)
Anhui, Bengbu	Anhui BBCA Biochemical Co.	Corn	362,880	399,168	Anhui, Shandong (7 Cities), Jiangsu (5 Cities) and Hebei (2 Cities)
Guangxi	Guangxi COFCO Bioenegry Co.	Cassava	108,864	181,440	Guangxi
Total			1,433,376	1,605,744	

Table 2. Current Bioethanol Production
Source: USDA-FAS (2009a).
Note: Rice is partly used for bioethanol production in Heilongjiang.

2.1.3 Production costs and subsidies

In China, the cost of corn-based bioethanol is 4,937 Yuan/ton and the feedstock cost of corn is 3,456 Yuan/ton (Table 3). The feedstock cost of cassava is 1,716 Yuan/ton and the cost of cassava-based bioethanol is 4,259 Yuan/ton. The feedstock cost of corn stover is 1,500 Yuan/ton and total cost is 5,800 Yuan/ton. The Chinese bioethanol production cost from corn is equivalent to 1.022 US$/ℓ[1], while the U.S. bioethanol production cost from corn was 0.492 US$/ℓ (F.O.Licht, 2008). The cost of Chinese bioethanol production from cassava is equivalent to 0.882 US$/ℓ[1], while Thailand's bioethanol production cost from cassava is 0.300 US$/ℓ (F.O.Licht, 2008). Thus, the cost of Chinese bioethanol production is much higher than that of the U.S. and Thailand.

Because of high feedstock prices, all bioethanol producers receive subsidies to cover operating losses. The government subsidy is necessary to produce bioethanol. The average subsidy for fuel bioethanol production set by the government reached 1,836 Yuan/ton in 2005, 1,625 Yuan/ton in 2006, 1,374 Yuan/ton in 2007, and 1,754 Yuan/ton in 2008[2]. The average subsidy decreased gradually between 2005 and 2007. However, it increased from 2007 to 2008 because of high feedstock prices resulting from soaring international grain prices at that time.

In addition, value-added tax (17%) of these plants has been removed (Wang, 2011), five percent consumption tax on bioethanol has been exempted, and approximately 100 Yuan in profit is guaranteed for each stock on a preferential basis. Stock grain subsidies are determined by referencing market prices in each relevant area. The government will cover any loss incurred as a result of adjustment, transportation, or sale of E10. The Ministry of Finance will provide a specified amount of compensation. It is estimated that the removal of Value Added Tax and Consumption Tax totaled 190 million Yuan (US$28 million), and the

---

[1] It is calculated that 1US$ is equivalent to 6.57 Yuan (2011.3).
[2] This bioethanol cost is estimated from USDA-FAS (2009a).

direct financial subsidy totaled 2 billion Yuan (US$294 million) for grain-based bioethanol plants from 2002 to 2008 (Lang et al., 2009).
All supporting policies are directed toward state-owned enterprises, whereas only a few of them are accessible by private enterprises. Currently, five licenses have been issued in China. In some cases, the lack of supporting policy is the main reason for the failure of private enterprise investment in biofuel plants (Wang, 2011).

Feedstock	Feedstock cost (Yuan/ton)	Production Cost (Yuan/ton)	Location
Corn	3,456	4,937	Jilin
Cassava	1,716	4,259	Guanxi
Sweet Potato	2,240	3,200	Henan
Potato	3,735	5,335	Yunnan
Jerusalem artichoke	2,292	3,274	Shandong
Sugarcane	2,295	3,278	Guanxi
Sweat Sorgam	2,000	4,400	Shandong
Sugarbeet	3,675	5,250	Xinjiang
Corn Stober	1,500	5,800	Henan

Table 3. Bioethanol production cost in China
Source: Song et al., (2008) and, Huang and Yabe (2010).

2.1.4 Feedstock for bioethanol production

The Chinese bioethanol industry used corn as a feedstock for 80 percent of its 2005 production. The government limited the use of inferior agricultural products as feedstock for bioethanol to mitigate the impact on the agricultural market at the first stage of operation. The government prohibited the use of standard corn, traditionally used for feed, food, and other industrial materials[3], as a feedstock for bioethanol. Inferior corn[4] for bioethanol can come from reserve stocks after a period of two to three years. The supply of this inferior corn and wheat has been decreasing since 2001, because of decreased production. In addition, the government has promoted effective food marketing systems and tried to reduce these inferior agricultural foods since 2001. In the mid 2000s there was not enough inferior corn to meet bioethanol demand in China. All bioethanol facilities in Heilongjiang and Jilin have used standard corn as a feedstock for the production of bioethanol, because they can't get enough inferior corn to produce it. Wheat is the main feedstock for bioethanol at the Henan plant. However, wheat is a staple food in China and has a high domestic consumption. The government policies shy away from the use of grain-based feedstock materials for bioethanol production, and the government will not expand bioethanol production from wheat.
Guangxi, Guangdong, Hainan, Fujian, Yunnan, Hunan, Sichuan, Guizhou, Jiangxi, and nine other provinces are suitable for cassava growth. In 2007, total output of cassava in China was about 7 million tons (Wang, 2011). Cassava-based bioethanol plants are operating in the Guangxi in Southern China. Its production capacity in 2009 was 181.4 thousand tons (USDA-FAS, 2009a). In addition to these crops, bioethanol productions from sweet sorghum, crop stalks and straw, sugarcane, sweet potatoes, rice, sugar beet, woody biomass, and others are at an experimental stage.

---

[3] Other industrial feedstocks are used for adhesives, gummed tape, polished goods, and other products.
[4] Inferior corn is unsuitable for food use and is delivered from reserved stock to the market after a 2-3 year reserved period.

2.1.5 Developments and perspectives of the Chinese biofuel program

The utilization and development of renewable energy in China is a very crucial national program that not only contributes to energy security and improves environmental problems, but also develops rural areas, promoting new industries and technical innovation. In January 2006 the government enacted the "Renewable Energy Law" to promote renewable energy utilization and production. The government promotes biomass energy policy, which is divided into four categories: biofuel, rural biomass, biogas, and bioelectricity. The national bioethanol program was started in 2001, and the government strongly promoted the bioethanol program to provide an alternative fuel for gasoline. It is assumed the government will promote the bioethanol program in the future, because of the increasing gravity of the energy security problem and the air pollution problem.

Corn is the main feedstock for bioethanol production in China. Chinese corn consumption for feed and starch use has increased since 1990 and the domestic corn price has also increased since December 2004. Chinese corn ending stocks decreased dramatically from 123,799 thousand tons in 1999/2000 to 36,602 thousand tons in 2006/07 (Figure 1). When the government started to expand the corn-based bioethanol program, corn ending stocks were abundant and the government tried to manage the decrease in these stocks.

In China, the domestic corn wholesale price increased from 1,190 Yuan/ton in February 2005 to 1,547 Yuan/ton in September 2006[5], because the Chinese corn supply and demand situation was very tight. Corn consumption for bioethanol was competing with corn consumption for feed, food, and other industries. In this regard, the NDRC started to regulate corn-based bioethanol expansion on December 21, 2006. This regulation allowed the current bioethanol production level in Heilongjiang and Jilin, but limited further expansion of corn-based bioethanol production. This regulation will apply to wheat-based bioethanol production as well.

Instead of expanding corn-based bioethanol production, the government wants to diversify bioethanol production, especially from cassava. Cassava-based bioethanol production was 108.9 thousand tons in 2008 and in 2009 production capacity was 181.4 thousand tons. Total cassava production in China was 3.9 million tons in 2009, which is much smaller than cassava production in Thailand (22.8 million tons in 2008[6]). Although Guangxi is trying to increase cassava production, it is assumed that it is difficult to produce enough cassava in China to meet domestic consumption for bioethanol production. If China is to expand bioethanol production from cassava, it will have to rely on cassava imports from Thailand. China has mastered cassava-based bioethanol technology by constructing a demonstration project in Guangxi, but with regard to liquefaction, saccharification, fermentation, separation process, and sterilization devices, it still lags behind advanced international levels (Wang, 2011). A key to success for developing cassava-based bioethanol production in China is technical innovation for mass production.

Sweet sorghum can grow under dry conditions in saline alkaline soil. Although a number of provinces are trying to increase sweet sorghum production, its production is much lower than corn[7]. In addition, Chinese sweet sorghum-based bioethanol production has a technical

[5] It was derived from Institute of Agricultural Economics, Chinese Academy of Agricultural Science (2007.10).

[6] This data was derived from FAOSTAT Data (FAO, 2011).

[7] In 2010/11, sorghum production is 1.5 million and corn production is 28.6 million tons (USDA-FAS, 2011).

problem. It is technically immature and bioethanol content is so low (20%) that it cannot be used as fuel (Wang, 2011). At present, biofuel productions from non-food resources such as cassava and sweet sorghum are still in the pilot scale project stage in China.

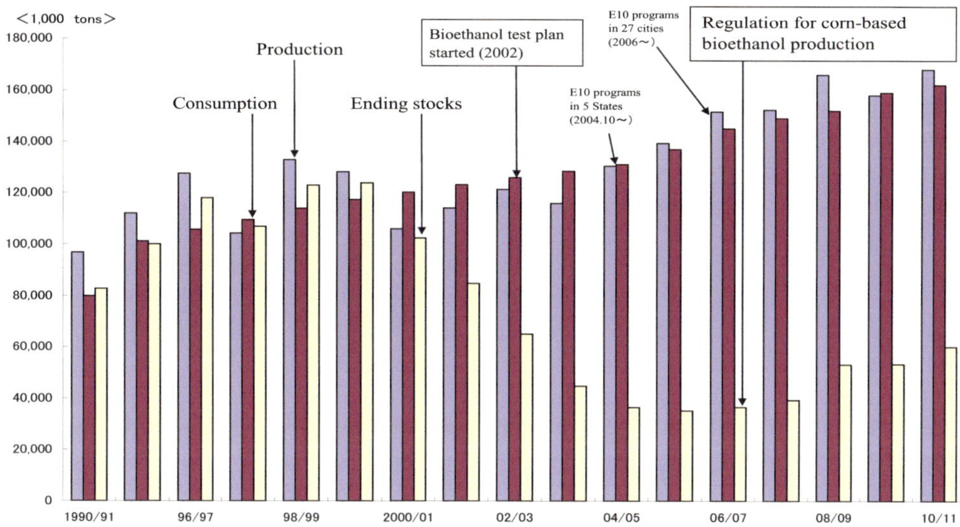

Fig. 1. Chinese corn ending stocks: production and consumption
Source: Data were derived from USDA-FAS (2011)

The NDRC provided a mid- to long-term plan for renewable energy in September 2007. This plan indicated that hydroelectric power generation would increase from 190 million kW in 2010 to 380 million kW in 2020, wind-power generation would increase from10 million kW in 2010 to 150 million kW in 2020, biomass generation would increase from 5.5 million kW in 2010 to 30 million kW in 2020, and solar energy generation would increase from 0.3 million kW in 2010 to 20 million kW in 2020. The plan indicated that bioethanol from non-food grade would be 2 million tons in 2010 and 10 million tons in 2020. The plan also indicated that biodiesel production would be 0.2 million tons in 2010 and 2 million tons in 2020. The Chinese government will promote the expansion of biofuel production from non-food grade in the future. In this plan, the government will promote agricultural resources that can be grown in waste land. In the long term, the National Energy Research Institute has projected that renewable energy will dominate more than 30% of the total primary energy supply in 2030 and 50% in 2050 (Kaku, 2011). This projection indicates that renewable energy will become a leading factor in the Chinese energy supply.

2.2 Japanese biofuel production and programs

2.2.1 The history of Japanese biofuel production and biomass storage

The history of bioethanol production in Japan dates back to 1889, when a factory was built in Hokkaido to produce bioethanol using potatoes as feedstock through malt saccharification.[8] After that, the office of the governor general of Taiwan (during Japan's colonial rule from 1895 to 1945) took the lead in developing bioethanol technologies. In 1937,

[8] As for Japanese biofuel production and programs, it depends on Koizumi (2011).

an alcohol monopoly system was launched to produce bioethanol from potatoes to meet military demand, and by 1944 Japan produced 170 thousand kℓ of bioethanol per year (Daishyo and Mitsui & Co., 2008). During World War II, bioethanol-blended fuel was used for airplanes as an alternative to gasoline, and a significant quantity of bioethanol-blended fuels was used for fighter-attack and trained airplanes at the end of WWII. It is estimated that bioethanol constituted 26.7% of total liquid fuels in 1945[9] because petroleum import lines from the Pacific area were broken at the end of the war. Biodiesel from soybean oil was also produced and used for naval fleets, mainly destroyers. Jatropha curcas-based biodiesel was developed by former army-related petroleum refiners and used for tank fuel and lamps. Japan's biofuel resources were developed as emergency alternative fuel for gasoline and diesel during WWII. The quality and production cost of biofuel were not suitable for commercial use after WWII. Most of these technologies were abandoned and forgotten after that. After WW II, Japan continued to produce bioethanol from imported molasses. However, the two oil crises in the 1970s shifted the focus of Japan's energy policy to energy savings and to reducing the country's reliance on oil,[10] with the result that the adoption of biofuel was not considered until recently. However, under the Kyoto Protocol, Japan was committed to cutting greenhouse gas emissions by 6% from 1990 levels before the end of the first commitment period (2008-2012). The decision to promote the recycling of various types of resources, including biomass, was enacted as the "Basic Law on Promoting the Formation of a Recycling-Oriented Society" in 2001. The first time the government announced a plan to promote biofuel production and utilization of biofuel was in the Biomass Nippon Strategy[11], which the Cabinet adopted in December 2002.

The Kyoto Protocol Target Achievement Plan, adopted by the Cabinet in April 2005, calculated that the new energy input in 2010FY[12] resulting from the implementation of the new energy countermeasures would be equivalent to 19.1 million kℓ of crude oil, which was projected to result in a reduction of 46.9 million tons of CO_2 emissions. The goal was to achieve a reduction in CO_2 equivalent to 500 thousand kℓ of crude oil[13]. When the Kyoto Protocol came into force in April 2005, Japan determined that, to meet its targets, it would be necessary to convert biomass energy into useful forms of energy, such as transportation fuels, and to draw a roadmap for the adoption of domestically produced biomass as transportation fuel. In March 2006, the Cabinet adopted the revised Biomass Nippon Strategy, the most striking features of which were that biofuel became the main force among various biomass products.

The Biomass Nippon Strategy categorizes biomass into three types: waste biomass, unused biomass, and energy crops. Based on data as of 2008, Japan stored 298 million tons of waste biomass and 17.4 million tons of unused biomass. The provisional estimate for the energy potential of unused biomass is approximately 14 million kℓ in crude oil, and the provisional estimate for the energy potential of energy crops is approximately 6.2 million kℓ in crude oil (Ministry of Agriculture, Forestry and Fisheries, 2010). Thus, there is potential to expand the production of biofuel in Japan.

---

[9] This figure is estimated from Miwa (2004).

[10] Japan relied on oil for 77.4% of energy consumption in 1973, and 71.5% in 1979, but this dropped down to 49.4% in 2001 (Ministry of Economy, Trade and Industry, 2009).

[11] Nippon means Japan in Japanese.

[12] FY means fiscal year from April to March of next year.

[13] 500 thousand kℓ of crude oil is equivalent to 800 thousand kℓ of bioethanol.

2.2.2 Developments and perspectives of the Japanese biofuel program

The Japanese government has been promoting bioethanol production and its use for automobiles since 2003. The Japanese bioethanol production level was estimated at 200 kℓ in March 2009 (Ministry of Agriculture, Forestry and Fisheries, 2010). At present, verification tests and large-scale projects for bioethanol production have been launched at ten locations in Japan. Demonstration projects include large-scale projects that began in 2007 to collect data for domestic transportation biofuel and to support a model project for the local utilization of biomass. The Ministry of Economy, Trade and Industry is promoting biofuel programs from an energy security incentive, while the Ministry of Agriculture, Forestry and Fisheries is promoting it mainly from the perspective of rural development, and the Ministry of Environment is promoting it for environmental reasons.

Hokkaido Bioethanol Co. Ltd in Shimizu Town, Hokkaido, produces bioethanol from surplus sugar beets and substandard wheat. Its facility's capacity is 15 thousand kℓ/year. Oenon Holdings, in Tomakomai City, Hokkaido, produces bioethanol from nonfood rice, and its facility's capacity is 15 thousand kℓ/year. JA Agricultural Cooperatives in Niigata City, in Niigatas Prefecture, produces bioethanol from nonfood rice with a capacity of 1.0 thousand kℓ/year (Ministry of Agriculture, Forestry and Fisheries, 2010). In addition to these projects, the soft cellulose-based bioethanol project has been promoted since 2008 to use rice straw and wheat straw to produce bioethanol. Rice and wheat straw-based bioethanol is produced at 3.7ℓ/day in Hokkaido, and rice straw and rice husk-based bioethanol is produced at 200ℓ/day in Akita Prefecture. Rice straw and other cellulose material-based bioethanol is produced at 100ℓ/day in Chiba Prefecture, and rice straw and wheat straw-based bioethanol is produced at 16ℓ/day in Hyogo Prefecture (Ministry of Agriculture, Forestry and Fisheries, 2010).

The municipal government and non-governmental organizations are promoting the production of biodiesel from used cooking oil blended with diesel used for public buses, official cars, and municipal garbage trucks. The total amount of biodiesel production was estimated at 10,000 kℓ as of March 2008 (Ministry of Agriculture, Forestry and Fisheries, 2010). Most of their biodiesel production levels are smaller than those of the bioethanol facilities since NGOs and local governments produce biodiesel in small plants using recycled rapeseed oil as the main feedstock. Twenty biodiesel fuel projects have started since 2007.

In February 2007, seven ministries and the cabinet office released a "roadmap" to expand biofuel. The goal was to produce 50 thousand kℓ of biofuel domestically per annum by 2011 FY. If appropriate technical development is achieved, such as reducing the costs of collection and transportation, developing resource crops, and improving bioethanol conversion efficiency, a significant increase in the production of domestic biofuel can be feasible by around 2030[14]. The budget in 2008 FY to enlarge Japanese biofuel production was 8 billion JPY. These measures included developing technologies for low-cost and highly efficient biofuel production, demonstrating the efficient collection and transportation of rice straws, and establishing technologies to manufacture biofuel from cellulose materials. The budget in 2009 FY to increase Japanese biofuel production was 20.3 billion JPY. To promote bioethanol production and utilization, a tax privilege for bioethanol production and utilization was also established in 2008. First, a 50% reduction in fixed assets tax for biofuel manufacturing

[14] The Ministry of Agriculture, Forestry and Fisheries calculated the production of domestic biofuel at 6 million kℓ to the year 2030.

facilities was applied for three years. Second, a tax reduction was established for the portion of bioethanol in bioethanol-blended gasoline; in the case of 3% bioethanol blended in gasoline, 1.6JPY/ℓ is tax exempted.
In 2009 the Ministry of Economy, Trade and Industry and the Ministry of Agriculture, Forestry and Fisheries set up a study panel for cellulose-based biofuel production to the year 2020. The panel released its estimates of biofuel production potential using Japanese technology in 2009: domestic cellulose-based bioethanol can be produced at about 330 thousand kℓ (crude oil equivalent); starch and glucose-based bioethanol can be produced at about 30 thousand kℓ; and biodiesel can be produced at about 50 thousand kℓ. Thus, domestic biofuel can be produced at about 400 thousand kℓ. The panel defined imported biofuel developed in Asian countries as "quasi domestic biofuel," which can be produced at about 100 thousand kℓ in 2020, based on their refineries' technologies and production scale.
In 2010, the Ministry of Economy, Trade and Industry set up the target amount of bioethanol utilization for oil refineries based on Notification No.242 of the Ministry of Economy, Trade and Industry. The target amount will be 210 thousand kℓ in 2011 increasing to 500 thousand kℓ in 2017.

2.2.3 Cost of bioethanol production and securing feedstock

The domestic costs of bioethanol are much higher than those of gasoline and imported bioethanol because of expensive land usage. The feedstock cost of sugarcane molasses is 7 JPY/ℓ, the processing cost is 83.4 JPY/ℓ, and gasoline tax is applied at the rate of 52.2 JPY/ℓ [15](Figure 1). The cost of sugarcane molasses-based bioethanol is 142.6 JPY/ℓ, and the production cost of rice from bioethanol use is 146.2 JPY/ℓ. There are two types of bioethanol utilization in Japan: a direct 3% blend with gasoline and ETBE (Ethyl Tertiary-Butyl Ether)[16] use. Bioethanol from sugarcane molasses and rice for bioethanol use in Niigata are used for direct blending with gasoline. The direct-blended gasoline has to be sold at the same price as standard gasoline to compete. The gasoline wholesale price is 59.6 JPY/ℓ, and gasoline tax is applied to 53.8 JPY/ℓ, so the total gasoline price is 113.4 JPY/ℓ. The price difference between sugarcane molasses for bioethanol use and the gasoline price is 29.2 JPY/ℓ, and the price difference between rice for bioethanol use and the gasoline price is 32.8 JPY/ℓ.
The production cost of bioethanol from non-food-grade wheat is 150.2 JPY/ℓ. This type of bioethanol is used in Hokkaido for ETBE production. The price of bioethanol for ETBE use is based on the imported Brazilian bioethanol price, determined by the Petroleum Association of Japan (PAJ). The total price of bioethanol from Brazil is 127.3 JPY/ℓ, and the price difference between that of non-food wheat and the Brazilian bioethanol price is 22.9 JPY/ℓ. Food-based biofuel is not produced in Japan, so these biofuel production costs are theoretical figures (Fig.2). It is not realistic to produce bioethanol from food use grains in Japan, because production costs are high.
These price differences present crucial challenges to the goal of expanding biofuel production in Japan. At present, bioethanol producers are bearing the price deficiencies using subsidies. However, these subsidies have been limited to between 3-5 years, and at present no bioethanol producers can operate their production facilities without subsidies.

[15] The tax reduction was established for the portion of bioethanol out of bioethanol-blended gasoline in February 2009. In the case of 3% bioethanol blended in gasoline, 1.6JPY/ℓ is tax exempted.
[16] ETBE (Ethyl Tertiary-Butyl Ether) is made from bioethanol and isobutylene.

Reducing the cost of producing bioethanol is the key to increasing its domestic production, but it will be difficult to reduce the domestic bioethanol cost to the level of gasoline prices and imported bioethanol prices in a short period. If the government wants to maintain domestic bioethanol production levels, policy measures to diminish their price deficiencies will be necessary, at least in the short term.

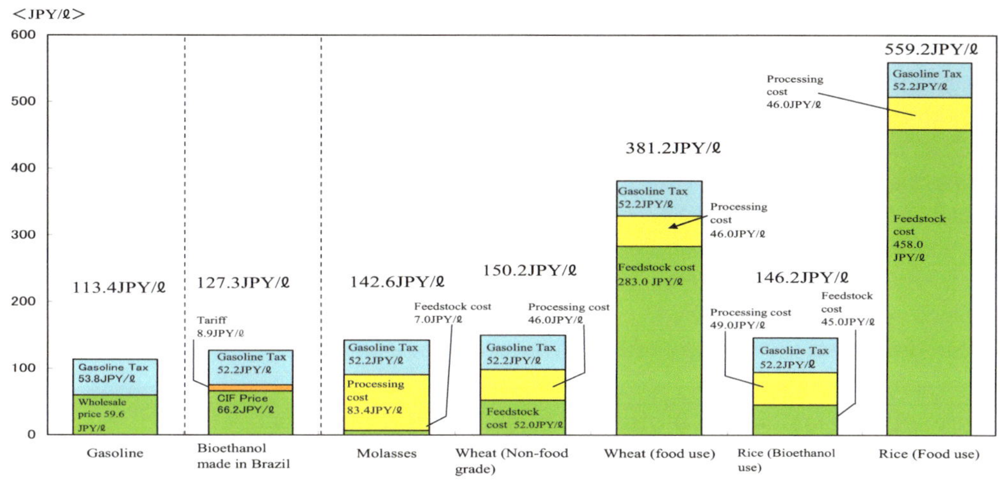

Fig. 2. Japanese bioethanol production cost

Note:

1. Production cost includes capital cost and variable cost. Retail price includes transportation cost and consumption tax. These data are based on Ministry of Agriculture, Forestry, and Fisheries of Japan (2010).
2. The wholesale price of gasoline is the average March 2010 price from the the Oil Information Center of Japan.
3. The Brazilian bioethanol CIF price is the average March 2010 price from trade statistics. The custom tariff is 13.4%

At present, ten bioethanol production projects are operating. It is difficult for most of these facilities to increase their production levels because of limited feedstock. In addition, agricultural products are strongly influenced by the weather, and Japan is a net food-importing country. What's more, there is strong critical opinion that food-based biofuel may damage domestic and world food availability. Thus, in order to increase the volume of domestically produced bioethanol in Japan, it is necessary to produce biofuel from cellulose materials and unused resources.

2.3 Other countries and regions

The government of Korea promotes biofuel utilization to eliminate GHG emission. The presidential committee for green growth has released a plan to cut GHG emissions by 4% until 2020, compared with the 2005 level. The Korean government strongly promotes a national renewable energy program. At present, the biodiesel program is the leading project in the program.

The Korean biodiesel production level was 300 thousand kℓ in 2009. Of that amount, 75-80 percent was imported soybean oil and palm oil, while the remainder was mainly domestically used cooking oil (USDA-FAS, 2010). The Korean government has set the biodiesel targeted blend ratio at 2.0% but plans to increase this to 3.0% in 2012. To meet biodiesel demand, Korea will have to increase biodiesel production in the future. The government is exploring research for alternative feedstock for biodiesel, such as rapeseed oil, animal fats, and other sources. However, it is difficult to increase the production and yield of rapeseed, and further R&D is needed for animal fats-based biodiesel. Ensuring feedstock is a crucial problem in expanding biodiesel production and utilization in Korea.

The government of Taiwan has promoted the B1 (1% biodiesel blend to diesel) mandate program since 2008. The main incentive for promoting the biodiesel program in Taiwan is to cut GHG emission. Although Taiwan is not a member of the Kyoto Protocol, it has tried to pursue the global trend of cutting GHG emission. Biodiesel production in Taiwan was estimated at 36 thousand Kℓ in 2009 (F.O.Licht, 2010). The feedstock of biodiesel production is used cooking oil. Taiwan's demand for biodiesel is estimated at 45 thousand kℓ per year (USDA-FAS, 2009b). The gap between domestic demand and supply depends on biodiesel imports from the EU. The government plans to increase the biodiesel blend ratio in the future.

3. Impacts of East Asian biofuel policies on food markets

3.1 Impacts of Chinese bioethanol imports on world sugar markets

3.1.1 Methodology and baseline projection

This study examines the impacts Chinese bioethanol import expansion from Brazil would have on Brazilian and international sugar markets by applying the World Sugar Market Model[17]. This model was developed in order to analyze how bioethanol, energy, or environmental policies in major sugar-producing countries affect not only domestic and world bioethanol markets but also corresponding sugar markets. The model was developed as a dynamic partial equilibrium model that extends to the world sugar and bioethanol markets. The world sugar market consists of 11 major sugar-producing countries, namely: Brazil, the U.S., the EU27, Australia, Mexico, Japan, India, China, Thailand, the former USSR, and the rest of the world. The Brazilian bioethanol market is involved in the model.

Brazil is the world's largest producer of sugarcane and sugarcane-based bioethanol. More than half of the sugarcane produced in Brazil goes towards bioethanol production, and the remainder goes to the bioethanol market, meaning developments in Brazil have considerable implications for global sugar and bioethanol markets. In the model, these two markets are inter-linked through the Brazilian sugar and bioethanol markets. In the Brazilian market, a "sugarcane allocation ratio variable" is defined as the relative proportions of sugarcane going to bioethanol production and sugar production respectively. Each country market consists of production, consumption, exports, imports, and ending stocks activities up to the year 2020/21. The sugar market activities are defined on a raw sugar equivalent basis. The baseline projection is based on a series of assumptions about the general economy, agricultural policies and technological changes in exporting and importing countries during the projection period. It is assumed that the Chinese

---

[17] As for the World Sugar Market Model, refer to Koizumi and Yanagishima (2005).

government doesn't import bioethanol from Brazil. Based on these assumptions, world sugar production is projected to increase by 2.0% and its consumption is projected to increase by 2.5% per annum from 2006/07 to 2020/21, while world sugar exports and imports are projected to increase by 1.8% per annum during this period.

3.1.2 Impacts of Chinese bioethanol imports on world sugar markets

The bioethanol mid-to long-term plan for renewable energy indicated that bioethanol production from non-food grade would be 2 million tons in 2010 and 10 million tons in 2020 (Table 4). According to this plan, bioethanol is not produced from corn and wheat, and produced from non-food grade feedstock. However, it is assumed to be difficult to expand bioethanol from non-food grade feedstock in China. In this scenario, it is hypothesized that during the projection period technological innovation for bioethanol production will not be developed and non-food grade feedstock for bioethanol supply will not expand. Thus, it was assumed bioethanol production from non-food grade would not expand from 2007/08 in this scenario.

The Chinese bioethanol production cost was 0.827US$/ℓ in 2007, while the Brazilian bioethanol production cost was 0.30 US$/ℓ in 2006/07 (F.O.Licht, 2008). The CIF price of bioethanol landed in China is estimated at 0.63 US$/ℓ[18], which is lower than the domestic production cost. The Chinese bioethanol production cost is higher than that of Brazil, which has a large capacity for exporting bioethanol. If the Chinese government promotes the utilization of alternative fuels, it may consider importing Brazilian bioethanol in the future. It is assumed that both bioethanol trades will expand in the future. The Chinese government will import bioethanol from Brazil as a mid-to long-term goal to address the deficiency in domestic production. As a result, bioethanol imports will total 1,700 thousand tons in 2010/11 and 9,700 thousand tons in 2020/21.

As a result of Chinese bioethanol imports from Brazil from 2010/11, the Brazilian sugar price (Domestic crystal sugar price) is predicted to increase by 24.8% in 2020/21 and the world raw sugar price (New York No.11) is predicted to increase by 15.9% in 2020/21 (Table 5). This can be concluded from analysis using the econometric model, that expanded bioethanol imports from China to Brazil would have an impact not only on the Brazilian sugar market, but also on world sugar markets. A higher world raw sugar price will also benefit other sugar-exporting countries. Other sugarcane-based sugar exporters are expected to materialize benefits with a two-year time lag, because of the agricultural conditions associated with the growth of sugarcane. Brazilian bioethanol and sugar producers are assumed to materialize benefits from relatively higher domestic bioethanol and sugar prices, because more than 60% of Usina (local sugar producers) have both bioethanol and sugar facilities in Brazil.

However, some developing countries may decrease their imports and consumption due to the relatively high sugar price. The expansion of Chinese bioethanol imports from Brazil can have a negative impact on some countries, due to the higher sugar prices[19]. In addition, the expansion of Chinese bioethanol imports from Brazil can cause an increase in the volatility of the world sugar price.

---

[18] Freight from Brazil to China, including insurance, is 0.21US$/ℓ, estimated from Sao Paulo Esalq and 1.9 DT Chemical tanker. The tariff equivalent is 0.1235 US$/ℓ (Tariff rate 2207.1 0-1 90).

[19] For detailed model simulation, please refer to Koizumi (2009).

	Feedstock	2008 Production (tons/year)	2009 Production Capacity (tons/year)	2010 Target (tons/year)	2020 Target (tons/year)
Heilongjiang	Corn	180,000	180,000	0	0
Jilin	Corn	470,000	500,000	0	0
Henan	Wheat	410,000	450,000	0	0
Anhui	Corn	400,000	440,000	0	0
Guangxi	Cassava	120,000	200,000	200,000	200,000
Hubei	Inferior grains	0	0	100,000	100,000
Total (1)		1,580,000	1,770,000	300,000	300,000
National Target (2)		–	–	2,000,000	10,000,000
Domestic defficiency (3)=(2)-(1)		–	–	1,700,000	9,700,000

Table 4. Chinese mid- to long-term plan and bioethanol production (Scenario)
Source: NDRC, Mid-long term plan of renewable energy (September 2007) and author's estimation

	2020/21
World raw sugar price (New York, No.11)	15.9%
Brazil crystal sugar price	24.8%
World white sugar price (London, No.5)	15.9%

Table 5. Impact on sugar prices (Scenario/baseline)
Source: Koizumi (2009)

3.2 Impacts of the biofuel and feedstock import on world agricultural markets in other countries and region

It is estimated that Japan will import bioethanol from Brazil to meet its goal. It is hypothesized that Japan will start the E3 (3% of bioethanol blend in gasoline) program in 2012 and will depend on imported bioethanol from Brazil. As a result of the E3 program in all areas of Japan from 2012, the Brazilian sugar price (Domestic crystal sugar price) is predicted to increase by 1.5% and the world raw sugar price (New York No. 11) is predicted to increase by 1.4% in 2015 (Koizumi, 2007). In addition to this analysis, it is hypothesized that Japan will import 3 million kℓ of Brazilian bioethanol starting in 2010[20]. As a result of the 3 million kℓ of bioethanol imported from Japan to Brazil, the Brazilian sugar price is predicted to increase by 4.4% and the world raw sugar price is predicted to increase by 3.1% in 2015 (Koizumi, 2007). As a result of the analysis using the econometric model, it is concluded that an expansion of bioethanol exports from Brazil to Japan would have an impact not only on the Brazilian sugar market, but also on world sugar markets[21].

Korea imports soybean oil as feedstock for biodiesel use from Argentina and Brazil, and imports palm oil as feedstock for biodiesel from Malaysia and Indonesia. Taiwan imports biodiesel from the EU. It is estimated that Korean soybean oil imports from Argentina and

[20] It is hypothesized that Japan will import 3 million kℓ of Brazilian bioethanol for thermal power generation if technical and transportation problems are resolved via cooperation between Japan and Brazil.
[21] For this model simulation, refer to Koizumi (2007).

Brazil can impact the soybean and soybean products markets in these countries; Korean palm oil imports from Malaysia and Indonesia can impact their palm oil markets; and Taiwan's biodiesel imports from the EU can impact biodiesel and oilseed markets in the EU. However, the amount of their imports is very small; it is estimated that their impacts on world vegetable oil and related markets are quite small and limited.

4. Future directions for the biofuel program in East Asia

4.1 Securing biofuel production and R&D for second-generation biofuel production

Governments in Asian countries and the region are promoting biofuel programs to deal with energy security, environmental problems, and agricultural problems. Securing feedstock for biofuel is the most crucial problem in expanding biofuel production in East Asia. In addition, high production costs and an unstable production system caused by a lack of feedstock supply are also obstacles to the expansion of biofuel production in East Asia.

At present, it is difficult to expand food-based biofuel production in East Asia. In the future, the most crucial factors for promoting biofuel production will be technological innovation in producing biofuel from rice straw or wooden biomass efficiently, and the development of crops that can produce bioethanol in large quantities. The R&D of second-generation biofuel that is developing in Japan includes improving varieties of energy resource crops, developing technologies for manufacturing biofuel, and developing cultivation methods. The sugar and corn starch yield of genetically engineered varieties is higher than that of conventional varieties. In addition, technologies have been developed to manufacture bioethanol more efficiently from non-food resources, such as woody biomass, rice straws, and energy crops. Japanese research institutes are also working on increasing the efficiency of cellulose-based bioethanol production[22]. Some Japanese universities and private companies are researching the production of biodiesel from algae, such as *pseudochoricystis ellipsdoidea* and *Botryococcus braunii* for automobile fuel and jet fuel.

In the future, China will have to diversify feedstock for biofuel production. China has switched from grain-based biofuel to non-food grade biofuel, such as sweet sorghum and cassava. However, biofuel production from non-food resources such as cassava and sweet sorghum are still in the pilot scale project at present and it is difficult to expand bioethanol from cassava and sweet sorghum, because of the difficulty securing feedstock. In addition, China is exploring second-generation biofuel production from corn stalk and algae. The Tianguan Group Co. Ltd., has constructed a pilot cellulose bioethanol production line, with a capacity of 300 tons/year. China National Cereals, Oils and Foodstuffs Corp. (COFCO), Sinopec, and Novozymes signed a new agreement to advance cellulose bioethanol technology in 2009 (Wang, 2011).

At present, high enzyme cost is one of the problems in expanding cellulose-based bioethanol production around the world. As for cellulose-based bioethanol production, the main research area is reducing the cost of enzymes in cooperation with the U.S. and private European companies. China is conducting R&D for biofuel production from algae in collaboration with private U.S. companies and government. While Chinese R&D for second-

[22] In 2006 RITE (Research Institute of Innovation Technology for the Earth) and Honda R&D Co., LTD. developed the *RITE strain*, which substantially reduces the harmful influence of fermentation inhibitors. RITE is also developing high STY (Space Time Yield), which promotes productivity in a unit of reaction volume per hour and simultaneous utilization of C6 and C5 sugars.

generation biofuel production has just begun, its R&D can be active in the future. Korea is conducting researches into producing biofuel from seaweed and Taiwan is conducting research into cellulose and agricultural waste-based bioethanol production. However, their researches are also at an experimental stage.

4.2 GHG reduction from domestic bioethanol production

Japanese bioethanol production is in an experimental stage, and is not mature enough to decide on a default ratio for LCA analysis of GHG emissions from domestic bioethanol production[23]. However, the Japanese government has released reference LCA results for domestic bioethanol production to introduce sustainable criteria for biofuel[24]: High-yield rice with changes for water management emitted 91 gCO2/MJ; high-yield rice without changes for water management emitted 57 gCO2/MJ; minimum-access rice emitted 60 gCO2/MJ; non-food grade wheat emitted 44 gCO2/MJ; surplus sugar beets emitted 39 gCO2/MJ; sugar beets for bioethanol use emitted 60 gCO2/MJ; wasted wood emitted 8 gCO2/MJ; and sugarcane molasses emitted 55 gCO2/MJ (Table 6). It is estimated that Japanese gasoline emitted 81.7 gCO2/MJ. As for the GHG elimination ratio compared with gasoline, the ratio of domestic bioethanol to gasoline ranges widely from -11% to 90%. The ratio of high-yield rice with changes for water management is -11%, and the ratio of wasted woods is 90%. It will be necessary to examine these LCA analyses again whenever bioethanol-related technological developments occur, because Japanese biofuel production is in an experimental stage.

(gCO2/MJ)

	Feedstock production	Feedstock transportation	Biofuel production	Biofuel transportation	Total	GHG elimination ratio compared with gasoline
High yield rice with change for water management	53	1	33	4	91	-11%
High yield rice without change for water management	19	1	33	4	57	30%
Minimum-Access Rice	21	1	33	4	60	27%
Wheat (Non-food grade)	7	1	32	4	44	46%
Surplused sugar beet	7	5	24	4	39	51%
Sugar beet for bioethanol use	28	5	24	4	60	27%
Wasted woods	0	1	3	4	8	90%
Sugarcane molasses	0	0	51	4	55	33%

Table 6. Reference study results of GHG emission and reduction for Japanese bioethanol
Source: Data were derived from the Ministry of Economy, Trade and Industry, Japan (2010)

4.3 Establishing sustainability criteria for biofuel

East Asian countries and the region are importing or will import biofuel and feedstock for biofuel from other countries. To ensure the sustainability of biofuel not only in their

[23] In this study, GHG covers CO2, CH4 and N20. The GHG emission was equivalent to CO2 emission. The GWP (Global Warming Potential) is 21(CH4) and 310(N20).

[24] The Japanese Government didn't release the default ratio for LCA analysis of GHG emissions from domestic biodiesel production. The governments of China, Korea, and Taiwan didn't release the reference and default ratio for LCA analysis of GHG emissions from domestic biofuel production.

countries and region but also on a global scale, they have to take care of the environment, food availability, and the social consequences among their trading partner countries. Thus, establishing sustainable criteria is crucial in promoting biofuel utilization and production in these countries and the region. The Sophisticated Methods of Energy Supply Structures Law, enacted in July 2009, required oil refiners (petroleum and gas enterprises) to use biofuel and biogas. To decide sustainable criteria for the use of biofuel in Japan, the government organized a study panel to discuss the introduction of the criteria in 2009, and in November 2010 the criteria were finally stipulated in Notification No. 242 of the Ministry of Economy, Trade and Industry.

The criteria included several issues: First, the biofuel should eliminate 50% of GHG, compared to gasoline or diesel. Second, oil refiners should pay attention to ensure food availability, and not to impair such availability in the course of promoting biofuel utilization.[25] Third, oil refiners should recognize the impact of biofuel production on biodiversity and obey domestic laws and regulations related to these areas.[26] Fourth, oil refiners should promote cellulose-based and algae-based biofuel R&D and utilization. These sustainable criteria took into account not only domestic biofuel production, but also imported biofuel. At present, most of the domestic biofuel production does not satisfy the criteria (50% GHG reduction), with the exception of waste woods and sugar beet for bioethanol use[27]. However, these criteria are applied to each project if the project is fairly evaluated as a demonstration project. This means these criteria will not apply for most of the domestic project for the time being, because Japanese biofuel production is in an experimental stage. The notification recognized it would be necessary to examine domestic criteria for these LCA analyses, whenever bioethanol-related technological developments occur.

The government of Japan decided on mandatory sustainable criteria for biofuel. The criteria cover the limitation of GHG emission, while paying attention to biodiversity and food availability. However, the criteria do not cover social consequences and other environmental issues, such as air quality, water availability, and others. On this account, Japan has been contributing to discussions in the Global Bioenergy Partnership (GBEP) to establish international guidelines for sustainable criteria for biofuel with the Food and Agricultural Organization of the United Nations (FAO) and other countries since 2007. The category of proposed sustainable criteria in the GBEP are much wider than those of Japan. [28] Although China, Korea, and Taiwan have not introduced sustainability criteria, it has been strongly

[25] If they are concerned that bioethanol production of the trading partner country will dramatically decrease, oil refiners should report their situation to the Japanese Government.

[26] If they are concerned that the biodiversity of the biofuel trading partner country will be damaged dramatically, oil refiners should report their situation to the government.

[27] In the case of Brazil, the panel reported that bioethanol production from existing crop land could eliminate 60% more GHG emissions than gasoline. It means that Brazilian bioethanol production from existing crops land can pass the draft criteria. The panel also reported that bioethanol production from converted pasture land could increase GHG emissions 8% over those of gasoline.

[28] GPEP brings together public, private, and civil society stakeholders in a joint commitment to promote bioenergy for sustainable development. The proposed criteria covers environmental (GHG emissions, productivity capacity of the land and ecosystems, air quality, water availability, use efficiency and quality, biological diversity, land-use change, including indirect effects), social, economic and energy related security (March 2011).

recommended they do so to comply with international sustainable criteria for GBEP, when the GBEP criteria are finalized and released.[29]

5. Conclusion

The governments of East Asian countries and the region are promoting biofuel programs to address energy security and environmental problems, and problems related to agriculture and rural development. Their main incentives to promote biofuel are different and produce various resources. Their feedstock for biofuel production includes various agricultural products. In China, energy security is the main incentive to promote its biofuel program. The enforcement of the Kyoto Protocol influenced the start of the biofuel programs in Korea and Taiwan and required Japan to start a biofuel program.

Verification tests and large-scale projects for biofuel production have been launched in China and Japan, but current biofuel production have experienced some problems because of high production costs and securing feedstock. In particular, securing feedstock for biofuel is a crucial problem, because this feedstock comprises various agricultural products that are used as food sources. Increasing biofuel consumption is exacerbating this problem. The gap between domestic demand and supply of biofuel has created a reliance on imported biofuel. Although bioethanol imports from Brazil will have an impact on the world sugar price, this impact differs from the impact of grain and staple food.

To ensure energy security, biofuel should be produced domestically in the long term. The governments of East Asian countries and the region are working on biofuel programs that will not compete with food availability. It is expected that the introduction and development of second-generation biofuel can mitigate the competition between food and energy. Japan, China, Korea, and Taiwan are promoting the production of biofuel from cellulose and unused resources, and Japanese R&D for second-generation biofuel production is very active. These countries and the region will have to continue to assist on these research projects in the mid to long term, so these governments can increase their domestic biofuel production and imports in the future. There are no international frameworks for the R&D of second-generation biofuel in East Asia and other regions. Establishing international cooperation to develop second-generation biofuel is needed in East Asia and other region.

However, there is still uncertainty about whether second-generation biofuel production can be economically viable. For the time being, some countries and regions may have to depend on imported biofuel and its feedstock from other regions to meet their national goals. East Asian countries and the region are importing biofuel or feedstock for biofuel from other countries and region. Because the biofuel program was introduced from environmental incentives in East Asian countries and the region, the introduction of these biofuels should improve environmental conditions not only in their countries and the region, but also globally. When countries promote a biofuel program, they have to pursue sustainability simultaneously. To pursue the sustainability of biofuel, they have to take care of the environment, food availability, and the social consequences in their trading partner countries. Thus, establishing sustainable criteria for biofuel, which determine the limitations of GHG emissions, and paying close attention to biodiversity, food availability, and social

[29] At present (March 2011), GBEP doesn't decided final guidelines for sustainable criteria for biofuel.

consequences, are needed for East Asian countries and the region. Japan decided the mandated criteria for oil refiners in 2010. However, further researches and dialogue with related countries will be required to realize the sustainability of biofuel. East Asian countries and the region, especially China and Japan, have put emphasis on promoting the expansion the amount of biofuel production until now. It is time to change this emphasis to pursuing the sustainability of biofuel, rather than expanding the amount of production. International cooperation in the region is needed to realize the sustainability of biofuel.

6. References

BP (2009), BP Statistical review of world energy, 2.3.2011, Available from http://www.bp.com/productlanding.do?categoryId=6929&contentId=7044622.

Chavez, E. Liu,D.H. and Zhao, XB. (2010). Biofuels Production Development and Prospects in China, *Journal of biobased materials and Bioenergy*, 4(3), pp.221-242, ISSN 15566560.

Daishyo, Y. and Mitsui & Co., LTD,(2008). *Bioethanol Frontline*, Kougyo Chosakai, ISBN9784769371632, Tokyo, Japan.

F.O.Licht (2008), Ethanol Production Costs a Worldwide Survey, F.O.Licht, ISSN14785765, Ratzeburg, Germany.

F.O.Licht.(2010). F.O.Licht World Ethanol & Biofuels Report,Vol.7,No18, F.O.Licht. ISSN14785765, Ratzeburg, Germany.

FAO (2011), FAOSTAT, 4.3.2011, Available from http://faostat.fao.org/site/567/default.aspx#ancor.

Huang, B and Yabe, M. (2010), Chinese bioethanol production and agricultural Policy, *Biofuel of Rice and Rural Development*, Yabe B and Morozumi K, pp229-253, Tsukuba Shobou, ISBN 9784811903590, Tokyo, Japan.

Institute of Chinese Affairs (2010). *Chinese Yearbook 2010*. Institute of Chinese Affairs, ISBN 9784620907000, Tokyo, Japan.

International Energy Agency (2008), *World Energy Outlook*, International Energy Agency, ISBN 9789264045606, Paris, France.

Kaku, S. (2011), *Chinese Energy*, Iwanami Shoten, ISBN 9784003412895, Tokyo, Japan.

Koizumi, T. (2008). Biofuel Policies in Asia, *Proceedings of FAO Expert Meetings*, 5&6, Rome, Italy, Feb.2008.

Koizumi, T. (2007). Impacts of the Japanese Bioethanol Import Expansion on the World Sugar Markets; Econometric Simulation Approach, *Proceedings of 5th FANEA Meeting*, Beijing, China, October 2007.

Koizumi, T. and Yanagishima, K. (2005). Impacts of the Brazilian Ethanol Program on the World Ethanol and Sugar Markets: An Econometric Simulation Approach, *The Japanese Journal of Rural Economics*, 7, 2005, pp61-77.

Koizumi, T. and Ohga, K. (2007). Biofuels Policies in Asian Countries: Impacts of the Expanded Biofuels Programs on World Agricultural Markets, *Journal of Agricultural & Food Industrial Organization, Special Issue: Explorations in Biofules Economics, Policy and History*, Volume 5, 2007, Article 8, pp.1-20, ISSN 15420485.

Koizumi, T. (2009). Biofuel and World Food Markets, Nourin Toukei Kyokai, ISBN 9784541036674, Tokyo, Japan.

Koizumi,T.(2011). The Japanese biofuel program-developments and perspectives, *Journal of Cleaner Production*, JCLP 2497, ISSN 0959-6526.

Lang, X. Zheng, F. and Cui, H. (2009). Evolution of fuel ethanol policy in China, *Forestry Economics*,pp.29-33.

Matsumoto, N. Sano, D. and Elder,D. (2009), Biofuel initiatives in Japan: Strategies, policies, and future potential, *Applied Energy*, 86:pp.69-76, ISSN 03062619.

Ministry of Agriculture, Forestry and Fisheries, Japan. (2010). New Era for Domestic Biofuel Production, proceedings of Biofuel seminar, Tokyo, Japan, September 2010.

Ministry of Economy, Trade and Industry, Japan. (2009). General Energy Statistics 2008FY, Ministry of Economy, Trade and Industry, 12.01.2010, Available from http://www.enecho.meti.go.jp/info/statistics/jukyu/index.htm.

Ministry of Economy, Trade and Industry, Japan.(2010). Notificication No.242 of Ministry of Economy, Trade and Industry, Japan.

Miwa, S. (2004). *Pacific War and Petroleum*, Nihon Keizai Hyouronsha Ltd., ISBN 4818815640, Tokyo, Japan.

National Bureau of Statistics of China. (2009). China Statistical Yearbook, China Statistics Press, ISBN978750375800, Beijing, China.

Song, A. Pei, G. Wang F. Wan, D. and Feng, C. (2008). Survey for Fuel Biofuel Feedstock Multiple Production, Academic Report of Agricultural Process, Vol.24(3), pp.302-307.

U.S. Department of Agriculture, Foreign Agricultural Service (USDA-FAS)(2009a). China-Peoples Republic of Biofuel Annual, CH9059, USDA-FAS, 01.03.2011, Available from http://gain.fas.usda.gov/Recent%20GAIN%20Publications/BIOFUELS%20ANNUAL_Beijing_China%20-%20P eoples%20Republic%20of_2009-7-17.doc.pdf.

U.S. Department of Agriculture, Foreign Agricultural Service (USDA-FAS)(2009b). Taiwan-Biofuel Annual, TW9033, USDA-FAS, 01.03.2011, Available from http://gain.fas.usda.gov/Recent%20GAIN%20Publications/General%20Report_Taipei_Taiwan_5-19-2009.pdf.

U.S. Department of Agriculture, Foreign Agricultural Service (USDA-FAS)(2010). Korea-Republic of, Bio-Fuels Production, KS1001, USDA-FAS. 01.03.2011, Available from http://gain.fas.usda.gov/Recent%20GAIN%20Publications/Bio-Fuels%20Production_Seoul_Korea%20-%20Republic%20of_2-10-2010.pdf

U.S. Department of Agriculture, Foreign Agricultural Service (USDA-FAS)(2011). PS&D, USDA-FAS, 01.03.2011, Available from http://www.fas.usda.gov/psdonline/psdQuery.aspx.

Wang, F. Xiong, XR. and Liu, CZ. (2009). Biofuel in China: Opportunities and challenges, *IN Vitro cellular & Developmental biology plant*, 45(3), pp.342-349. ISSN 10545476.

Wang, QA. (2011). Time for commercializing non-food biofuels in China, *Renewable & Sustainable Energy Reviews*, 15(1), pp.621-629, ISSN 13640321.

Zhou, A. Thomson, E. (2009). The development of biofuels in Asia, *Applied Energy*; 2009; 86(Suppl.1):S.11-20, ISSN 03062619.

12

Air Quality and Biofuels

S. Prasad and M.S. Dhanya
Division of Environmental Sciences,
Indian Agricultural Research Institute, New Delhi
India

1. Introduction

The energy sector has played a crucial role in the context of the global economy as well as the socio-economic development. The world energy consumption is growing at the rate of 2.3% per year. The Energy Information Administration estimated that the primary sources of energy consisted of petroleum 36.0%, coal 27.4% and natural gas 23.0% amounting to 86.4% share for fossil fuels in primary energy consumption in the world (EIA, 2010). Fossil fuel consumption is the largest contributor to air pollution, greenhouse gas emissions and the environmental impacts with a large endowment of coal and has an energy system that is highly carbon intensive. The combustion of fossil fuel releases VOCs, nitrogen oxides (NOx), carbon monoxide (CO) and particulate matter (PM). The combination of VOCs and NOx with sunlight further results in the formation of tropospheric ozone, the main component of smog. The burning of fossil fuels produces around 21.3 billion tonnes (21.3 giga tonnes) of carbon dioxide (CO_2) per year and the natural processes can only absorb about half of that amount, so there is a net increase of 10.65 billion tonnes of atmospheric carbon dioxide (USDoE, 2007). Coal combustion also leads to sulphur dioxide (SO_2) emissions with serious implications for local pollution (Shukla, 1997). Biomass burning is also recognized as a significant global source of emissions contributing as much as 40% of gross carbon dioxide and 38% of tropospheric ozone (Levine, 1991). Besides, 1.4 million tonnes of methane (CH_4) emissions are also reported from burning traditional biomass fuels. Apart from these emissions, there are a number of other environmental problems associated with energy use. Thus, the energy system is turning out to be 'unsustainable' in the 21st century.

In recent years, researchers have recognized the importance of holistic thinking. Current Kyoto-based approaches to reduce the earth's greenhouse gas involve seeking ways to reduce emissions. Biofuels have emerged as one of the most strategically important alternative fuel sources and are considered as an important way of progress for limiting greenhouse gas emissions, improving air quality and finding new energetic resources (Delfort et al., 2008). A fuel is considered as biofuel if it is derived from biomass such as agricultural products or residues, industrial and urban residues, wood residuals and forest products, either as liquid or as gas (Granda et al., 2007; Prasad et al., 2007a). It encompasses mainly bioethanol, biodiesel, biogas and biohydrogen (NREL, 2006). Ideally a biofuel should be carbon neutral and should therefore not contribute to the overall accumulation of carbon in the atmosphere (Oliveira et al., 2005). Carbon in crops is the result of the photosynthetic conversion of carbon dioxide in the atmosphere (capturing CO_2) into dry matter determined

by solar radiation during the growing season (Tilman et al., 2006) and by natural resources (e.g. climate, water) and external inputs (e.g. fertilizers, pesticides). Biofuel is thus considered an important component of the global strategy to reduce green house gas emission, improve air quality and to increase energy security by providing an alternative to fossil fuels (Farrell et al., 2006; Larson, 2006; Prasad et al., 2007b).
The worldwide investment in new biofuels production capacity has also been growing rapidly and was expected to exceed $4 billion in 2007. The value of biofuels production plants under construction and announced construction plans through 2009 exceeded $4 billion in the United States, $4 billion in Brazil and $2 billion in France (REN21, 2008). Biofuels production technologies, despite their techno economic potential have found meagre deployment due to myriad barriers. Recent developments in global climate change negotiations which culminated in the Kyoto Protocol are likely to remove some of the vital barriers to RETs which permitted fossil fuels to externalize the environmental costs. If biofuels want to be part of the solution they must accept a degree of scrutiny unprecedented in the development of a new industry. That is because sustainability deals explicitly with the role of biofuels in ensuring the well-being of our planet, our economy, and our society both today and in the future (Sheehan, 2009).
There are three key arguments for the commercial use of biofuels:
a. Economic-driven rise in consumption, resulting in higher prices for fossil fuels;
b. Energy security and geo-political dependence of regions with a high volatility;
c. Anthropogenic-based CO_2 emissions and climate change.

2. Global biofuel scenario

Soaring prices of fossil-fuels and environmental pollution associated with their use has resulted in increased worldwide interest in the production and use of biofuels. Both developed and developing countries have made mix of policies which have triggered public and private investments in biofuel crop research and development and biofuels production (EPA, 2009; REN21, 2008). Biofuels already constitute the major source of energy for over half of the world's population, accounting for more than 90% of the energy consumption in poor developing countries (FAO, 2005). Presently, biofuels production is expanding, especially in Brazil, the USA and South-East Asia, where sugar cane, maize and palm oil are converted into ethanol or biodiesel (Anonymous, 2008). Over the next several decades, the most certain increase in demand for biofuels is going to focus on displacing liquid fuels for transport, mostly in the form of ethanol which currently supplies over 95% of the biofuels for transportation (Fulton et al., 2004). The world's top ethanol fuel producers in 2009 were the United States with 10.75 billion US liquid gallons (bg) and Brazil (6.58 bg), accounting for 89% of world production of 19.53 billion US gallons (73.9 billion liters or 58.3 million metric tonnes according to data assembled by F.O. Licht. The Global Renewable Fuels Alliance (GRFA) is an international federation representing more than 65 per cent of the world's renewable fuels production from 30 countries. The GRFA predicts global production will reach 85.9 billion litres in 2010 growing by 16.2 per cent from 2009 production (Enagri, 2010). Global production of biodiesel has grown rapidly as well, although starting from a much smaller base. Biodiesel output expanded from 0.23 billion gallons in 2000 to 3.9 billion gallons in 2008 (EPA, 2009). The European Union produces nearly 80 percent of the world's biodiesel, largely from rapeseed; Germany is the single largest biodiesel producer, followed by the United States which produces the fuel mainly from soybeans (Nicholas, 2007). According to Pitkanen et al. (2003) lignocellulosic materials could support the sustainable production of liquid transportation fuels. The 73.9 Tg dry wasted crop material worldwide could potentially produce 49.1 GL $year^{-1}$ of bioethanol (Kim & Dale, 2004), about 16 times higher than the current world ethanol production. The potential bioethanol production could replace 353 GL of gasoline i.e. 32% of the global gasoline consumption (Prasad et al., 2007a).

2.1 Biofuel policies in different countries

Policy choices are instrumental in determining the direction of national as well as global biofuels development. Around the world, governments are considering a number of biofuel policy options. The biofuel policy aims to promote the use in transport of fuels made from biomass as well as other renewable fuels. The central policy of biofuel concerns job creation, greater efficiency in the general business environment and protection of the environment. A range of policies are currently being implemented to promote renewable bioenergy in United States, including the Energy Policy Act of 2005, the Energy Independence and Security Act of 2007, the 2002 Farm Bill and the Biomass Research and Development Act of 2000 (FAO, 2008).

Policy targets for renewable energy exist in at least 66 countries worldwide, including all 27 European Union countries, 29 U.S. states (and D.C.) and 9 Canadian provinces. Most targets aim for the 2010–2012 timeframe, although an increasing number of targets aim for 2020. There is now an EU-wide target of 20 percent and a Chinese target of 15 percent of primary energy by 2020. Besides China, several other developing countries adopted or upgraded targets during 2006/2007 (REN21, 2008). China finalized targets for the equivalent of 13 billion liters of ethanol and 2.3 billion liters of biodiesel per year by 2020. The directive sets a European target of 2% substitution of conventional transport fuels by biofuels by December 2005 and a further 5.75% substitution by December 2010. Moreover, the European Commission is committed to encourage the production and use of biofuels by proposing to set a binding minimum target for renewable energy sources of 10% of final energy use in the transport sector by 2020 and is also working on changing fuel specifications to allow higher than 5% blends of biofuel (Lechon, 2009; FAO, 2008).

New U.S. renewable fuels standard requires fuel distributors to increase the annual volume of biofuels blended to 36 billion gallons (136 billion liters) by 2022. The new standard implies that 20 percent of gasoline for road transport would be biofuels by 2022. Several states within the U.S. have also taken steps to promote development and increased use of biofuels. Under the Energy Policy Act of 2005, U.S. renewable transportation fuels are scheduled to reach 7.5 billion gallons by 2012. The 2007 Energy Independence and Security Act require 36 billion gallons of ethanol by 2020, with 21 billion gallons coming from advanced biofuels such as cellulose-based ethanol (Hoekman, 2009). The United Kingdom has a similar renewable fuels obligation, targeting 5 percent by 2010. Japan's new strategy for long-term ethanol production targets 6 billion liters per year by 2030, representing 5 percent of transport energy (FAO, 2008).

Developing countries like India also started Biofuel mission in 2003 to cope with the global fuel crisis. Government of India through a notification in September 2002 made 5% ethanol blending mandatory in petrol in 9 states and 3 Union Territories. In the next phase, supply of ethanol-blended petrol would be extended to the whole country and efforts would be made to increase the percentage of ethanol mixture in petrol to 10 percent (Prasad et al.,

2007b). National Biofuel Policy drafted by the Ministry of New and Renewable Energy Sources (MNRE), assures that biofuel programme would not compete with food security and the fertile farm lands would not be diverted for plantation of biofuel crops. The policy deals with a number of issues like minimum support prices (MSPs) for biofuel crops, subsidies for growers of biofuel crops, marketing of oil-bearing seeds, subsidies and fiscal concessions for the biofuel industry, R&D, mandatory blending of auto-fuel with biofuel, quality norms, testing and certification of biofuels. An indicative target of 20% by 2017 for the blending of biofuels-bioethanol and biodiesel has been proposed in the National Biofuel Policy (Indian Express, 2008).

3. Biomass resource base and biofuel generation technology

Since the mid-1970s many research initiatives have focused on increasing the biomass resource base for production of biofuel. Several technologies used for the conversion of plant material into biofuels are available and depend on the type of feedstock; the conventional and new technologies can be classified into the following four groups:

3.1 First generation biofuel technology

In general, first generation biofuels are produced from cereal crops (e.g. wheat, maize), oil crops (e.g. rape, palm oil) and sugar crops (e.g. sugar beet, sugar cane) using established technology. Based on the conversion of sugars (sugar cane) and starch (potato, cassava, maize) or oil (oil palm, rapeseed) accumulated in food crops into ethanol and biodiesel respectively accounts the first generation biofuels (Cassman & Liska, 2007). Some have called for an integrated systems biology approach to define ideotypes that meet the requirements of feedstocks for biofuels. However, the scientific evidence that crop traits can be genetically modified to meet the requirements for fuel without any trade-off on the value as a food crop is absent. Alternatively, different varieties may be developed for food and fuel production.

3.2 Second generation biofuel technology

In general, second generation biofuels are produced from cellulosic materials (Somerville, 2006) and also based on the use of dedicated energy crops like switch grass *(Panicum virgatum)* grown with low external inputs and using conversion methods that result in high net energy efficiency (output/input). Conversion of cellulosic biomass, which is both abundant and renewable, is considered as a promising alternative for ethanol produced from starch or sugar. Plant triacylglycerols are another potential feedstock to produce biofuels, especially biodiesel. Most vegetable oils are derived from triacylglycerols stored in seeds. Novel energy crops may be developed that produce triacylglycerols in non-seed tissues (Durrett et al., 2008). To avoid competition with food crops there is a growing interest in woody/tree borne oil plants. Native energy oil plants are more frequently present in tropical and subtropical regions. Non-edible oils obtained from plant species such as *Jatropha curcas* (Ratanjyot), *Pongamia pinnata* (Karanj), *Calophyllum inophyllum* (Nagchampa), *Hevea brasiliensis* (Rubber) and other oil-based crops can be efficiently used for biodiesel production. *Jatropha curcas* is a drought resistant, perennial oil plant (ca. 40% oil content) with favourable traits to produce biodiesel in unfavourable regions of India, Sub-Sahelian Africa and Latin America (Kumar & Sharma, 2008).

3.3 Third generation biofuel technology

This is based on algae or cyanobacteria that contain a high oil mass fraction (up to 70%) and are grown in ponds. Microalgae are sunlight-driven cell factories that convert carbon dioxide to potential biofuels (Akkerman et al, 2002; Ghirardi et al., 2000). Oil content in microalgae can exceed 80% by weight of dry biomass (Metting, 1996). Depending on species, microalgae produce many different kinds of lipids, hydrocarbons and other complex oils (Guschina & Hardwood, 2006; Metzger & Largeau, 2005). Certain algae and cyanobacteria have high lipid contents (Spolaore et al., 2006) biodiesel derived from microalgal oil (Dunahay et al., 1996; Sheehan et al., 1998). Under proper conditions, these micro-organisms can produce lipids for biodiesel with yields per unit area that are 50-100% higher than those with any plant system (Chisti, 2008). Microalgae can also provide several different types of renewable biofuels which include methane produced by anaerobic digestion of the algal biomass (Gavrilescu & Chisti, 2005) and photobiologically produced biohydrogen (Fedorov et al., 2005; Kapdan & Kargi, 2006).

The microalgae appear to be the only source of renewable biodiesel that is capable of meeting the global demand for transport fuels which will not compromise production of food, fodder and other products derived from crops (Chisti, 2008; Hu et al., 2008). However, it is still not proven that this high efficiency can be maintained after scaling-up the technology to a large production plant. Furthermore, the feedstock is waste derived from plant material used for food and feed. Yet, we do not know what the trade-off is between maximizing the utilization of primary production for food and feed and the use of residues and waste to produce methane or hydrogen. Chemical composition of the residues and waste also will matter.

3.4 Fourth generation biofuel technology

It is based on biohydrogen production by embedding parts of the photosynthesis apparatus in artificial membranes (Kruse et al., 2005). These biofuels are derived from the bioconversion of living organisms (microorganisms and plants) using biotechnological tools. The mean conversion efficiency for the total solar spectrum amounts to ca. 20%, which is on average about 10 times higher than for annual crops. This high efficiency should be considered a potential level. The gap between the potential level and actual efficiency is still not known. Currently, this technology is still expensive and not yet ready for commercial exploitation.

4. Biofuels for improvement in air quality

The emissions from engines using gasoline have SO_2, CO_2, VOCs, nitrogen oxides (NOx), carbon monoxide (CO) and particulate matter (PM), which causes pollution (Gaffney & Marley, 2009). The VOCs emitted from gasoline-fueled vehicles arise from uncombusted or partially combusted fuel and typically include cyclohexane, octanes, and aromatics. The NOx and VOC emissions react in the presence of sunlight by way of a series of photochemical reactions involving hydroxyl, peroxy and alkoxy radicals, to form the secondary pollutant ozone (Finlayson-Pitts & Pitts, 2000). The emissions of NOx and total VOCs lead to the formation of ozone in the troposphere, the main component of smog. CO is a deadly poison and the inhalation of fine particulate matter (PM2.5) is a serious health concern (Peter et al., 2003).

The energy and environmental crisis which the world is experiencing forced to find alternative uses for renewable resources and using clean technologies. Feedstocks include agricultural and food processing wastes, trees, and various grasses that are converted to ultra-clean (minimal SOx and NOx pollutants) biofuel in elaborate biochemical or thermo-chemical steps. And depending on the choice of a microorganism the bio-conversion can yield cellulosic ethanol, biogas or biohydrogen. Biofuels has a number of health and environmental benefits including improvement in air quality by reducing pollutant gas emissions relative to fossil fuels (Vasudevan et al., 2005). Therefore, it is imperative to develop and promote alternative energy sources that can lead to sustainability of the energy system. Hall & House (1993) have examined the role of biomass in mitigating global warming and contributing to the development of future energy strategies and concluded that the use of biomass for fossil fuel substitution would be far more effective in reducing atmospheric CO than to simply sequester CO_2 in forests in most circumstances. Currently, the second generation biofuels are projected to reduce carbon emissions by 90%, and by 2040 these could potentially replace up to 40% of all conventional fuels (Krisztina et al., 2010).

4.1 Combustion profile of biofuels

The success of oxygenated gasoline has sparked interest in the use of oxygenated compounds as emissions reducing additives in diesel fuel. Oxygenated compounds used as diesel additives are structurally similar to diesel fuel but have one or more oxygen atoms bonded to the hydrocarbon chain. Numerous oxygenated compounds have been investigated as either diesel fuel additives or replacements and have shown emissions reducing properties.

4.1.1 Properties and combustion profile of ethanol

Although ethanol was always a good oxygenate candidate for gasoline, the compound first approved by Environmental Protection Agency was methyl tertiary butyl ether (MTBE), a petrochemical industry product (Gaffney & Marley, 2000). The introduction of MTBE in gasoline has been studied as a classic case of solving one problem (reducing vehicle carbon monoxide emissions) while causing a new problem (persistent contamination of water systems with MTBE). Use of MTBE increased until 1999, but reports then appeared of environmental pollution incidents caused by MTBE spillage; US bans on MTBE came into force during 2002. Presently, ethanol is prospective material for use in automobiles as an alternative to petroleum based fuels. The main reason for advocating ethanol is that it can be manufactured from natural products or waste materials, compared with gasoline, which is produced from non-renewable natural resources. Ethanol can be independently used as a transportation fuel together with additives (e.g. ignition improver, denaturing agents, etc.). In addition, instead of pure ethanol, a blend of ethanol and gasoline is a more attractive fuel with good anti-knock characteristics (Al-Hasan, 2003).

Ethanol contains 34.7% oxygen by weight, and adding oxygen to fuel results in more complete fuel combustion, and therefore contributes to a reduction in exhaust emission and petroleum use (Huang et al., 2008; Prasad et al., 2007b). Ethanol is a high octane fuel and its use displaces toxic octane boosters such as benzene, a carcinogen. Ethanol is a virtually sulfur free additive and is biodegradable. Thus, it's easy to see why many states use ethanol to reduce vehicular emissions. The physical and thermo-physical properties of ethanol compared to the other fuels (gasoline and diesel) indicates that ethanol is more suitable and environmentally safe fuel (Table 1) as its normal boiling point lies in between gasoline and diesel, while heating value, carbon and sulfur content are lower (Lynd, et al., 1991; Vaivads et al., 1995).

Properties	Ethanol	Gasoline	Diesel
Density (g cm^{-3})	0.785	0.737	0.856
Normal boiling point (ºC)	78.00	38-204	125-400
Lower heating value, LHV (kJ cm^{-3})	21.09	32.05	35.66
LHV (kJ g^{-1})	26.87	43.47	41.66
Energy (MJ l^{-1})	23.10	32.84	33.32
Energy (MJ kg^{-1})	29.40	47.46	46.94
Carbon content (%)	52.20	85.50	87.00
Sulfur content (ppm)	0.00	~200	~250

Table 1. Comparison of thermo-physical properties of ethanol, gasoline and diesel fuel

A comparison of flammability variables for neat diesel, ethanol and gasoline clearly showed that ethanol (Table 2) falls between diesel and gasoline in terms of flashpoint and flammability temperature limits (Battelle, 1998). In the engine durability tests conducted by Meiring and coworkers (1983), no abnormal deterioration of the engine or fuel injection system was detected after 1000 hrs of operation on a blend containing 30% dry ethanol, small amount of octyl nitrate ignition improver and ethyl acetate phase separation inhibitor and the remainder diesel fuel. The Chicago Transit Authority in the US monitored the condition and overall performance of a fleet of 30 buses, of which 15 were the control run on number one diesel. After completion of 434,500 km distance by the 15 buses running on the blend, no abnormal maintenance or fuel related problems were encountered (Marek & Evanoff, 2001).

Characteristics	Neat diesel	Neat ethanol	Neat gasoline
Vapour-pressure at 37.8 °C (kPa)	0.3	17	65
Flash point (°C)	64	13	-40
Auto-ignition temperature (°C)	230	366	300
Flammability limits (%)	0.6-5.6	3.3-19.0	1.4-7.6
Flammability limits (°C)	64-150	13-42	-40-18

Table 2. Approximate fuel ethanol characteristics related to flammability

Low-percentage ethanol-gasoline blends (5-10%) can be used in conventional spark-ignition engines with almost no technical change. New flex-fuel vehicles of which there are over 6 million running mainly in Brazil, United States and Sweden, can run on up to 85% ethanol blends that had modest changes made during production. Ethanol combustion offers fuel and emissions savings due to the high octane number, the high compression ratio and the combustion benefits from ethanol vapour cooling which partly offsets its lower energy content per liter (IEA-ETE, 2007).

4.1.2 Properties and combustion profile of biodiesel

Biodiesel is a mono-alkyl ester based oxygenated fuel made from vegetable oil or animal fats. It has properties similar to petroleum based diesel fuel and can be blended into conventional diesel fuel. This interest is based on a number of properties of biodiesel, non toxic and its potential to reduce exhaust emissions (Jha, 2009; Knothe et al., 2006). The advantages of biodiesel as diesel fuel are its portability, ready availability, renewability, higher combustion efficiency, lower sulfur and aromatic content (Knothe et al., 2006; Ma &

Hanna, 1999), higher cetane number, and higher biodegradability (Mudge & Pereira, 1999; Speidel et al., 2000; Zhang et al., 2003). Biodiesel is by nature is an oxygenated fuel with oxygen content of about 10%. This improves combustion and reduces CO, soot and unburnt hydrocarbon.

Biodiesel is non-flammable and, in contrast to petrodiesel, is non explosive. The flash point of biodiesel (>130 °C) is significantly higher than that of petroleum diesel (64 °C) or gasoline (−45 °C) (Anonymous, 2010a). Biodiesel has a density of ~0.88 g/cm^3, higher than petrodiesel (~0.85 g/cm^3). Biodiesel has better lubricating properties and much higher cetane ratings than today's lower sulfur diesel fuels (Knothe et al., 2005; Mittelbach & Remschmidt, 2004). Biodiesel addition reduces fuel system wear (Anonymous, 2010b) and in low levels in high pressure systems increases the life of the fuel injection equipment that relies on the fuel for its lubrication. The calorific value of biodiesel is about 37.27 MJ/L (Elsayed et al., 2003). Variations in biodiesel energy density are more dependent on the feedstock used than the production process and properties of biodiesel from different oils are shown in Table 3 (Chhang et al., 1996; Rao & Gopalakrishnan, 1991). Biodiesel has virtually no sulfur content, and it is often used as an additive to Ultra low sulphur diesel (ULSD) fuel to aid with lubrication, as the sulfur compounds in petrodiesel provide much of the lubricity.

Biodiesel from Vegetable oil	Kinematic Viscosity mm^2/s	Cetane no:	Heating value MJ/kg	Flash Point °C	Density kg/l
Peanut	4.9	54	33.6	176	0.883
Soybean	4.5	45	33.5	178	0.885
Babassu	3.6	63	31.8	127	0.875
Palm	5.7	62	33.5	164	0.880
Sunflower	4.6	49	33.5	183	0.860
Diesel	3.06	50	43.8	76	0.855
B20 (20%blend)	3.2	51	43.2	128	0.859

Table 3. Approximate fuel biodiesel characteristics related to flammability

Since the key properties of the biodiesel are comparable to those of diesel fuel, it can be used in all diesel engines with little modification or no modification either on its own or as a blend with conventional or low sulphur diesel (Ryan, 1999). The disadvantages of biodiesel are its higher viscosity, lower energy content, higher cloud point and pour point, higher nitrogen oxide (NOx) emissions, lower engine speed and power, injector coking, engine compatibility, high price and greater engine wear. The technical disadvantages of biodiesel fossil diesel blends include problems with fuel freezing in cold weather, reduced energy density and degradation of fuel under storage for prolonged periods. However there are solutions to this such as using a blend of biodiesel upto B20 which has a gelling point of -15 degrees F, adding a biodiesel additive such as Fuel Boost to the blend also lowers the gel point even further and useful in the winter (Petracek, 2011).

4.1.3 Properties and combustion profile of biogas

Biogas is a renewable fuel produced by anaerobic fermentation of organic material (Pathak et al., 2009). The value of a substrate in the biogas process depends on its potential as a high yield plant species and on the quality of the biogas produced such as the achievable methane content. The most suitable plant species for the production of biogas are those which are rich in degradable carbohydrates such as sugars, lipids and proteins, and poor in hemicelluloses and lignin, which have a low biodegradability (El Bassam, 1998). Its composition varies with the source, but usually it has 50-70% CH_4, 25-50% CO_2, 1-5% H_2, 0.3-3% N_2 and traces of H_2S (Bedoya, 2009). Methane is the only combustible constituent of biogas, which is utilized in different forms of energy. Biogas can be used for heating, lighting, transportation, small-scale power generation, and large gas turbines as a complementary fuel (e.g., to natural gas) (Bedoya, 2009). Constraints like cost of cleaning, upgrading (to remove CO_2) and transportation of biomass limit the use of biogas (Jahangirian et al., 2009).

Methane is very light fuel gas. If we increase the number of hydrogen and carbon atoms, we have got progressively heavier gases, releasing more heat, therefore more energy, when ignited. Specific gravity of methane is 55 which is less than petrol & LPG. This means that biogas will rise if escaping, thus dissipating from the site of a leak. This important characteristic makes biogas safer than other fuels. It does not contain any toxic component; therefore there is no health hazard in handling of fuel. The calorific value of biogas is 5000-7000 Kcal/m^3. In calorific value, one cubic meter of biogas is equivalent to 0.7 m^3 of natural gas, 0.7 kg of fuel oil and 4 kWh of electricity (Asankulova & Obozov, 2007).

Motive power can be generated by using biogas in dual fuel internal combustion (IC) engine. Air mixed with biogas is aspirated into the engine and the mixture is then compressed, raising its temperature to about 350°C, which is the self-ignition temperature of diesel. Biogas has a high (600°C) ignition temperature. Therefore, in order to initiate combustion of the charge, a small quantity of diesel is injected into the cylinder just before the end of compression. The charge is thus ignited and the process is continued smoothly. Converting a spark-ignition engine for biogas fueling requires replacement of the gasoline carburettor with a mixing valve (pressure-controlled venturi type or with throttle). A spark-ignition engine (gasoline engine) draws a mixture of fuel (gasoline or gas) and the required amount of combustion air. The charge is ignited by a spark plug at a comparably low compression ratio of between 8:1 and 12:1. Power control is affected by varying the mixture intake via a throttle (Biogas Digest, 2010). Biogas has very high octane number approximately 130. By comparison, gasoline is 90 to 94 & alcohol 105 at best. This means that a higher compression ratio engine can be used with biogas than petrol. Hence, cylinder head of the engine is faced so that clearance volume will be reduced and compression ratio can sufficiently increase. Thus volumetric efficiency and power output are increased.

4.2 Biofuels for GHGs emission reduction and air quality

Vehicular emissions from petroleum products in the form of CO, NOx, unburnt hydrocarbons and particulates are of high environmental concern especially in air pollution (Subramanian et al., 2005). Thermal power plants are a major source of SPM (suspended particulate matter) and solid waste. The inefficient burning of biomass causes exposure to various pollutants and is considered a major health hazard and has been shown to lead to lung and chest problems among women and children (Smith, 1987). Biofuels has a number of health and environmental benefits including improvement in air quality by reducing pollutant gas emissions relative to fossil fuels (Vasudevan et al., 2005). Therefore, it is imperative to develop and promote alternative energy sources that can lead to sustainability of the energy system. This would not only warrant major reforms in the energy policies and infrastructure, but also huge international investments.

4.2.1 Reduction in exhaust emission by ethanol

Ethanol is one of the best tools available today to reduce air pollution from vehicles. Ethanol-diesel emulsion gives beneficial results in terms of pollution emission reduction in engines (Jha, 2009; Knothe et al., 2006). It is found that a remarkable improvement in PM-NOx trade-off can be achieved by promoting the premixing based on the ethanol blend fuel having low evaporation temperature, large latent heat and low cetane number as well, in addition, based on a marked elongation of ignition delay due to the low cetane number fuel and the low oxygen intake charge (Ishida et al., 2010). As a result, very low levels of NOx and PM which satisfies the 2009 emission standards imposed on heavy duty diesel engines in Japan, were achieved without deterioration of brake thermal efficiency in the PCI engine fuelled with the 50% ethanol blend diesel fuel and the high exhaust gas recirculation (EGR) ratio. It is noticed that smoke can be reduced even by increasing the EGR ratio under the highly premixed condition (Ishida et al., 2010). A 41% reduction in particulate matter and 5% NOx and 27% CO emission has been observed with 15% ethanol blends. Emission tests conducted especially on ethanol-diesel blends (Table 4) confirm the effect of substantially reducing particulate matter (Prasad et al., 2007b).

Pollutant	Emission (%)		Emission (g/km)	
	10% ethanol	15 % ethanol	22% ethanol	100 % ethanol
Particulate matter	27	41	0.08	0.02
NOx	4	5	0.45	0.34
Carbon monoxide	20	27	0.76	0.65
Unburned hydrocarbons	-	-	0.004	0.02
Sulfur dioxide	-	-	0.064	0.0

Table 4. Reduction in pollution emission with different percentages of Ethanol blending

If blended at the refinery, as opposed to "splash blending" outside the refinery, ethanol-blended gasoline can reduce NOx emissions as well, thus further reducing the potential for smog. Compared with conventional unleaded gasoline, ethanol is a particulate-free burning fuel source that combusts with oxygen to form carbon dioxide, water and aldehydes. Gasoline produces 2.44 CO_2 equivalent kg/l and ethanol 1.94 (Popa, 2010). Since ethanol contains 2/3 of the energy per volume as gasoline, ethanol produces 19% more CO_2 than gasoline for the same energy. When compared to gasoline, depending on the production method, ethanol releases less green house gases and savings of GHG emissions from ethanol produced from various crops are seen (Wang et al., 2009). Ethanol could play an important role in reducing petroleum consumption by enabling a substantial increase in the fuel efficiency of gasoline engine vehicles. This ethanol boosted engine concept uses a small amount ethanol to increase the efficiency of use of a much larger amount of gasoline by approximately 30%. Gasoline consumption and the corresponding CO_2 emissions would thereby be reduced by approximately 25%. In combination with the additional reduction that results from the substitution of ethanol for gasoline as a fuel, the overall reduction in gasoline consumption and CO_2 emissions is greater than 30% (Cohn et al., 2005).

4.2.2 Atmospheric pollution reduction by biodiesel

Biodiesel is a clean-burning renewable fuel that is compatible with petroleum diesel and can be produced domestically. The biodiesel performs as well as diesel while reducing the emissions of particulate matter, carbon monoxide (CO), hydrocarbons, oxides of sulphur (SOx), particulate matter and smoke density (Ali et al., 1995; Bagley et al., 1998; Durbin et al., 2000; Koo & Leung, 2000). Biodiesel is considered as 'carbon neutral' because all the carbon dioxide (CO2) released during consumption had been sequestered from the atmosphere for the growth of vegetable oil crops (Barnwal and Sharma, 2005). Other environmental benefits of biodiesel include the fact that it is highly biodegradable and appear to reduce emissions of air toxics and carcinogens (relative to diesel). The benefits of 100% (B 100) and 20% (B 20) biodiesel blending, in terms of per cent pollutants emission reduction (Planning Commission of India, 2003) and reduction emission in g/km for 10 and 15 % blend (Vasudevan et al., 2005) is shown in Table 5. According to the EPA's Renewable Fuel Standards Program Regulatory Impact Analysis, released in February 2010, biodiesel from soy oil results an average of 57% reduction in greenhouse gases compared to fossil diesel, and biodiesel produced from waste grease results in an 86% reduction (Petracek, 2011).

Pollutant	Emissions reduction (%)		Emission (g/km)		
	B 100	B20	Diesel	B 10	B 15
Particulate matter	-30	-22	0.129	0.093	0.080
NOx	+13	+2	0.79	0.83	0.89
Carbon monoxide	-50	-20	0.77	0.65	0.62
Unburned hydrocarbons	-93	-30	0.37	0.22	0.16
Sulfur dioxide	-100	-20	--	---	---

*(-) and (+): Less and more % of pollutant emission from biodiesel in comparison to 100% diesel

Table 5. Reduction in pollution emission with different percentages of biodiesel blending

Biodiesel has higher cetane number, lower sulfur content and lower aromatics than that of conventional diesel fuel. It also reduces emissions due to presence of oxygen in the fuel (Subramanian et al., 2005). In addition, the exhaust emissions of sulfur oxides and sulfates (major components of acid rain) from biodiesel are essentially eliminated compared to diesel. Of the major exhaust pollutants, both unburned hydrocarbons and nitrogen oxides are ozone or smog forming precursors. The use of biodiesel results in a substantial reduction of unburned hydrocarbons. However, a marginal increase in NOx (1-6%) is reported (Table 5) for biodiesel use in many engines. Emissions of nitrogen oxides are either slightly reduced or slightly increased depending on the duty cycle of the engine and testing methods used. Based on engine testing, using the most stringent emissions testing protocols required by EPA for certification of fuels or fuel additives in the U.S., the overall ozone (smog) forming potential of the hydrocarbon exhaust emissions from biodiesel is nearly 50 percent less than that measured for diesel fuel (Petracek, 2011). The summary report given by NREL stated that the maximum estimated increase and decrease in daily maximum 1-hour or 8-hour ozone concentrations due to the use of either a 100% or 50% penetration of a B20 fuel in the HDDV fleet in any of the areas studied is +0.26 ppb and –1.20 ppb for 1-hour ozone and the 100% B20 fuel scenario. As the maximum ozone increase (+0.26 ppb) is well below 1 ppb, the use of biodiesel is estimated to have no measurable adverse impact on 1-hour or 8-hour ozone attainment in Southern California and the Eastern United States (Morris et al., 2003). The mass concentration of the particles/smoke decreased up to 33% when the engine burned 100% biodiesel as fuel, compared to the 100% petroleum diesel (Zou and Atkinson, 2003).

4.2.3 Atmospheric pollution reduction by biogas

The fossil fuels combustion leads to emission of air pollutants such as CO, NOx, SO_2, volatile organic compounds and particulates (Parashar et al., 2005). Biogas technology, besides supplying energy and manure, provides an excellent opportunity for reducing environmental hazards and pollution through substituting firewood for cooking, kerosene for lighting and cooking and chemical fertilizers (Pathak et al., 2009). The benefits of biogas are generally similar to those of natural gas. In addition, burning biogas reduces greenhouse gas (GHG) emissions; it reduces the net CO_2 release and prevents CH_4 release. Thus, biogas combustion is a potential means to satisfy various legislative and ecological constraints (Jahangirian et al., 2009). Borjesson & Berglund (2006) analyzed fuel-cycle emissions of CO2, CO, NOx, SO_2, hydrocarbons (HC), CH_4, and particles from a life-cycle perspective for biogas systems based on different digestion technologies and raw materials. They suggest that the overall environmental impact of biogas depends largely on the status of uncontrolled losses of CH_4, the end-use technology that is used, the raw material digested, and the energy efficiency in the biogas production chain.

Biogas is a smokeless fuel offering an excellent substitute for kerosene oil, cattle dung cake, agricultural residues and firewood which are used as fuel in most of the developing countries (MNES, 2006). Burning of kerosene, firewood and cattle dung cake as fuels emits 0.8 to 2.2, 0.7 to 4.0 g kg^{-1} NOx, and SO_2, respectively along with varying amounts of CO, volatile organic compounds, particulate matters, organic matter, black carbon and organic carbon (Table 6).

A family size biogas plant substitutes 316 L of kerosene, 5,535 kg firewood and 4,400 kg cattle dung cake per annum as fuels. Substitution of kerosene reduces emissions of NOx, SO_2 and CO by 0.7, 1.3, and 0.6 kg $year^{-1}$. Substitutions of firewood and cattle dung cake results in the reduction of 3.5 to 12.2, 3.9 to 6.2, 436.9 to 549.6 and 30.8 to 38.7 kg $year^{-1}$ NOx, SO_2, CO and volatile organic compounds, respectively. Total reductions of NOx, SO_2, CO and volatile organic compounds by a family size biogas plant are 16.4, 11.3, 987.0 and 69.7 kg $year^{-1}$ (Pathak et al., 2009).

Pollutants	Pollution reduction due to a biogas plant (kg $year^{-1}$)			
	Kerosene	Firewood	Dung cake	Total
Oxides of N (NOx)	0.7	12.2	3.5	16.4
Oxides of S (SO_x)	1.3	3.9	6.2	11.3
Carbon monoxide	0.6	549.6	436.9	987.1
Volatile organic compounds	0.2	38.7	30.8	69.7
Particulate $matter^{10}$	0.1	16.6	13.2	29.9
Particulate $matter^{<2.5}$	0.1	11.6	28.6	40.3
Organic matter	0.4	7.2	17.6	25.2
Black carbon	0.1	3.3	11.0	14.4
Organic carbon	0.1	19.4	55.4	74.9

Table 6. Pollution reductions due to use of biogas plant

The biogas used as vehicle fuel presents better characteristics than the natural gas (Table 7). Some disturbance still appears for the NOx emissions, but they stay below the EU norms.

Concerning CO_2, hydrocarbons and CO emissions, the biogas is far better than the Natural Gas used for Vehicles (NGV), (Traffic & Public Transport Authority, 2000).

Pollutant	Emission (g/km)		
	Diesel	Natural Gas	Biogas
Particulate matter	0.1	0.022	0.015
NOx	9.73	1.1	5.44
Carbon monoxide (CO)	0.2	0.4	0.08
Unburned Hydrocarbons (HC)	0.4	0.6	0.35
CO2	1053	524	223

Table 7. Pollution reductions due to biogas used as vehicle fuel

Methane has a greenhouse gas (GHG) heating factor 21 times higher than CO_2. Combustion of biogas converts methane into CO_2 and thereby reduces the GHG impact by over 20 times. Combustion of biogas reduces the flame temperature, which reduces NOx emissions since the main pathway for NOx formation is thermal (Lafay et al., 2007). The digester reduces emissions of methane, carbon dioxide and ammonia from manure while in the enclosed vessel. Combustion of the biogas releases some carbon dioxide and sulphur compounds back into the atmosphere. However this combustion process releases carbon dioxide, which was captured by plants in the last year by the crop fed to the animals in contrast to fossil fuels, which are releasing carbon from ancient biomass.

4.3 Effect of biofuels on health

The exhaust gases from transportation vehicles contain many types of gaseous and particulate air pollutants, including trace levels of some particulate polycyclic aromatic hydrocarbons (PAHs) which have adverse effects on human health (Prasad et al., 2007b; Subramanian et al., 2005). Burning of biomass or any solid fuel, most closely associated with air quality problems and has some negative impacts on health (Pathak et al., 2009), particularly when burned in household cooking/heating stoves where there is little or no ventilation. Exposure to particulates from biomass burning causes respiratory infections in children, and carbon monoxide is implicated with problems in pregnancy. Coal and biomass are also suspected of causing cancer, where exposure rates are high (Smith, 1993). Petroleum fuels produce aromatic compounds of a polycyclic nature which are responsible for producing cancer in humans. But increased levels of NOx and HC may effects the human health as these may contain carcinogenic HC as well. If these productions can be reduced then considerable reduction in cancer amongst human beings can be hoped for. So for all of these reasons and biofuel production should be increased to improve our environmental as well as physical health (Wang et al., 1997).
It is highly likely that the net public health impact of using biofuels is beneficial. This is likely true even if the alleged negative impacts of ethanol and biodiesel blending (NOx, permeation) are assumed to be true. This theory is supported by the fact that: (1) ethanol and biodiesel blending significantly reduces emissions of pollutants that are generally believed to pose the greatest public health threat (PM and Toxics i.e. Hazardous Air Pollutants or HAPs); and (2) the actual ozone impact of the alleged increases in NOx and permeation emissions, if assumed to be true, is negligible or extremely small (Coleman,

2011). Ozone levels are significantly increased, thereby increasing photochemical smog and aggravating medical problems such as asthma (Hulsey, 2006; Jacobson, 2007).

4.3.1 Bioethanol and human health

On the positive side, the use of alcohols and alcohol/petroleum blends in diesel engines has been shown to reduce emissions of the potentially carcinogenic carbonaceous soot particles (Gaffney et al., 1980; Wang et al., 1997). Dynamometer studies of the use of gasahol (10% ethanol in gasoline) in motor vehicles report an average decrease in total HC emissions of 5%, a decrease in CO emissions of 13% with an increase in NOx emissions of 5% (HEI, 1996). The same studies showed a decrease in the emissions of the air toxics, benzene and 1, 3-butadiene of 12% and 6%, while acetaldehyde emissions increased by 159%. Although the atmospheric reactivity of ethanol is much lower than that of gasoline, no significant change was reported in the overall atmospheric reactivity (Maximum Individual Risk, MIR) of the exhaust emissions from gasohol when the higher reactivity of acetaldehyde is included. In terms of the health-related PAH emissions, some marked reductions were demonstrated for less toxic gaseous PAHs such as naphthalene, but the particulate PAH emissions, which have more implications for adverse health effects, remaining virtually unchanged and did not show a statistically significant reduction (Zou & Atkinson, 2003).

4.3.2 Biodiesel and human health

The use of biodiesel in a conventional diesel engine results in a substantial reduction of unburned HC, CO and particulate matter compared to emissions from diesel fuel (Table 5). Biodiesel exhaust emission has been extensively characterized under field and laboratory conditions. Biodiesel reduces emissions of CO and CO_2 on a net lifecycle basis and contain fewer aromatic hydrocarbons. Biodiesel can also reduce the tailpipe emission of particulate matters. Vellguth (1983) proved that rapeseed oil methyl esters (RME) are an adequate substitute for fossil diesel fuel (DF). Bünger and his coworkers (1998) investigated the mutagenic and cytotoxic effects of diesel engine exhaust (DEE) from a modern passenger car using rapeseed oil methyl esters (RME) biodiesel as fuel and directly compared to DEE of DF derived from petroleum. The results indicated a higher mutagenic potency of DEE of DF compared to RME due to the lower content of polycyclic aromatic compounds (PAC) in RME exhaust. The existing engines can use 20% biodiesel blend without any modification and reduction in torque output (Vasudevan et al., 2005). The use of a B20 fuel in the HDDV fleet is estimated to reduce the per million risk of premature death due to exposure to air toxics in the SoCAB region of southern California by approximately 2% and 5% respectively (Table 8) for the 50% and 100% HDDV fleet penetration of B20 biodiesel in the HDDV fleet emission scenarios calculated with no indoor/outdoor (I/O) effects and accounting for I/O effects on an annual average and hourly basis, (Morris et al., 2003).

Scenario	Std Diesel	50% B20 Fuel		100% B20 Fuel	
	Risk	Risk	(%)	Risk	(%)
No I/O Effects	1950	1910	-2.1	1835	-5.9
Annual I/O Effects	1284	1261	-1.8	1216	-5.3
Hourly I/O Effects	1257	1235	-1.8	1191	-5.3

Table 8. Average risk (out of a million) of premature death for the standard diesel base case and the 50% and 100% penetration of B20 biodiesel in the HDDV fleet emission scenarios

Scientific research confirms that biodiesel exhaust has a less harmful impact on human health than petroleum diesel fuel. Pure biodiesel emissions have decreased levels of polycyclic aromatic hydrocarbons (PAH) and nitrited PAH compounds that have been identified as potential cancer causing compounds. Also, particulate matter, an emission linked to asthma and other diseases, is reduced by about 47 percent, and carbon monoxide, a poisonous gas, is reduced by about 48 percent (Sinobioenergy, 2011). Biodiesel is the only alternative fuel to have fully completed the health effects testing requirements of the 1990 Clean Air Act Amendments as biodiesel produces less sulfur emissions than regular diesel. The public health benefits of reduced particulate and HAP exposure from biofuels outweigh the negligible smog impact of any relative small NOx and permeation emissions increases from biofuels blends (Coleman, 2011).

4.3.3 Biogas and health benefits

Biogas can have significant health benefits especially in rural areas. According to the Integrated Environmental Impact Analysis carried out by Biogas Support Program for 600 biogas users and 600 non-users, four percent more non-biogas users have respiratory diseases (Tables 9) than those who own biogas plants (BSP, 2000).

Disease	Problems in the past (HHs)*		Present status of HHs	
	Yes	No	Improved	Remained same
Eye Infection	72	18	69	3
Cases of burning	29	71	28	1
Lung problem	38	62	33	5
Respiratory problems	42	58	34	8
Asthma	11	89	9	2
Dizziness/headache	27	93	16	11
Intestinal/diarrhea	58	42	14	44

Table 9. Health benefits of biogas

Qualitative information from various household surveys carried out by BSP has revealed that problems like respiratory illness, eye infection, asthma and lung problems have decreased after installing a biogas plant. According to the Biogas Users' Survey conducted in 2000 with 100 households (HHs*), biogas can have positive impacts on the health of its users. Out of 42 respondents who had respiratory problems in the past, it was reported that the problem has improved for 34 of them. Similarly, those who had problems like asthma, eye infections and lung problems found that their problems had decreased after displacing dirtier fuels with biogas. If parasitic diseases had previously been common, the improvement in hygiene also has economic benefits (reduced working time). The more fully the sludge is digested, the more pathogens are killed. High temperatures and long retention times are more hygienic. The following are the principal organisms killed in biogas plants: Typhoid, Paratyphoid, Cholera and dysentery bacteria (in one or two weeks), Hookworm and bilharzia (in three weeks), Tapeworm and roundworm die completely when the fermented slurry is dried in the sun. Biogas has a positive effect more on rural health conditions.

5. Fuel economy in biofuel blends engines

Ethanol (E100) consumption in an engine is approximately 51% higher than for gasoline since the energy per unit volume of ethanol is 34% lower than for gasoline (Chauhan et al., 2011). The higher compression ratios in an ethanol-only engine allow for increased power output and better fuel economy could be obtained with lower compression ratios than gasoline-powered engines. In flexible fuel vehicles, the lower compression ratio requires tunings that give the same output when using either gasoline or hydrated ethanol. A 2004 MIT study (Stauffe, 2006), and an earlier paper published by the Society of Automotive Engineers, identified a method to exploit the characteristics of fuel ethanol substantially better than mixing it with gasoline (Stokes et al., 2000). The improvement consists of using dual-fuel direct-injection of pure alcohol (or the azeotrope or E85) and gasoline, in any ratio up to 100% of either, in a turbocharged, high compression-ratio, small-displacement engine having performance similar to an engine having twice the displacement. Direct cylinder injection raises the already high octane rating of ethanol up to an effective 130 and resulted in over-all reduction of gasoline use and CO_2 emission of 30%.

Biodiesel blends can reduce emission levels of HC (hydrocarbons) and CO (carbon monoxide); however, biodiesel blends may somewhat increase emission levels of NOx (oxides of nitrogen) in some engines. Biodiesel blends, used in new, low emissions engines may not significantly affect emissions. B20 is most widely used by fleets in the United States, because B20 balances performance, EPA emission levels, costs, and availability. B20 is also the minimum blend level that qualifies as an alternative fuel, in compliance with the Energy Policy act of 1992. Blends lower than B20 are used regionally, depending on favorable tax incentives that vary from state to state. However, NOx and evaporative VOCs (permeation) are regulated to control ozone formation, and recent air shed model runs suggest that the use of ethanol (E10) and biodiesel (up to B20) do not measurably increase actual ozone levels. With regard to permeation emissions, it is useful to remember that permeation (an evaporative VOC) is a very small percentage of any state's overall gasoline hydrocarbon emissions inventory (e.g. ~ 4% in California), and that ethanol generally reduces tailpipe (i.e. non-evaporative) hydrocarbon emissions as (at least a partial) offset. Also, because ambient temperature is a primary catalyst for fuel permeation, states with colder climates than California will have much lower permeation rates (Coleman, 2011).

6. Further scope of biofuels on environmental benefits

Tackling air pollutions and climate change requires the simultaneous deployment of available commercial clean technologies, demonstration and commercialisation of technologies at the advanced research, development and demonstration stage and research into new technologies. So for centuries, biofuels has been playing a vital role in the provision of energy services at the household level. However, at the beginning of the 21st century large scale commercial use of biofuel is the most rapidly growing renewable energy source in the developed countries as well as developing countries. Several clean energy options are viable today and several others are likely to be so in the future, as technologies improve, costs are reduced, and the competitive landscape for biofuel technologies evolves.

The Intergovernmental Panel on Climate Change has considered a range of options for mitigating climate change and increased use of biomass for energy features in all of its scenarios. The biomass takes an increasing share of total energy over the next century, rising to 25±46% in 2100 in its five scenarios. In the biomass intensive energy scenario, with biomass providing for 46% of total energy in 2100, the target of stabilizing CO_2 in the atmosphere at present-day levels is approached. Annual CO_2 emissions fall from 6.2 Gt C in 1990 to 5.9 Gt C in 2025 and to 1.8 Gt C in 2100: this results in cumulative emissions of 448 Gt C between 1990 and 2100, compared to 1300 Gt C in their business-as-usual case (IPCC, 1996).

However, developing countries with tropical climates may have a comparative advantage in growing energy rich biomass and second generation technologies could enable expansion of the range of feedstock used from the traditional sugarcane, maize, and rapeseed to grasses and trees that can thrive in less fertile and more drought prone regions. Biodiesel production efforts are focused on using non-edible oil seeds from plants (*Jatropha curcas, Pongamia pinnata* and other tree borne oilseeds) and animal fats like fish oil. The focus is to encourage the use of wastelands and other unproductive land for the cultivation of these relatively hardy new biofuel crops so that biofuel feedstock crop cultivation does not compete with food crops for scarce agricultural land and water (Singh, 2009).

As ethanol yields improve or different feedstocks are introduced, ethanol production may become more economically feasible. Currently, research on improving ethanol yields from each unit of feedstock is underway using biotechnology. Also, as long as oil prices remain high, the economical use of other feedstocks such as cellulose, become viable. Environmental costs per unit of ethanol decline with higher biomass yield, lower fertilizer and fuel inputs into biomass production, and improvements in biomass to biofuel conversion efficiencies (Cassman and Liska, 2007). By-products such as straw or wood chips can be converted to ethanol. Fast growing species like switch grass can be grown on land not suitable for other cash crops and yield high levels of ethanol per unit area. The development of commercial cellulosic technology would allow agricultural residues to be used and increase ethanol yield per hectare.

The biogas plants may reduce the dependence on conventional sources of energy by the turn of the century, provided promotional efforts are continued. Although, cattle dung has been recognized as the chief raw material for bio-gas plants, other materials like night-soil, poultry litter and agricultural wastes are also used where they are socially acceptable.

Climate change, air quality and energy security will change the way energy is used and supplied over the next century. Supplying increasing amounts of clean and secure energy will be a challenge that will require a great deal of innovation and investment. There are plenty of biomass resource and technology options for biofuel productions that could lead to emissions reductions in the heat, transport and electricity sector, while improving energy security and air quality.

7. Conclusion

Biofuels are non-polluting, locally available, accessible and reliable fuels obtained from renewable sources. Biomass can act as a reservoir of carbon or as a direct substitute for fossil fuels with no net contribution to atmospheric CO_2 if produced and used sustainably. Fuel security and the reduction of air pollution are some of the fundamental gains of an expanded biofuels industry. When particularly favorable improvements in technology over the next decade are assumed, the costs of emissions from biofuel could be approximately equal to, but unlikely less than, those of conventional gasoline. Cellulosic ethanol holds the promise of yet greater environmental benefits, but economical ways of producing it must first be discovered. New biofuel feedstocks especially low input cultivation of non-food crops (e.g., *Jatropha*, hybrid poplar, new varieties of switchgrass, and better multispecies plant mixtures) and algal biodiesel production technology may also yield substantial improvements. Biofuel markets can serve as an opportunity to trigger additional investments that could lead to increased production of food as well as biofuel crops by small-scale farmers. Further research on the use of indigenous non-food crops should be encouraged. Conversely, other ways of increasing biofuel production may increase air pollutant emissions unless accompanied by simultaneous improvements in abatement technology. Consideration should also be given to improved emissions controls and increases in fuel efficiency and fuel conservation that would reduce the need for increased fuel imports. Thus biofuels provide lots of environmental benefits including reduction of greenhouse gas emissions, improvement in air quality, reduction of fossil fuel use, increased national energy security, increased rural development and a sustainable fuel supply for the future and it also requires careful assessment on its impact of the environment especially in lowering greenhouse emissions.

8. References

Akkerman, M.; Janssen, J.; Rocha & Wijffels, R.H. (2002). Photobiological hydrogen production: photochemical efficiency and bioreactor design, *International Journal of* Hydrogen Energy, Vol.27, pp. 1195–1208

Al-Hasan, M. (2003). Effect of ethanol–unleaded gasoline blends on engine performance and exhaust emission. *Energy Conversion and Management*, Vol.44, pp. 1547–1561

Ali, Y.; Hanna, M. A. & Lecviticus, L. I. (1995). Emissions and power characteristics of diesel engines on methyl soyate and diesel fuel blends. *Bioresource Technology,* Vol.52, pp. 185-195

Anonymous, (2008). Biofuels Research in the CGIAR: A Perspective from the Science Council. Position paper, p.27, Food and Aquiculture Organization of the United Nations FAO, Rome

Anonymous, (2010a). Biodiesel. Fuel fact sheets, 15.03.2010, Available from http://www.biodiesel.org/pdf_files/fuelfactsheets/Lubricity.pdf

Anonymous, (2010b). Generic biodiesel material safety data sheet, 15.03.2010, Available from http://www.biodiesel.org/pdf _files/fuelfactsheets/MSDS.pdf

Asankulova, A., & Obozov, A. D. (2007). Biogas in Kyrgyzstan. *Applied Solar Energy* Vol.43, No.4, pp. 262-265

Bagley, S.T.; Gratz, L.D.; Johnson, J.H. & McDonald, J. F. (1998). Effects of oxidation catalytic converter and a biodiesel fuel on the chemical, mutagenic and particle size characteristics of emissions from a diesel engine. *Environmental Sciences and Technology, Vol.*32, pp. 1183- 1191

Barnwal, B.K. & Sharma, M.P. (2005). Prospects of biodiesel production from vegetable oils in India. *Renewable and Sustainable Energy Review*, Vol.9, pp. 363–78

Battelle, 1998. Flammability limits for ethanol/diesel blends. Final Report prepared by Battelle, Columbus, OH, USA.

Bedoya, I.D.; Arrieta, A.A. & Cadavid, F.J. (2009). Effects of mixing system and pilot fuel quality on diesel–biogas dual fuel engine performance. *Bioresource Technology,* Vol.100, pp. 6624–6629

Biogas Digest, (2010). Basics Information and Advisory Service on Appropriate Technology, *Volume I.* 15.11.2010, available from http://www.gtz.de/de/dokumente/en-biogas-volume1.pdf

Borjesson, P. & Berglund, M. (2006). Environmental systems analysis of biogas systems-Part I: Fuel-cycle emissions. *Biomass and Bioenergy,* Vol.30, No. 5, p. 469

BSP, (2000). Biogas Support Program (BSP). Winrock International, Nepal, Biogas Support Program Nepal http://practicalaction.org/docs/energy/docs50/bp50-nepal-biogas.pdf

Budny, D. & Sotero, P. (2007). Brazil Institute Special Report: The Global Dynamics of Biofuels, Brazil Institute of the Woodrow Wilson Center, 03.05.2008, Available from http://www.wilsoncenter.org/topics/pubs/Brazil_SR_e3.pdf

Bünger, J.; Krahl, J.; Franke , H., Munack, A., Hallier, E. (1998). Mutagenic and cytotoxic effects of exhaust particulate matter of biodiesel compared to fossil diesel fuel. *Mutation Research*, Vol. 415, pp.13–23

Cassman, K.G. & Liska, A. (2007). Food and fuel for all: realistic or foolish? *Biofuels, Bioproducts and Biorefining,* Vol.1, pp. 18-23.

Chauhan, B.S.; Kumar, N.; Pal, S.S. & Yong, D.J. (2011). Experimental studies on fumigation of ethanol in a small capacity Diesel engine. Energy, Vol.36, pp. 1030-1038

Chhang, D.Y.Z.; Van Gerpen, J.H., Lee, I.; Johnson, L.A.; Hammond, E.G. & Marley, S.J. (1996). Fuel properties and emissions of soybean oil esters as diesel fuel. *J American oil and Chemical Society,* Vol.73, p.1549

Chisti, Y., (2008). Biodiesel from microalgae beats bioethanol. *Trends in Biotechnology,* Vol.26, pp. 126-131

Cohn, D.R.; Bromberg, L. & Heywood, J.B. (2005). Direct Injection Ethanol Boosted Gasoline Engines: Biofuel Leveraging For Cost Effective Reduction of Oil Dependence and CO_2 Emissions. April 20, 2005, available from http://ethanolboost.com/LFEE-2005-01.pdf Massachusetts Institute of Technology, Cambridge, MA 02139

Coleman, Brooke (2011). A brief Summary of Air Quality and Impacts of Biofuels. Available from http://www.nebiofuels.org/pdfs/AQ_Summary.pdf

Delfort, B.; Durand, I.; Hillion, G.; Jaecker-Voirol, A. & Montagne, X. (2008). Glycerin for new biodiesel formulation. *Oil Gas Science and Technology Reveiw IFP,* Vol.63, No.4, pp. 395-404

Duailibi, J. (2008). Ele é o falso vilão (in Portuguese). Veja Magazine, 03.05.2008, Available from http://veja.abril.com.br/300408/p_058.shtml

Dunahay, T.G.; Jarvis, E.E.; Dais S.S. & Roessler, (1996). P.G. Manipulation of microalgal lipid production using genetic engineering, *Applied Biochemistry and Biotechnology,* Vol.57–58, pp. 223–231

Durbin, T. D.; Collins, J. R.; Norbeck, J. M. & Smith, M. R. (2000). Effects of biodiesel, biodiesel blends, and a synthetic diesel on emissions from light heavy-duty diesel vehicles. *Environmental Science and Technology,* Vol 34, pp. 349-355

Durrett, T.P.; Bennmg, C. & Ohlrogge, J. (2008). Plant triaglycerols as feedstocks for the production of biofuels. *The Plant Journal* Vol.54, pp. 593-607

EIA, (2010). Energy Information and Administration, International Energy Statistics. 12.01.2010, Available from http://tonto.eia.doe.gov/cfapps/ipdbproject/IED Index3.cfm

El Bassam, N. (1998). Energy Plant Species-Their Use and Impact on Environment, p.321, London, James & James, Science Publishers Ltd

Elsayed, M.A; Matthews, R. & Mortimer, N.D. (2003). Carbon and Energy Balances for a Range of Biofuels Options. Resources Research Unit, Sheffield Hallam University

Enagri, (2010). Global ethanol production to reach 85.9 bn litres. *Enagri eMagazine*, Vol.47, pp. 20, ISSN 1750-6972, Lincolnshire, UK

EPA, (2009). U.S. Environmental Protection Agency, EPA Proposes New Regulations for the National Renewable Fuel Standard Program for 2010 and beyond, fact sheet, p.3, May 2009, Washington, DC

FAO, (2005). Bioenergy. Sustainable Development Department, FAO, Rome, Italy. 11.3.2006, Available from http://www.fao.org/sd/dim_en2/en2_050402_en.htm

FAO, (2008). FAO, The State of Food and Agriculture, Biofuels: Prospects, Risks and Opportunities, ISBN 978-92-5-105980-7, Chapter 3, p.30-32.

Farrell, A.E., Plevin, R.J., Turner, B.T., Jones, A.D., O'Hare, M. & Kammen, D.M. (2006). *Science*, Vol.311, pp.506–508

Fedorov, A.S.; Kosourov, S., Ghirardi M.L. & Seibert, M. (2005). Continuous H2 photoproduction by Chlamydomonas reinhardtii using a novel two-stage, sulfate-limited chemostat system, *Applied Biochemistry and Biotechnology*, Vol.121-124, pp. 403–412

Finlayson-Pitts, B.J.; Pitts Jr., J.N., (2000). Chemistry of the Upper and Lower Atmosphere: Theory, Experiments, and Applications. Academic Press, San Diego

Fulton, L.; Howes, T. & Hardy, J. (2004). Biofuels for Transport: An International Perspective. *International Energy Agency*, Paris

Gaffney, J. S. & Marley, N.A. (2009). The impacts of combustion emissions on air quality and climate - From coal to biofuels and beyond, *Atmospheric Environment*, Vol.43, pp.23–36

Gaffney, J.S. & Marley, N.A. (2000). Alternative fuels. In: Brimblecombe, P., Maynard, R. (Eds.), Air Pollution Reviews. *The Urban Air Atmosphere and Its Effects*, vol.1, Chapter 6, pp. 195–246, Imperial College Press, London, UK

Gaffney, J.S.; Sapienza, R.; Butcher, T.; Krishna, C.; Marlow, W. & O'Hare, T. (1980). Soot reduction in diesel engines: a chemical approach. *Combustion Science and Technology*, Vol.24, pp. 89–92

Gavrilescu, M. & Chisti, Y. (2005). Biotechnology- a sustainable alternative for chemical industry, *Biotechnology Advances*, Vol.23, pp. 471–499

Ghirardi, M.L.; Zhang, J.P.; Lee, J.W.; Flynn, T.; Seibert, M. & Greenbaum, E. (2000). Microalgae: a green source of renewable H_2. *Trends in Biotechnology*, Vol.18, pp. 506–511

Goettemoeller, J. & Goettemoeller, A. (2007). Sustainable Ethanol: Biofuels, Biorefineries, Cellulosic Biomass, *Flex-Fuel Vehicles, and Sustainable Farming for Energy Independence*, p. 42, Prairie Oak Publishing, Maryville, Missouri

Granda, C.B.; Li Zhu & Holtzapple, M.T. (2007). Sustainable liquid biofuels and their environmental Impact. *Environmental Progress*, Vol.26, pp. 233-250

Guschina, I.A. & Harwood, J.L. (2006). Lipids and lipid metabolism in eukaryotic algae, Progress in Lipid Research, Vol.45, pp. 160–186

Hall, D.O. & House, J.I. (1993). Reducing atmospheric CO, using biomass energy and photobiology. *Energy Conversion Management*, Vol.34, pp.889-896

HEI, 1996. The potential health effects of oxygenates added to gasoline: a review of the current literature. Health Effects Institute, Cambridge, MA. 01.01.1996, Available from http://www.healtheffects.org/Pubs/oxysum.htm

Hoekman, S. Kent (2009). Biofuels in the U.S.-Challenges and Opportunities. *Renewable Energy*, Vol.34, pp.14–22

Hu, Q.; Sommerfeld, M.; Jarvis, E.; Ghirardi, M.; Posewitz, M.; Seibert, M. & Darzins, A. (2008). Microalgal triacylglycerols as feedstocks for biofuel production: perspectives and advances. *Plant Journal*, Vol.54, pp. 621-639

Huang, H.J.; Ramaswamy, S.; Tschirner, U.W. & Ramarao, B.V. (2008). A review of separation technologies in current and future biorefineries. *Separation and Purification Technology*, Vol.62, pp. 1–21

Hulsey, B. (2006). Cleaning the Air with Ethanol, Reports Ethanol Today, Available from www.betterenvironmentalsolutions.com/reports/EthanolToday.pdf, pp.56-57

IEA-ETE, (2007). IEA-Energy Technology Essentials-Biofuel Production (ETE02), January. 2007, pp.1-4, Available from http://www.iea.org/techno/essentials2.pdf

Indian express (2008). Petrol with 20% biofuel to be mandatory by 2017. http:/www.indian express.com/news/petrol-with-20-biofuel-to be mandatory-by/333541/

IPCC, (1996). Climate Change 1995: Impacts, adaptations and mitigation of climate change: Scienti®c-technical analysis, *Intergovernmental Panel on Climate Change Working Group* II report, Cambridge University Press, Cambridge, 1996

Ishida, M.; Yamamoto, S.; Ueki, H. & Sakaguchi, D. (2010). Remarkable Improvement of NOx-PM trade-off in a Diesel Engine by Means of Bioethanol and EGR. *Energy*, Vol.35, No.12, pp. 4572-4581

Jacobson, M. Z. (2007-03-14). Effects of Ethanol (E85) vs. Gasoline Vehicles on Cancer and Mortality in the United States. ACS Publications, 14.01.2008, available from http://pubs.acs.org/cgi-bin/sample.cgi/esthag/asap/html/es062085v.html

Jahangirian, S.; Engeda, A. & Wichman, I.S. (2009). Thermal and Chemical Structure of Biogas Counter flow Diffusion Flames. *Energy and Fuels*, Vol.23, pp. 5312–5321

Jha, S.K.; Fernando, S.; Columbus, E. & Willcutt, H. (2009). A Comparative Study of Exhaust Emissions Using Diesel-Biodiesel-Ethanol Blends in New and Used Engines. *Transactions of the American Society for Agricultural and Biological Engineering*, Vol.52, No.2, pp. 375-381

John, A.M. (2008). Carbon-negative biofuels. *Energy Policy*, Vol.36, pp. 940–945

Kapdan, I.K. & Kargi, F. (2006). Bio-hydrogen production from waste materials. *Enzyme and Microbial Technology*, Vol.38 pp. 569–582

Keay, D. (2007). Study warns of health risk from ethanol, San Francisco Chronicle, 18. 04. 2007, Available from http://sfgate.com/cgi-bin/article.cgi?file

Kim, S. & Dale, B.E. (2004). Global potential bioethanol production from wasted crops and crop residues. *Biomass and Bioenergy* Vol.26, pp. 361–375

Knothe, G. & Steidley, K.R. (2005). Lubricity of components of biodiesel and petrodiesel: The origin of biodiesel lubricity. *Energy and Fuels*, Vol.19, pp. 1192-1200

Knothe, G. (2006). Biodiesel and vegetable oil fuels: Then and now. Paper presented at the *97th American Oil Chemists Society Annual Meeting*, St. Louis, MO

Koo, B.C.P. & Leung, D.Y.C. (2000). Emission testing on a biodiesel produced from animal fats. In: *Proceedings of 3rd APCSEET*, pp. 242-246, ISBN 981-02-4549-1. World Scientific Publishing, Singapore, Hong Kong

Krisztina, U.; Scarpete D.; Panait, T. & Marcel, D. (2010). Thermo economical Performance Criteria in Using Biofuels for Internal Combustion Engines. *Advances in Energy Planning*, pp. 81-86

Kruse, 0.; Rupprecht, J.; Mussgnug, J.H.; Dismukes G.C. & Hankamer, B. (2005). Photosynthesis: a blueprint for solar energy capture and biohydrogen production technologies. *Photochemical and Photobiological Science*, Vol.4, pp. 957-969

Kumar, A. & Sharma, S. (2008). An evaluation of multipurpose Oil seed crop for industrial uses *Jatropha curcas* L.): a review. *Industrial Crops and Products* Vol.28, pp. 1-10

Lafay, Y.; Taupin, B.; Martins, G.; Cabot, G.; Renou, B. & Boukhalfa, A. (2007). Experimental study of biogas combustion using a gas turbine configuration. *Experimental Fluids*, Vol.43, No.2, p. 395

Larson, E.D. (2006). A review of life-cycle analysis studies on liquid biofuel systems for the transport sector, *Energy and Sustainable Development*, Vol.10, No.2, pp.109–126.

Lechon, Y.; Cabal, H.; de la Rua, C.; Caldes, N.; Santamarıa, M. & Saez, R. (2009). Energy and greenhouse gas emission savings of biofuels in Spain's transport fuel. The adoption of the EU policy on Biofuels. *Biomass and Bioenergy*, Vol.33, pp. 920-932

Levine, J.S. (1991). Global Biomass Burning: Atmospheric, Climatic, and Biospheric Implications. The MIT Press, pp.25-30, Cambridge, MA

Lynd, L.R.; Cushaman, J.H.; Nichols, R.J. & Wyman, C.E. (1991). Fuel ethanol from cellulosic biomass. *Science*, Vol.251, p. 1318

Ma, F. & Hanna, M. A. (1999). Biodiesel production: a review. *Bioresource Technology*, Vol.70, pp. 1-15

Marek, N. & Evanoff, J. (2001). The use of ethanol blended diesel fuel in unmodified, compression ignition engines: an interim case study. In: Proceedings of the Air and Waste Management Association, *94th Annual Conference and Exhibition*, Orlando, FL

Meiring, P.; Hansen, A.C.; Vosloo, A.P. & Lyne, P.W.L. (1983). High concentration ethanol-diesel blends for compression-ignition engines. SAE Technical Paper No.831360, Society of Automotive Engineers, Warrendale, PA

Metting, F.B. (1996). Biodiversity and application of microalgae, *Journal of Industrial Microbiology*, Vol.17, pp. 477–489

Metzger, P. & Largeau, C. (2005). Botryococcus braunii: a rich source for hydrocarbons and related ether lipids, *Applied Microbiology Biotechnology*, Vol.66, pp. 486–496

Mittelbach, M. & Remschmidt, C. (2004). Biodiesel, *The comprehensive handbook*, pp. 27-35, *Boersedruck Ges*, M.B.H., Vienna, Austria

MNES (2006). Renewable energy for rural applications. Ministry of Non-conventional Energy Sources, Government of India

Morris, R.E.; Pollack, A. K.; Mansell, G. E.; Lindhjem, C.; Jia, Y. & Wilson, G. (2003). Impact of Biodiesel Fuels on Air Quality and Human Health Summary Report September 16, 1999–January 31, 2003. NREL/SR-540-33793

Mudge, S.M. & Pereira, G. (1999). Stimulating the biodegradation of crude oil with biodiesel preliminary results. *Spill Science and Technology Bulletin*, Vol.5, pp. 353–355

Nicholas, Z. (2007). Coproducts Energy Value is rising, Ethanol Producer Magazine, October 2007

NREL, (2006). From biomass to biofuels. National Renewable Energy Laboratory, NREL/BR-510-39436

Oliveira, M.E.D. de; Vaughan, B.E. & Rykiel Jr., E.J. (2005) Ethanol as Fuel: Energy, Carbon Dioxide Balances, and Ecological Footprint, *BioScience*, Vol.55, No.7, pp. 593-602

Parashar, D.C.; Gadi, R.; Mandal, T.K. & Mitra, A. P. (2005). Carbonaceous aerosol emissions from India. *Atmospheric Environment*, Vol.39, pp. 7861–7871

Pathak, H.; Jain, N.; Bhatia A.; Mohanty, S. & Gupta, N. (2009). Global warming mitigation potential of biogas plants in India, *Environmental Monitoring and Assessment*, Vol.157, pp. 407–418

Peter, J.G., David, J.G & John N.S. (2003). Wood-ethanol for climate change mitigation in Canada. *Applied Biochemistry and Biotechnology*, Vol.105, No.1-3, pp. 231-242

Petracek, R. (2011). The Advantages of Using Biodiesel Blends. 26.02.2011, Available from http://www.buzzle.com/articles/the-advantages-of-using-biodiesel-blends.html

Pitkanen, J.; Aristidou, A.; Salusjarvi, L.; Ruohonen, L. & Penttila, M. (2003). Metabolic flux analysis of xylose metabolism in recombinant *Saccharomyces cerevisiae* using continuous culture. *Metabolic* Engineering, Vol.5, pp.16–31

Planning Commission, (2003). Report of Committee on Development of Biofuel. (2003). Planning Commission, Government of India, New Delhi, India

Popa, B. (2010). Emissions: Gasoline vs. Diesel vs. Bioethanol, 27.12.2010, Available from http://www.autoevolution.com/news/emissions-gasoline-vs-diesel-vs-bioethanol-3657.html, autoevolution.com

Prasad, S.; Singh, A. & Joshi, H.C. (2007a). Ethanol as an alternative fuel from agricultural, industrial and urban residues. *Resources Conservation and Recycling*, Vol.50, pp. 1–39

Prasad, S.; Singh, A.; Jain, N. & Joshi, H.C. (2007b). Ethanol production from sweet sorghum syrup for utilization as automotive fuel in India. *Energy and Fuels*, Vol.21, No.4, pp. 2415 -2420

Rao, P.S. & Gopalakrishnan, K.V. (1991). Vegetable oils and their methylesters as fuels for diesel engines. *Indian Journal of Technology*, Vol.29, pp. 292-297

REN21. (2008). RE Policy Network for the 21st Century Global Status Report, Available from http://www.martinot.info/RE2007_Global_Status_Report.pdf

Ryan, T.W. (1999). Characterization of vegetable oils for use as fuels in Diesel engines ASAE, 4/99, 1999

Sanderson, K. (2006). A field in ferment, Business feature, *Nature*, Vol.444, pp. 673-676

Sheehan, J.; Dunahay, T.; Benemann, J. & Roessler, P. (1998). A look back at the U.S. Department of Energy's Aquatic Species Program-biodiesel from algae, National Renewable Energy Laboratory, Golden, CO Report NREL/TP-580–24190

Sheehan, John J. (2009). Biofuels and the conundrum of sustainability. *Current Opinion in Biotechnology*, Vol.20, pp. 318–324

Shukla, P.R. (1997). Energy Strategies and Greenhouse Gas Mitigation: Models and Policy Analysis for India. *Allied Publishers*, New Delhi

Singh, S. (2009). India Biofuels. Annual Report No. IN9080. USDA, Foreign Agricultural Service. Global Agri Info Network, Available from http://gain.fas.usda.gov/

Sinobioenergy, 2011, Approval for use in USA, Executive Summary 1. 26.02.11 available from http://www.sinobioenergy.com/ enArticleView.aspx?id=99v

Smith, K. R. (1993). Fuel combustion, air pollution exposure, and health: The situation in developing countries. *Annual Review of Energy and Environment*, Vol.18, pp. 529-566

Smith, K.S. (1987). Biofuels, Air Pollution, and Health, *Plenum Publishers*, New York.

Somerville, C. (2006). The billion ton biofuels vision. *Science*, Vol.312, pp. 1277

Speidel, H.K.; Lightner, R.I & Ahmed, I. (2000). Biodegradability of new engineered fuels compared to econventional petroleum yuels and alternative fuels in current use. *Applied Biochemistry and Biotechnology*, Vol.86, pp. 879-897

Spolaore, P.; Joannis, C.; Duran, E. & Isambert, A. (2006). Commercial applications of microalgae. *Journal of Biosciences and Bioengineering*, Vol.101, pp. 87–96

Stauffe, N. (2006). Pint-sized engine promises high efficiency. MIT Tech Talk Vol.51, No.6, pp. 1-8. Available from http://web.mit.edu/newsoffice/2006/techtalk51-6.pdf

Stokes, J.; Lake, T.H. & Osborne, R.J. (2000). A Gasoline Engine Concept for Improved Fuel Economy–The Lean Boost System. (Technical Paper) SAE paper 2001-01-2902

Subramanian, K.A.; Singal, S.K.; Saxena, M. & Singhal, S. (2005). Utilization of liquid biofuels in automotive diesel engines: An Indian Perspective. *Biomass and Bioenergy*, Vol.29, pp. 65-72

Tachinardi, M.H. (2008). Por que a cana é melhor que o milho, (in Portuguese). *Época Magazine*, pp.73, 06.08.2008, Available from http://revistaepoca.globo.com/Revista/Epoca/0,,EMI5865-15273.html

Tilman, D.; Hill, J. & Lehman, C. (2006). Carbon-negative biofuels from low-input high diversity grassland biomass. *Science*, Vol.314, pp. 1598–1600

Traffic & Public Transport Authority, (2000). Technology and biogas use in Sweden, November 2000, City of Gothenburg, Sweden

USDoE, (2007). US department of energy on greenhouse gases. 09.09.2007, Available http://www.eia.doe.gov/oiaf/1605/ggccebro/chapter1.html

Vaivads, R.H.; Bardon, M.F.; Rao, V.K. & Battista, V. (1995). Flammability tests of alcohol/gasoline vapours. SAE Technical Paper 950401

Vasudevan, P.; Sharma, S. & Kumar, A. (2005). Liquid fuel from biomass: an Overview, *Journal of Scientific and Industrial Research*, Vol.64, pp. 822-831

Vellguth, G. (1983). Performance of vegetable oils and their monoesters as fuels for diesel engines, Society of Automotive Engineers, *SAE Technical paper series No. 831358*, Warrendale, PA, USA

Wang, M.; Saricks, C.; Santini, D. (2009). Effects of Fuel Ethanol Use on Fuel-Cycle Energy and Greenhouse Gas Emissions, Argonne National Laboratory. 07. 07. 2009, Available from http://www.transportation.anl.gov/pdfs/TA/58.pdf

Wang, W.G.; Clark, N.N.; Lyons, D.W.; Yang, R.M.; Gautam, M.; Bata, M. & Loth, J.L., (1997). Emissions comparisons from alternate fuel busses and diesel busses with a chassis dynamometer testing facility. *Environment Science and Technology*, Vol.31, pp. 3132–3137

Zhang, Y.; Dube, M. A.; McLean, D. D. & Kates, M. (2003). Biodiesel production from waste cooking oil: 2. Economic assessment and sensitivity analysis. *Bioresource Technology*, Vol. 90, pp. 229-240

Zou, L. & Atkinson, S. (2003). Characterizing vehicle emissions from the burning of biodiesel made from vegetable oil. *Environmental Technology*, Vol.24, No.10, pp. 1253-1260

13

Identification of Work Tasks Causing High Occupational Exposure to Bioaerosols at Biofuel Plants Converting Straw or Wood Chips

Anne Mette Madsen
National Research Centre for the Working Environment Lersø Parkallé 105, Copenhagen Denmark

1. Introduction

A bioaerosol is a suspension of airborne dust that contains living organisms or was released from living organisms such as fungi, bacteria, actinomycetes, pollen and other plant material. Exposure to bioaerosols containing high concentrations of fungi, actinomycetes, or endotoxin from bacteria may cause various deleterious health effects mainly on the airways (Douwes et al. 2003). The respiratory disorders caused by bioaerosol components can be dependent on the exposure levels (Rylander et al. 1985; Eduard et al. 2001). For example, a 60 spore m^{-3} increment in concentration of the fungus *Epicoccum* has been found to be associated with increased incidence of morning cough (Neas et al. 1996). At biofuel plants, large quantities of biofuel are handled and exposure to bioaerosol components and other particles occurs (Madsen 2006) which may cause respiratory disorders. An epidemiological study shows that the exposure level to microorganisms has an impact on the occurrence of respiratory symptoms among biofuel workers (Schlünssen et al. 2010). Inflammation in relation to respiratory disorders has been evaluated in mice exposed to airborne dust collected in two working areas at a biofuel plant. The study indicates that dust from a biofuel plant, at doses corresponding to two weeks of observed human endotoxin exposure, results in a strong inflammatory response. The airborne dust from the straw storage hall at the biofuel plant induced a stronger inflammatory response than dust from the boiler room and had the highest concentration of most microbial components (Madsen et al. 2008). In contrast, airborne dust collected from a boiler room at a straw plant were more toxic in terms of mutagenicity, PAH (polycyclic aromatic hydrocarbons) concentration and ability to generate reactive oxygen species than dust generated from straw and wood chips and than airborne dust sampled in a straw storage hall (Cohn et al. 2010). Also particles from combustion of dried animal drugs are described to be highly oxidative (Mudway et al. 2005).

Occupational exposure to bioaerosol components and the inflammatory potential of these bioaerosols are different in different working areas at the plants (Madsen 2006; Timm et al. 2009). Furthermore the microbial dustiness of different biofuels (straw, wood chips, wood pellets and wood briquettes) differs (Madsen et al. 2004) and also depends on the storage method and period (Sebastian et al. 2006). Some people at biofuel plants work for a whole day in the straw storage and this can cause a high exposure. The straw shredding area has also been identified as a high risk exposure area (Madsen 2006). The aim of this chapter is to identify factors influencing exposure to bioaerosols in straw storage halls and to reveal the impact on the exposure of different attempts to reduce exposure, e.g. sealing of a straw shredder. Empirical data showing the influence of opening outdoor gates while straw is unloaded are presented. Furthermore the impact of the quality of the biofuel handled in the straw reception on the human exposure is studied as well as the impact on the exposure of the water content of the handled straw.

2. Methods

2.1 The biofuel plants

The study included 18 biofuel plants situated all over Denmark. To make this study comparable with earlier publications of studies on the same plants, the same names as used in these previous papers have been used. Thus 13 plants are called a number between 4 and 24 as in another study (Madsen and Nielsen 2010), and five other plants are called plant A,B,C,D and E, also as in another study (Madsen 2006). The plants generated energy using straw or wood chips as the fuel. Airborne dust was sampled in working areas in combined straw receiving and storage halls, which in the following are called straw storage halls. At plants A and E, airborne dust was sampled in areas where work with wood chips was performed and at plants B, C and D dust was sampled where work with straw was performed. At 11 of the plants straw was received on both days of sampling; up to 36 trucks arrived per day with straw. On receipt, the water content in the received straw was measured using a straw bale moisture probe by the people working at the plants. Results varied between 8.1 and 24.0 percentage by dry weight and averages at each plant and each day varied between 10.2 and 15.2 (Madsen and Nielsen 2010). During unloading of straw the gates in the straw storage halls were sometimes open, allowing outdoor air in, and sometimes they were closed. After unloading the straw, the truck body was usually cleaned using a vacuum cleaner or brooms.

2.2 Sampling of airborne dust at the biofuel plants

Measurements were performed in the early spring, late autumn and winter season in 2000 to 2006 during two to four working days. The stationary sampling and the measurement of concentrations and aerodynamic diameters (d_{ae}) of particles were performed 1.5 m above floor level. 'Total dust' has been defined as the dust collected by a sampler with an entry velocity of 1.25 m/s (Kenny and Ogden 2000); 'total dust' was sampled at plant numbers 4 to 24 using 25 mm closed-face cassettes (Millipore holder; Millipore, Bedford, MA, USA, with an inlet velocity of 1.25 m/s). The samplers were fitted with Teflon filters (pore size 1.0 µm) for endotoxin, pH and gravimetric analysis and with polycarbonate filters (pore size 1.0 µm, GE Water & Process Technologies) for other analysis.

Personal dust monitoring at all plants and stationary sampling at plants A to E was conducted using GSP inhalable samplers (CIS by BGI, INC Waltham, MA) as described in (Madsen 2006). The samplers were mounted with Teflon filters (pore size 1.0 µm) for endotoxin and gravimetric analysis and with polycarbonate filters (pore size 1.0 µm) for other analysis.

After sampling, the filters were transported carefully to the laboratory, and different microbial analyses were performed (**Table 1**). All results are presented as time-weighted averages.

An APS (APS-3321; TSI Inc., USA) or a particle counter (GRIMM model 1200) measured the number concentration of particles from 0.75 to 19.8 µm (aerodynamic diameter abbreviated d_{ae}) over one minute intervals in straw storage halls. Data are included in this chapter for measures at plants 14, 15, 16 and 18. The theoretical aspiration of the APS is near 100% for particles as large as 20 µm (Peters et al. 2006). These particle data are used to show the variation in particle concentration as a function of work task and to study the effect of open versus closed gates during unloading of straw. Arrows are drawn in the figures pointing at the time where a certain task starts or occurs.

2.3 Dustiness of biofuel collected at the plants

To measure the microbial dustiness of biofuels handled at biofuel plants in autumn and spring, biofuels were sampled at plants A, B, C, D and E in autumn 2000 and spring 2001. The wood chips were sampled from chips craves and the straw carefully sampled from the floor in the straw storage hall immediately after it fell from the bales during unloading from trucks. Consequently one straw sample represents many straw bales. Subsequently the biofuel samples were stored at 9-15°C for 15 hours before the microbial dustiness was studied. The study was performed in triplicate.

A rotating drum was used to generate airborne dust. The dust generator was a rotating drum with horizontal axis and a volume of 3.3 m^3 as described previously (Breum et al. 1999; Madsen et al. 2004). The biofuel (3.0 kg) was loaded into the bottom of the drum, which was then rotated (7 rpm, 5 min). A vacuum pump attached downstream of the drum maintained an airflow of 420 l min^{-1} through the drum; excess HEPA-filtered replacement air was supplied at the opposite end of the drum, ensuring ambient pressure inside the drum. Dust for microbial analysis was sampled on filter cassettes with teflon filters in closed-faced field monitors (25 mm dia., 8 µm; Millipore, Bedford, USA) with a 5.6 mm inlet at an airflow of 1.9 l min^{-1} (1.25 m s^{-1} inlet velocity), and with polycarbonate filters (25 mm dia., 0.4 µm, Nucleopore, Cambridge, MA, USA) with a 4.4 mm inlet at an airflow of 1.9 l min^{-1} (2.07 m s^{-1} inlet velocity) in closed-faced field monitors .

The data on microbial dustiness was used to study the impact of quality of biofuels on the exposure measured at biofuel plants.

2.4 Gravimetric analysis and extraction of dust

The mass of the dust collected on the Teflon filters was determined by weighing the filters before and after dust sampling. Before weighing, the filters were equilibrated at constant air temperature and humidity for 20-24 hours. The dust on the Teflon filters was extracted in 10.0 ml pyrogen-free water with 0.05% Tween 20 by orbital shaking (300 rpm) at room temperature for 60 min and centrifuging (1000g) for 15 min. The dust on polycarbonate filters was extracted in 10.0 ml sterile 0.05 % Tween 80 and 0.85 % NaCl aqueous solution by shaking for a 15 min period (500 rpm) at room temperature.

2.5 Determination of endotoxin, NAGase activity and pH

The supernatant from the Teflon filters was analysed (in duplicate) for endotoxin using the kinetic Limulus Amoebocyte Lysate test (Kinetic-QCL endotoxin kit, BioWhittaker, Walkersville, Maryland, USA) as earlier described (Madsen 2006). A standard curve obtained from an *Escherichia coli* O55:B5 reference endotoxin was used to determine the concentrations in terms of endotoxin units (EU) (10.0 EU≈1.0 ng). pH was measured in the

supernatant from the dust suspensions from the Teflon filters using a pH meter (PHM220 LABpHmeter, Meterlab).
To quantify the activity of NAGase (EC3.2.1.30) in the supernatant from the polycarbonate filters, the release of *p*-nitrophenol from the substrate *p*-nitrophenol-N-acetyl-β-D-glucosaminide (Sigma Chemical Co. USA) was estimated (Madsen and Neergaard 1999). Activities are expressed as pmol sec^{-1} per m^3 air.

Measured component	Unit	Description
Bacteria:		
Bacteria	cfu (colony forming units)	Bacteria able to grow on an agar medium
Mesophilic actinomycetes	cfu	A group of bacteria (Gram positive) able to grow on an agar medium at 25°C
Thermophilic actinomycetes	cfu	A group of bacteria (Gram positive) able to grow on an agar medium at 55°C
'Total bacteria'	Number	Living and dead bacteria counted by microscopy
Endotoxin	EU (Endotoxin units)	Endotoxin is a cell wall component from Gram negative bacteria
Fungi:		
Fungi	cfu	Fungi (moulds) able to grow on an agar medium
'Total fungi'	Number	Living and dead fungal spores counted by microscopy
Aspergillus fumigatus	cfu	A living thermotolerant fungal species (mould), able to grow at 45 °C
NAGase	pmol/sec	An enzyme (a chitinase) mainly produced by fungi

Table 1. Measured microbial components

2.6 Quantification of microorganisms (CAMNEA)

Microorganisms were quantified using a modified CAMNEA method (Palmgren et al. 1986). The number of fungi cultivable on Dichloran Glycerol agar (DG 18 agar, Oxoid, Basingstoke, England) at 25 °C was counted. In addition, DG 18 agar plates were incubated at 45 °C to quantify cultivable *Aspergillus fumigatus*. Estimates were made, firstly of the number of bacteria cultivable at 25 °C on Nutrient agar (Oxoid, Basingstoke, England) with actidione (cycloheximide; 50 mg l^{-1}) and secondly of the number of mesophilic actinomycetes and thermophilic actinomycetes (55 °C) cultivable on respectively 10% and 100% Nutrient agar with actidione (cycloheximide; 50 mg/ l). The numbers of microorganisms are expressed as cfu (colony forming units) per m^3 air.
The total numbers of fungal spores and bacteria were determined after staining with 20 ppm acridine orange (Merck) in acetate buffer for 30 sec with subsequent filtration through a polycarbonate filter (25 mm, 0.4μm; Nuclepore, Cambridge, MA, USA). Fungi and bacteria were counted at a magnification of x1250 using epi-fluorescence microscopy (Orthoplan;

Leitz Wetzlar). The numbers of fungi were determined in forty randomly chosen fields or until at least 400 cells were counted and are presented as number per m^3.

2.7 Treatment of data

The influence of using a broom versus a central vacuum cleaner (plants 6 and 15), the influence of water content in straw (plants 4, 6, 7, 9, 11, 12, 15, 20, 21, 23 and 24), the influence of sealing a straw shredder (plant 18) and the influence of open versus closed gates (plant 18) on exposure was compared inside the plants. The influence of quality of biofuel (plants A, B, C, D, and E) was studied with plants as random effect. All analyses were performed in SAS 9.1.

Different numbers of trucks with straw arrived and unloaded straw at the straw storage halls over the two days of sampling at 11 biofuel plants. To be able to compare the exposure level on two days of sampling at the same plant, we balanced the exposure level with the number of trucks arriving with straw. Subsequently, the effect of water content in the handled straw on the exposure to 'total dust', *Aspergillus fumigatus*, thermophilic and mesophilic actinomycetes was calculated on the log-transformed data using Proc Mixed, with the biofuel plants as the random effect.

Pearson's correlation coefficients were calculated for the log-transformed data of concentrations measured at the biofuel plants and compared with the microbial dustiness of biofuels measured using the rotating drum. The effect of microbial dustiness of biofuels, kind of biofuel and season on the exposure to 'total dust', endotoxin, fungi and bacteria was calculated on the log-transformed data using Proc Mixed, with the biofuel plants as the random effect. The effect of kind of biofuel and season on the microbial dustiness of biofuels in terms of 'total dust', endotoxin, fungi and bacteria was calculated on the log-transformed data using Proc Mixed, also with the biofuel plants as the random effect.

The number of airborne particles measured during straw unloading with open versus closed gates and data concerning cleaning using a broom versus a vacuum cleaner were compared using Proc Anova. Data on exposure as affected by sealing a straw shredder were analysed using Proc GLM with pair-wise comparisons.

3. Results and discussion

3.1 Variation in particle exposure through day and night

Particle concentration was measured over three-and-a-half days in March 2006 in a straw storage hall. Results showed an increasing concentration in the morning after the start of work and a decreasing concentration in the afternoon after the end of the working day (about 16:00) (**Figure 1**). **Figures 2** and **3** also show low particle concentrations in the morning before working hours start between 6:30 and 7:00. The last day of exposure measured at **Figure 1** is a Friday, when people at the plant stopped working earlier (about 12:00), and the particle concentration also decreased earlier. During the night, particles were also aerosolised due to the automatic straw feeding (**Figure 1**). In the figure only particles with a d_{ae} between 0.97 and 7.7 µm are shown, as fungi is typically present in the air as particles with a d_{ae} between 2 and 5 µm, and bacteria as particles with a d_{ae} between 1 and 8 µm (Madsen et al. 2009). Many particles had an d_{ae} between 0.54 and 0.97µm but particles with this d_{ae} and d_{ae} between 0.97µm and 7.7 µm mainly followed the same pattern (**Figure**

3). However in periods with low activity such as before 7:00 and between 12:15 and 13:00, there was a high number of particles with a d_{ae} between 1.0 and 7.7 μm compared to particles with a d_{ae} 0.54 between 0.97μm.

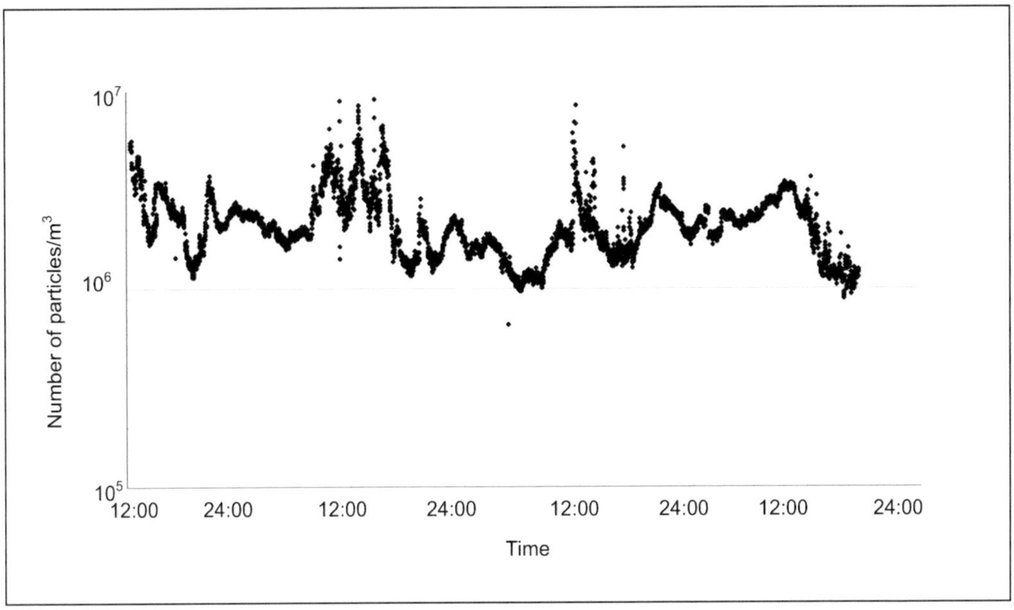

Fig. 1. Concentrations of airborne particles ($0.97 < d_{ae} < 7.7$ μm) in a straw storage hall at plant 14 as a function of time of the day. The measurements have been performed using an APS and the period shown is from a Tuesday in March 2006 at 12:00 and until midnight the following Friday

3.2 Unloading straw

At 92% of the biofuel plants, the engine of the trucks or tractors was shut off immediately after entering the straw storage hall. The first step in the unloading of straw was at most plants to remove a net covering the straw. Removal of the net caused an increase in particle exposure (example in **Figure 4**). Next the straw was removed using forklifts, cranes at the plant or, more rarely, cranes on the truck. Unloading of straw causes an increase in concentration of airborne particles (example in **Figures 2, 3** and **4**). At some plants the straw was unloaded and placed in the right place in one step (as in **Figures 1** and **3**), in some other plants it was done in more than one step (example in **Figures 2** and **4**). The extra reorganising of bales of straw can cause an extra exposure period which can cause a more than ten-fold increase in particle concentration, lasting for up to an hour. Based on these measurements it is suggested to explore the possibilities of reducing exposure by organising the unloading of straw and the subsequent straw feeding so that it is not necessary to move the straw bales once they have been unloaded.

3.3 Exposure as affected by open or closed gates

To assess the influence of open versus closed gates during unloading of straw, particle concentrations were measured in a period of four minutes before unloading the straw and during the first four minutes of unloading, when a big gate to the outdoor environment was either closed or open. When the gate was closed during unloading at plant 18, the particle concentration increased during the first four minutes of straw unloading by a factor of 2.9 to 4.4 (dependent on the particle size). When the gate was open, the concentration only increased by a factor 1.5 to 2.7 (**Table 2**). At plant 15 the highest increase in particle concentration (7.5 times) was found during unloading of the first load of straw in the morning and with closed gates (**Figure 3**).

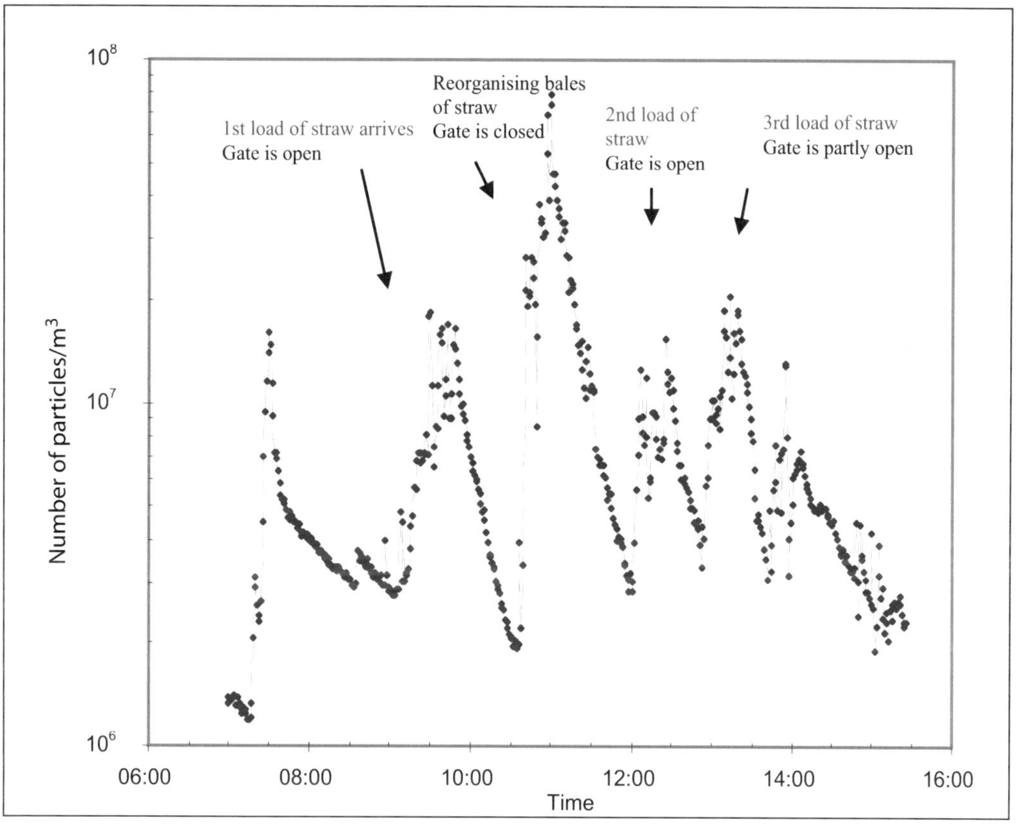

Fig. 2. Concentration of airborne particles ($1.0<d_{ae}<7.5\mu m$) in a straw storage hall at plant 18 as a function of time of the day. In the period measured, the three loads of straw were received and the gate was sometimes open and sometimes closed. The measurement was performed using a Grimm particle counter between 7:00 and 15:30

The half life period is the period from termination of unloading of straw and until the particle concentration has fallen by 50% of the difference between the peak and the level before unloading commenced. The clearance period is the period from termination of unloading of straw and until the particle concentration is at the same level as it was before unloading the straw. The half-life period and clearance period were lower when the outdoor gate was open than when it was closed (**Table 3**). The difference was significant for particles with d_{ae} of]0.75-1.0] ($p=0.041$),]1.0-2.0] ($p=0.044$),]5.0-7.5] ($p=0.047$) and]7.5-10.0] ($p=0.0108$) but not for particles with d_{ae} of]2.0-3.5] ($p=0.121$) and]3.5-5.0] ($p=0.64$).

Particle sizes* d_{ae} in µm	Number x10³/m³		Increase-factor	
	Closed	Open	Closed	Open
]0.75-1.0]	5600	2900	2.9	1.5
]1.0-2.0]	4900	1700	3.3	1.5
]2.0-3.5]	1200	430	3.7	1.5
]3.5-5.0]	3800	1100	4.3	1.8
]5.0-7.5]	550	120	4.2	2.6
]7.5-10.0]	36	8.4	4.2	2.2

*Measured using a Grimm particle counter

Table 2. Effect of open versus closed gates during unloading of straw at plant 18. Median concentration of particles during the first four minutes of unloading of straw and increase-factor in particle concentration in these four minutes of unloading relative to the preceding period

These data show that when opening the gates to the outdoor air, a dilution of the indoor bioaerosols occurs rather than an aerosolisation of settled dust or of particles on biofuels. The concentrations of bioaerosol components in the outdoor air in other industrial or urban areas (Nikkels et al. 1996; Nielsen et al. 2000; Park et al. 2000; Madsen 2006) are also described to be much lower than inside the biofuel plants. Opening gates could therefore be an obvious measure to reduce bioaerosol exposure.

	Half-life period				Clearance period			
Particle sizes*	Closed		Open		Closed		Open	
d_{ae} in µm	Minutes	SD	Minutes	SD	Minutes	SD	Minutes	SD
]0.75-1.0]	51	12.7	14	1.4	>70	-	22	2.5
]1.0-2.0]	51	12.9	14	0.71	>70	-	24	2.1
]2.0-3.5]	46	12.8	12	0.71	>70	-	27	1.8
]3.5-5.0]	29	4.2	8	1.1	>70	-	31	2.1
]5.0-7.5]	22	4.6	8	1.4	69	4.3	32	1.8
]7.5-10.0]	19	4.1	7	1.8	65	4.8	34	2.5

*Measured using a Grimm particle counter

Table 3. Half-life period and clearance period for concentrations of airborne particles from termination of unloading of straw at plant 18 when gates were closed (n=4), or when gates were open (n=2)

3.3 Cleaning

During or after the removal of the straw the body of the truck was cleaned. During unloading of bales of straw, pieces of straw were dropped on the floor, and the floor was sometimes cleaned using vacuum cleaners or brooms or other methods. In the example in **Figure 3,** straw is unloaded using forklifts and the truck body and floor are cleaned using a vacuum cleaner. During the cleaning of the floor the particle, concentration increased when using vacuum cleaners, brooms and compressors (**Figures 3** and **4**).

At two biofuel plants exposure to bioaerosol components was measured during the cleaning of the truck body using either brooms or central vacuum cleaners. The exposure levels to the different bioaerosol components were different at the two plants and the levels are presented separately in **Tables 4** and **5**. The personal exposure to different bioaerosol components was higher when cleaning the truck body using a broom than when using a vacuum cleaner (**Table 4** and **5**).

Bioaerosol components	Fraction (%)	Average exposure/m^{3a}
Endotoxin	77	147 EU
Inhalable dust	80	0.21 mg
'Total number of fungal spores'	29*	2.5×10^5 number
Aspergillus fumigatus	30*	738 cfu
NAGase	58*	0.38 pmol/sek
'Total number of bacteria'	20*	5.5×10^5 number
Mesophilic actinomycetes	56*	1377 cfu
pH	77	4.78 no unit
Particles d_{ae}]075-1.0]	28*	3.3×10^7 number
Particles d_{ae}]1.0-5.0]	34*	8.6×10^6 number
Particles d_{ae}]5.0-7.5]	85	4.8×10^5 number
Particles d_{ae}]7.5-10]	34*	7.8×10^4 number

[a]Exposure when the vacuum cleaner and not the broom was used. The exposure was measured for two persons during 2x2 days. Figures marked by an asterisk (*) were significantly different using a broom compared with a central vacuum cleaner

Table 4. Fraction (%) of personal exposure to bioaerosol components and particles in the straw storage hall at plant 15 using a broom for cleaning compared with using a central vacuum cleaner

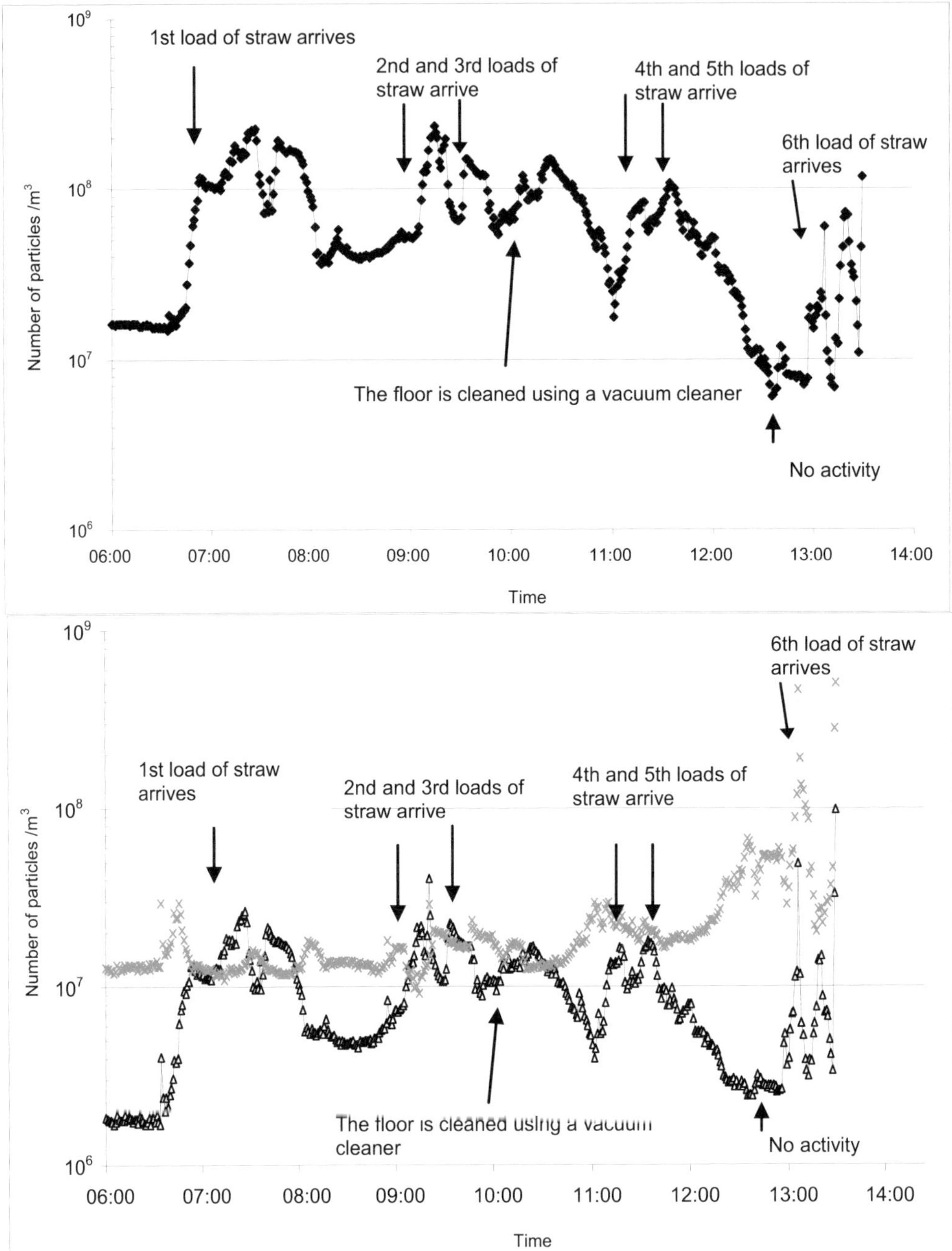

Fig. 3. Concentration of airborne particles ($0.54<d_{ae}<7.7\mu m$, top figure and $0.97<d_{ae}<7.7\mu m$, bottom figure, black symbols) in a straw storage hall as a function of time of the day. The grey symbols are the relation between large and small particles (($0.97<d_{ae}<7.7\mu m/0.54<d_{ae}<0.97\mu m$)*100,000). Six loads of straw were received and the gate was mainly closed. Unloading took between 15 and 20 minutes, and floor cleaning 25 minutes. The measurement was performed at plant 15 in autumn using an APS

When using a broom to clean at plant 15 the personal exposure was above a calculated "no effect level" (NOEL) of 150 EU/m^3 (Smid et al. 1992a; Smid 1993) and when using a central vacuum cleaner it was below the proposed NOEL (**Table 4**). At plant 6 the exposure to endotoxin was low compared to what has been found in straw storages earlier (Madsen 2006) and compared to plant 15 in this study, and it was lower than the suggested NOEL both when using a vacuum cleaner and brooms. Dust exposure was below the Danish Occupational Exposure limit (OEL) of 3 mg/m^3 (Danish Working Environment Authority (Arbejdstilsynet) 2007) both with and without use of the vacuum cleaner. When using the vacuum cleaner, the exposure to dust was reduced to 80% and 23% respectively of what it was when using brooms at the two plants. A study of sawmill workers has indicated that the lowest exposure causing symptoms in the throat is 3×10^5 fungal spores/m^3 (Alwis et al. 1999; Eduard, 2009). Exposure to fungi was reduced at both plants by using the central vacuum cleaners, but it still reached or exceeded this level. Exposure to the fungus *Aspergillus fumigatus* was not higher than a NOEL (Fogelmark et al. 1991) in both situations. Exposures larger than $2x10^4$ cfu of thermophilic actinomycetes m^{-3} have been suggested as a TLV (threshold limit value) (Dutkiewicz et al. 1994). This value was exceeded when using the broom but not when the central vacuum cleaner was used (**Table 5**). The pH of the dust suspensions seems to be affected by the presence of microorganisms - with a higher pH when more microorganisms were present. Mouldy hay causing farmers lung disease has earlier been described to be less acid than non-problematic hay (Gregory and Lacey 1963).

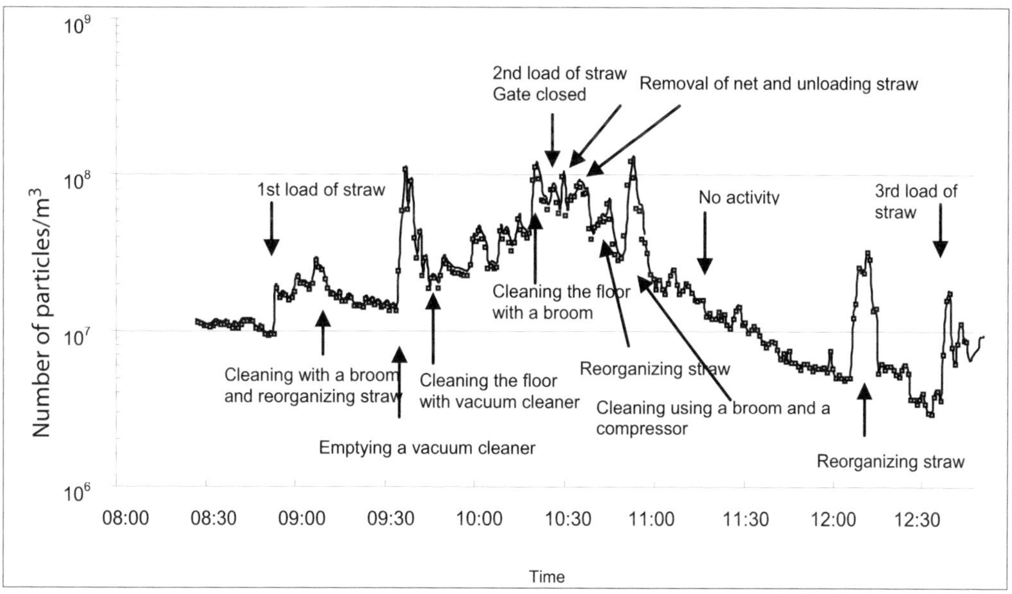

Fig. 4. Concentration of airborne particles ($0.54<d_{ae}<7.7\mu m$) in a straw storage hall at plant 16 as a function of time of the day. In the period measured, three loads of straw were received. The unloading was performed using forklifts and it took between 6 and 10 minutes. After unloading the straw, the bales of straw were reorganized. Between 9:48 and 10:25 cleaning activities were performed. The measurement was performed in spring 2004 using an APS

Bioaerosol components	Fraction (%)	Average exposure/m3a	
Endotoxin	133	29	EU
Inhalable dust	23*	0.32	mg
'Total number of fungal spores'	11*	2.4 x10^5	number
Aspergillus fumigatus	55*	1452	cfu
NAGase	9.3*	0.42	pmol/sek
'Total number of bacteria'	14*	7.4x10^5	number
Mesophilic actinomycetes	3.1*	2807	cfu
Thermophilic actinomycetes	1.2*	3608	cfu
pH	79	4.57	no unit

[a]Exposure when the vacuum cleaner and not the broom was used. The exposure was measured during 2x2 days. Figures marked by an asterisk (*) were significantly different using a broom compared with a central vacuum cleaner.

Table 5. Personal exposure to bioaerosol components in the straw storage hall at plant 6 using a broom for cleaning compared with using a central vacuum cleaner

The concentration of stationary measured particles of different sizes was also higher when cleaning the truck body using a broom than when using a vacuum cleaner (**Table 4**). Together the particle and bioaerosol exposure suggest that the exposure in the straw storage hall can be reduced by using a vacuum cleaner rather than a broom.

3.4 Exposure as affected by quality of the biofuel

To study the impact of the quality of biofuels on the exposure, exposure levels were compared with microbial dustiness of biofuels collected at biofuel plants. Correlation coefficients (r) between exposure in a working area and the microbial dustiness of the biofuel handled in the same area were 0.88 ($p<0.0001$), 0.77 ($p<0.0021$), 0.66 ($p<0.0001$) and 0.68 ($p<0.024$) for respectively endotoxin, cfu of bacteria, inhalable dust, and cfu of fungi (**Figure 5**). Statistical analysis showed that the quality of the biofuel when measured as dustiness in terms of endotoxin ($p<0.0001$), bacteria ($p<0.0001$), fungi ($p<0.0001$) and dust ($p<0.0001$) all had a significant effect on the exposure level. Also the season had a significant effect on the exposure to bacteria ($p=0.0003$), fungi ($p<0.0001$) and dust ($p<0.0001$), but not to endotoxin ($p=0.19$). In contrast the kind of biofuel handled (wood chips or straw) had no significant effect on exposure.

When the effect of season and kind of biofuel on the microbial dustiness of biofuels was studied separately (with plant as a random effect), significant effects of season on dustiness in terms of fungi ($p=0.011$) and dust ($p=0.0093$) but not of bacteria ($p=0.19$) and endotoxin ($p=0.79$) were found. The kind of fuel (straw versus wood chips) had a significant effect on dustiness in terms of bacteria ($p=0.0014$), endotoxin ($p<0.0001$) and dust ($p<0.0001$) but not of fungi ($p=0.10$). This higher dustiness of straw than of wood chips in terms of bacteria, endotoxin and dust supports earlier work (Madsen et al. 2004).

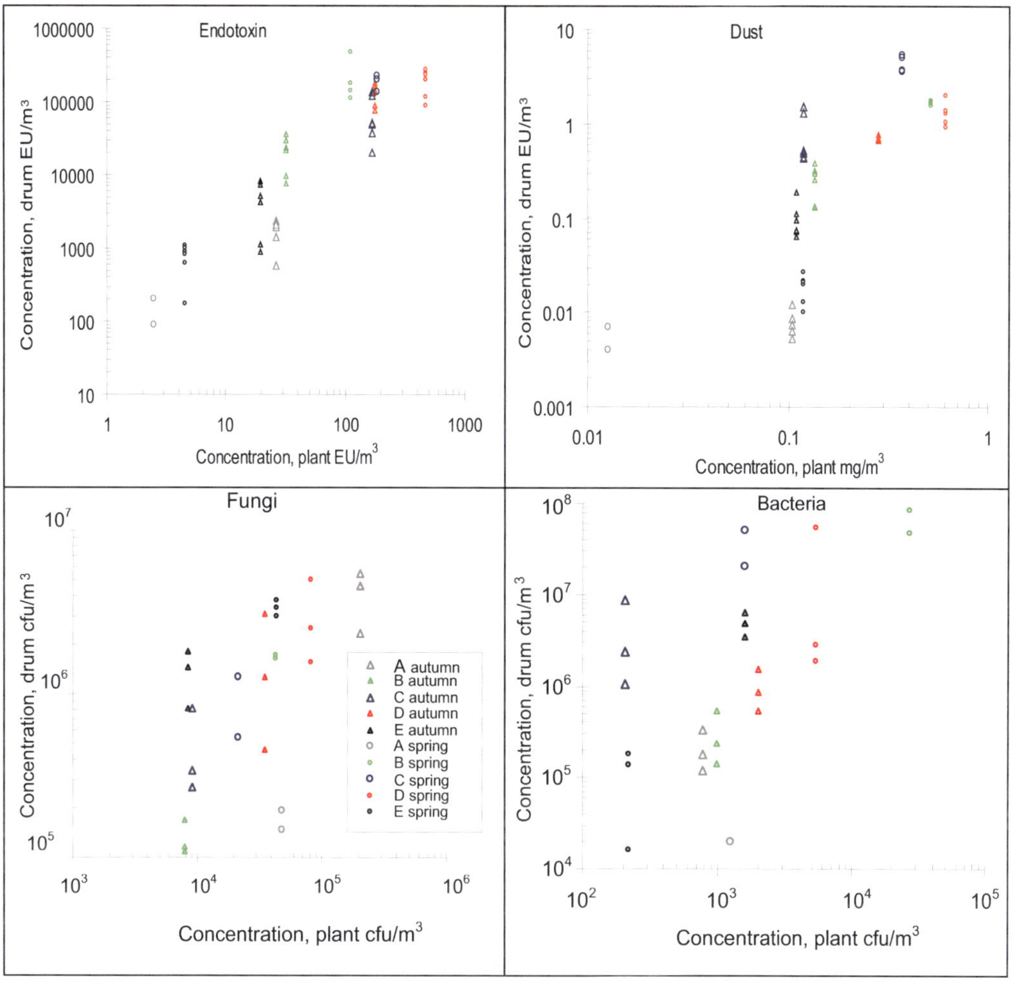

Fig. 5. Concentration (units/m³) of endotoxin, dust, fungi and bacteria in the air at five biofuel plants (A, B, C, D and E) versus concentration of these components released from straw or wood chips in a rotating drum. At plants B, C and D measurements were performed in a straw storage hall and at plants A and E measurements were performed where wood chips were unloaded

These positive associations between microbial dustiness on the exposure and the plant show an impact of the quality of the biofuel handled on the personal exposure. Checking the quality of straw and wood chips and rejecting problematic biofuel could thus be a measure to reduce exposure. There is however no easy way to evaluate the quality of biofuels regarding microbial dustiness, but the 'history' of the biofuel may give a hint about the quality of the biofuel. Thus 'storage history' may give a hint about the quality, as storing biofuels over summer outdoors increases their microbial dustiness (Sebastian et al. 2006). In a straw storage hall, higher exposure to dust, fungi, actinomycetes and bacteria is found in spring than in autumn (Madsen 2006); and as this study shows, there is a higher dustiness of biofuels in terms of fungi and dust in spring than in autumn. Furthermore the location where the biofuel sample is taken should also be considered, as samples taken from the inner part of a biofuel pile are dustier than samples taken from the surface (Sebastian et al. 2006). The kind of biofuel handled (e.g. wood chips, bark chips, straw or wood pellets) (Thörnqvist and Lundström 1982; Madsen et al. 2004; Madsen 2006) and the size of wood chips (Pellikka and Kotimaa 1983) should also be considered, as these factors have been shown to affect the microbial dustiness or the exposure. Furthermore storage of wood for chips as log stacks, rather than as wood chips, also affects the microbial dustiness (Thörnqvist and Lundström 1982) and could thus be considered when predicting the potential microbial dustiness of a material.

In relation to storage of biofuels, microorganisms and CO_2 formation should also be considered. Transport of logs and wood chips in confined spaces can result in rapid and severe oxygen depletion and CO_2 formation, possibly caused by microbial activity (Svedberg et al. 2009).

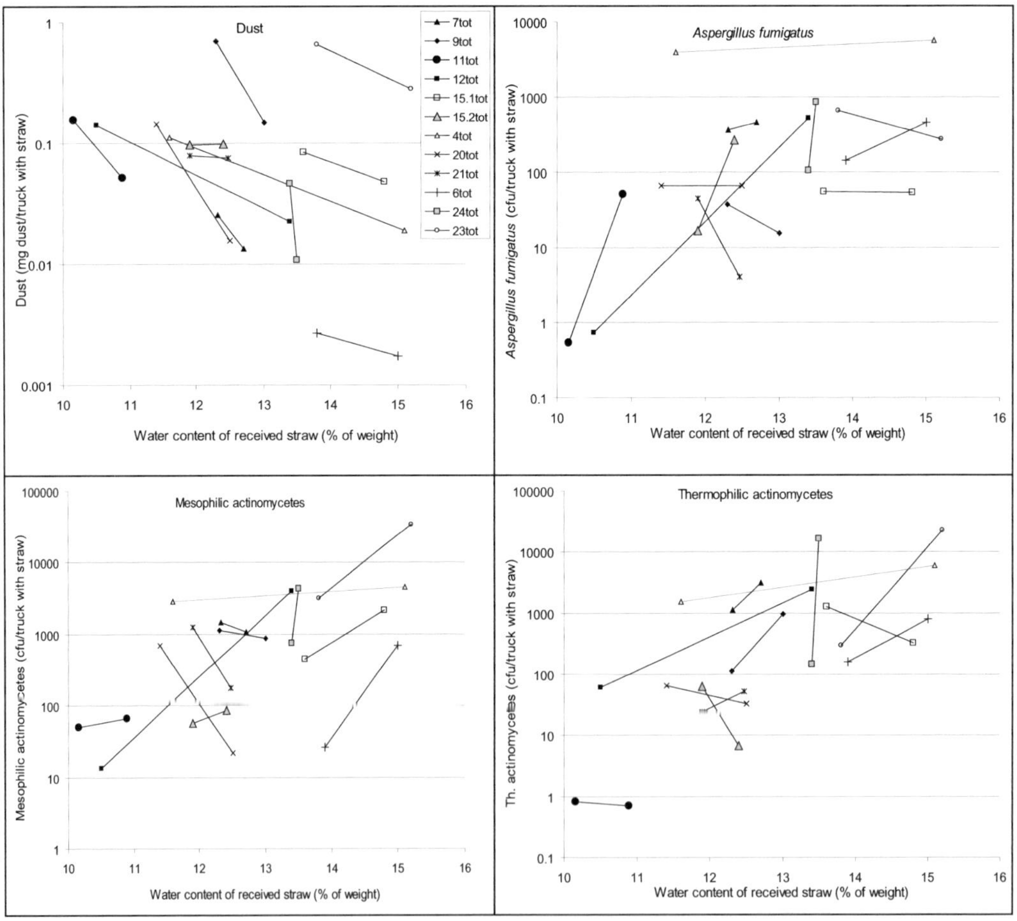

Fig. 6. Exposure to 'total dust' (mg/m^3/number of trucks with straw) and *Aspergillus fumigatus*, mesophilic and thermophilic actinomycetes (cfu/m^3/number of trucks with straw) as a function of water content (%) in the straw received during the two days of bioaerosol sampling at straw storage halls at plants 7, 9, 11, 12, 15, 4 , 20, 21, 6, 24 and 23

3.5 Water content in straw as an indicator of subsequent exposure

Water content in straw is usually measured by people working at the plants using straw bale moisture probes at reception of straw. Therefore whether water content of straw can be used as an indicator of a subsequent exposure level when working with the straw has been investigated.

At some plants (measurements from 24 days at 11 plants), straw was received on all days of bioaerosol sampling at each plant. The exposure to 'total dust', *Aspergillus fumigatus*, thermophilic and mesophilic actinomycetes per number of trucks with straw arriving at the straw storage hall was measured. Furthermore water content in the straw received at each plant was measured. The exposure to 'total dust' (p=0.0137) was lower on the days when the water content of the straw received was highest (**Figure 6**). Hence small increases in water content in the straw caused a lower exposure to dust. For *Aspergillus fumigatus* (p=0.0112) and mesophilic actinomycetes (p=0.0427) a significant effect of water content on exposure was also seen, although this association was opposite, with increasing water content associated with increasing exposure (**Figure 6**). For themorphilic actinomycetes (p=0.0536) no significant association was seen between exposure and water content of straw.

As for 'total dust' a higher water content in straw is also seen to cause a lower exposure to endotoxin. Water content in straw is seen to affect both the concentration, exposure level and size distribution of endotoxin-containing particles (Madsen and Nielsen 2010). The microorganisms measured, *Aspergillus fumigatus*, thermophilic actinomycetes and mesophilic actinomycetes, are living microorganisms, while endotoxin is from both living and dead Gram negative bacteria, and dust contains both living and dead microorganisms and other particles. The water content of an organic material may both affect the particle release and growth or sporulation of microorganisms. The effect of water content on dustiness of some materials, such as coal, is reviewed by (Hjemsted and Schneider 1996). Previous studies have shown that 'total dust' and endotoxin on the one side, and *Aspergillus fumigatus*, thermophilic actinomycetes and mesophilic actinomycetes on the other side are differently associated with biofuel (straw and wood chips), while actinomycetes and fungi seem to be more easily released from biofuel than other bacteria and endotoxin (Madsen et al. 2006). This may partly explain why *Aspergillus fumigatus* and actinomycetes were also easily released from the more wet straw.

Water content of straw is affected by the relative air humidity (rh); straw incubated at 20 °C and an rh of 54.4% has been shown to obtain a content of 11.8 % water, while straw stored at a rh of 81.3% has been shown to obtain a content of 17.7 % water (Lawrence et al. 2009). The water activity (a_w) level that limits the growth of the majority of bacteria is below 0.90 a_w and for fungi below 0.70 a_w. A water activity of 0.7 corresponds to a moisture content of 13%-15% in straw (Summers et al. 2003). Thus the water content in the bales of straw with the highest water content may have supported growth of some actinomycetes and fungi.

The average water content in the straw at the 11 biofuel plants was between 10.2 and 15.2% and none of the bales of straw was discarded or rejected because of high water content. This and the former study show that increasing water content may cause a higher exposure to both mesophilic and thermophilic actinomycetes and *Aspergillus fumigatus* and at the same time a lower exposure to dust and endotoxin.

3.6 Exposure before and after sealing a straw shredder

The concentration of airborne endotoxin (p=0.049), 'total number of microorganisms' (p=0.016) and NAGase (p=0.026) in the straw shredder room was significantly higher before than after sealing a straw shredder (**Figure 7**). The concentration of airborne dust (p=0.061) and 'total number of fungi' (p=0.065) tended to be higher in the straw shredder room before than after sealing the straw shredder.

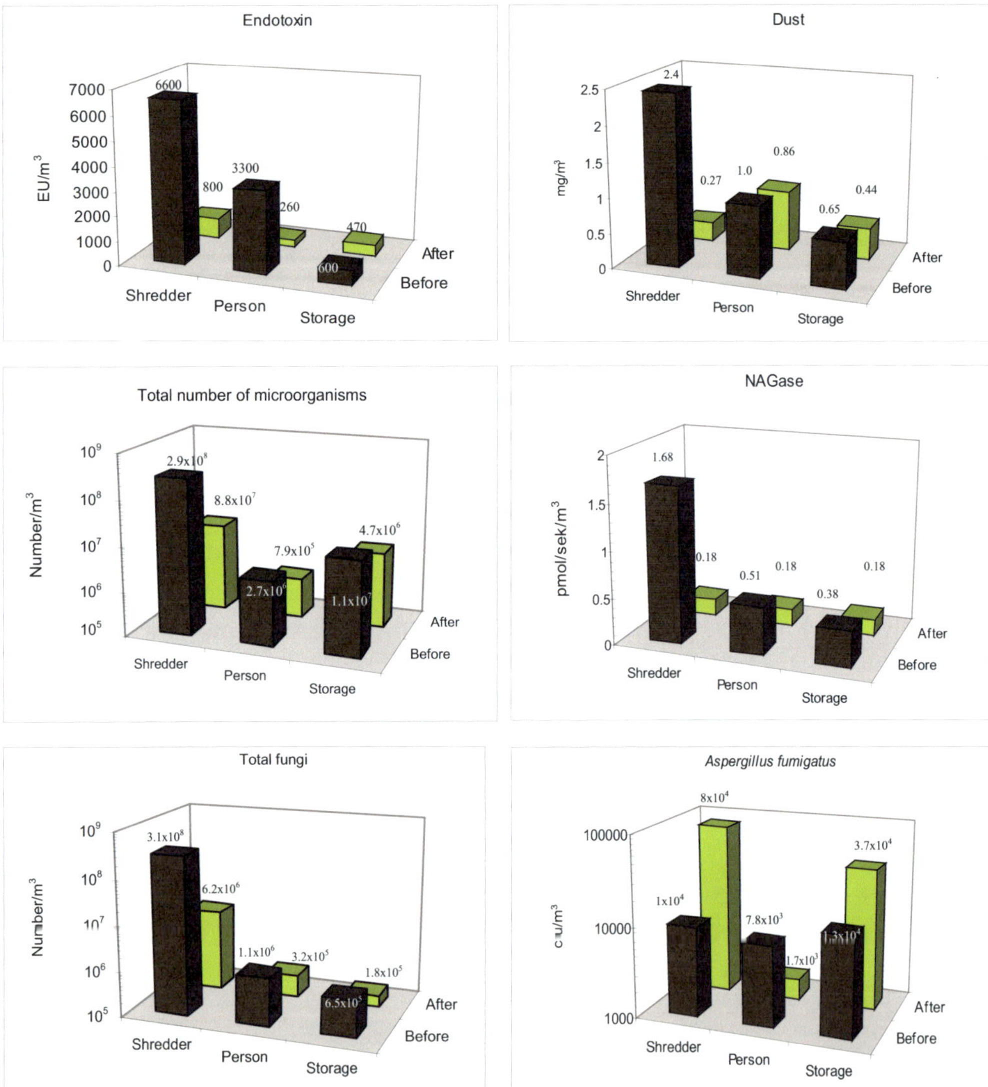

Fig. 7. Exposure to bioaerosol components before and after sealing a straw shredder at plant 18. 'Shredder' is stationary measurements in the straw shredder room; 'Person' is a personal exposure measurement of a person working in the straw storage hall and in the straw shredder room; 'Storage' is a stationary measurement in a straw storage hall next to the straw shredder room

Also the personal measured exposure and the concentration in the adjacent room - the straw storage hall - was affected positively by sealing the straw shredder.

Both before and after sealing the straw shredder, the concentration of endotoxin in the straw shredder room was considerably higher than the calculated NOEL of 150 EU/m^3. The personal exposure to endotoxin was also considerably lower after sealing but it was still higher than the NOEL. Also the exposure to dust was reduced significantly after sealing the straw shredder, and after sealing the dust concentration in the shredder room was lower than the Danish OEL.

In contrast to the other bioaerosol components, the concentration of *Aspergillus fumigatus* was significantly higher in the straw shredder room ($p=0.0045$) after sealing than before sealing. This may reflect differences in the quality of the straw in the two periods of exposure measure, because *Aspergillus fumigatus* is not always present in straw as it is a thermotolerant fungus, which is only predominant when heat is developed in a stored material like straw.

4. Conclusion

By measuring exposure to bioaerosol components using personal and stationary samplers and particle counters repeatedly at the same plant, it was possible to identify factors affecting the exposure level. Variations in concentrations of airborne particles were found through a day at biofuel plants. At some plants the straw was unloaded and placed in the right place in one step, in other plants this was done in more steps. The extra reorganising of bales of straw caused an extra increase in particle concentration lasting for up to an hour. It is suggested to explore the possibilities of reducing exposure by organising the unloading of straw and the following straw feeding so that reorganising the straw bales is not necessary.

In straw storage halls, unloading straw caused increased particle exposure. Using a broom to clean a truck body during and/or after unloading straw caused a higher exposure than cleaning using a central vacuum cleaner. Cleaning the straw storage hall caused a high exposure and cleaning using a compressor caused a peak exposure. It is recommended to investigate whether cleaning in the straw storage hall during the day between unloading deliveries of straw causes higher exposure than cleaning at the end of the day.

Open versus closed gates during straw unloading also affected the exposure significantly. Open gates caused a lower exposure, and from the data in this study it is suggested to open the gates while unloading straw. The water content in straw also influences the exposure level. While increasing water content causes a decreasing dustiness, the concentration of mesophilic actinomycetes and *Aspergillus fumigatus* in the dust increased, causing an increasing exposure to these living microorganisms.

The quality of biofuel, measured as microbial dustiness, had a significant effect on the exposure, with increasing microbial dustiness causing higher exposure. Consequently exposure may be reduced by using biofuel of high quality. The history of the biofuel may give information about its quality because quality is affected by the season and period and method of storage. Thus, higher dustiness, in terms of fungi and dust, is found in spring than in autumn. Furthermore straw has a higher dustiness, in terms of endotoxin, bacteria and dust, than wood chips.

Sealing a straw shredder caused a significantly lower exposure to bioaerosol components and can thus be recommended if a high exposure is found in this area.

5. Acknowledgements

Signe H. Nielsen, Margit W. Frederiksen and Tina T. Olsen are acknowledged for skilful technical assistance. We are particularly grateful to PSO- ELTRA (grant 4774 and 5786) for financial support. The workers at the biofuel plants are also greatly acknowledged for their involvement as well as Lars Lærkedahl (DONG Energy), Tove Kjær Hansen (DONG Energy), Mette Hansen (Dansk Fjernvarme) and Helle Mose Iversen (Vattenfall).

6. References

Arbejdstilsynet. (2007) At-vejledning. *Grænseværdier for stoffer og materialer* [The Danish Working Authority. Work Place Exposure Limits]. In . p. 1-85. ISBN .

Alwis KU, Mandryk J, Hoching AD (1999) Exposure to Biohazards in wood dust: bacteria, fungi, endotoxins, and (1,3) b-D-glucans. *Appl Occup Environ Hyg*; 14:598-608.

Breum NO, Würtz H, Ebbehøj N, Midtgård U (1999) Dustiness and bioaerosol exposure in sorting recyclable paper. *Waste Manage Res*; 17:100-108.

Cohn CA, Lemieux CL, Long AS, Kystol J, Vogel U, White PA, Madsen AM (2010) Physical-chemical and microbiological characterization, and mutagenic activity of airborne PM sampled in a biomass-fueled electrical production facility. *Environ Mol Mutagen*. DOI: 10.1002/em.20628

Douwes J, Thorne P, Pearce N, Heederik D (2003) Bioaerosols health effects and exposure assessment: progress and prospects. *Ann Occup Hyg*; 47:187-200.

Dutkiewicz J, Pomorski ZJH, Sitkowska J, Krysinska-Traczyk E, Skorska C, Prazmo Z, Cholewa G, Wojtowicz H (1994) Airborne microorganisms and endotoxin in animal houses. *Grana*; 33:85-90.

Eduard W, Douwes J, Mehl R, Heederik D, Melbostad E (2001) Short term exposure to airborne microbial agents during farm work: exposure-response relations with eye and respiratory symptoms. *Occup Environ Med*; 58:113-118.

Eduard, W (2009) Fungal spores: A critical review of the toxicological and epidemiological evidence as a basis for occupational exposure limit setting. *Critical Reviews in Toxicology* 39: 799-864.

Fogelmark B, Lacey J, Rylander R (1991) Experimental allergic alveolitis after exposure to different microorganisms. *Int J Exp Path*; 72:395.

Gregory PH, Lacey ME (1963) Mycological examination of dust from mouldy hay associated with farmer´s lung disease. *J gen microbiol*; 30:75-88.

Hjemsted K, Schneider T (1996) Documentation of a dustiness drum test. *Ann Occup Hyg*; 40:627-643.

Kenny LC, Ogden TL (2000) Twenty-five years of inhalable dust. *Ann Occup Hyg*; 44:561-563.

Lawrence M, Heath A, Walker P (2009) Determining moisture levels in straw bale construction. *Construction and Building Materials*; 23:2763-2768.

Madsen AM (2006) Airborne endotoxin in different background environments and seasons. *Ann Agric Environ Med*; 13:81-86.

Madsen AM (2006) Exposure to airborne microbial components in autumn and spring during work at Danish biofuel plants. *Ann Occup Hyg*; 50:821-831.

Madsen AM, Kruse P, Schneider T (2006) Characterization of microbial particle release from biomass and building material surfaces for inhalation exposure risk assessment. *Ann Occup Hyg*; 50:175-187.

Madsen AM, Mårtensson L, Schneider T, Larsson L (2004) Microbial dustiness and particle release of different biofuels. *Ann Occup Hyg*; 48:327-338.

Madsen AM, Neergaard Ed (1999) Interactions between *Pythium oligandrum* and sclerotia of the plant pathogen *Sclerotinia sclerotiorum*. *Eur J pl path*; 105:761-768.

Madsen AM, Nielsen SH (2010) Airborne endotoxin associated with particles of different sizes and affected by water content in handled straw. *Int J Hyg Environ Health*.

Madsen AM, Saber AT, Nordly P, Sharma AK, Wallin H, Vogel U (2008) Inflammation but no DNA (deoxyribonucleic acid) damage in mice exposed to airborne dust from a biofuel plant. Scand J work Environ Health; 34:278-287.

Madsen AM, Schlünssen V, Olsen TT, Sigsgaard T, Avci H (2009) Airborne fungal and bacterial components in PM_1 dust from biofuel plants. *Ann Occup Hyg*; 53:749-757.

Mudway IS, Duggan ST, Venkataraman C, Habib G, Kelly FJ, Grigg J (2005) Combustion of dried animal dung as biofuel results in the generation of highly redox active fine particulates. *Part Fibre Toxicol*; 2:6.

Neas LM, Dockery DW, Burge H, Koutrakis P, Speizer FE (1996) Fungus spores, air pollutants, and other determinants of peak expiratory flow rate in children. *Am J Epidemiol*; 143:797-807.

Nielsen BH, Nielsen EM, Breum NO (2000) Seasonal variation in bio-aerosol exposure during bio-waste collection and measurements of leaked percolate. *Waste Manage Res*; 18:64-72.

Nikkels AH, Terstegge P, Spieksma FThM (1996) Ten types of microscopcally identifiable fungal spores at Leiden, The Netherlands. *Aerobiologia*; 12:107-112.

Palmgren U, Ström G, Blomquist G, Malmberg P (1986) Collection of airborne micro-organisms on Nuclepore filters, estimation and analysis-CAMNEA method. *J Appl Bacteriol*; 61:401-406.

Park JH, Spiegelman DL, Burge HA, Gold DR, Chew GL, Milton DK (2000) Longitudinal study of dust and airborne endotoxin in the home. *Environ Health Perspect*; 108:1023-1028.

Pellikka M, Kotimaa M (1983) The mould dust concentration caused by the handling of fuel chips and its modifying factors. *Folia Forestalia*1-18.

Peters TM, Ott D, O'Shaughnessy PT (2006) Comparison of the Grimm 1.108 and 1.109 portable aerosol spectrometer to the TSI 3321 aerodynamic particle sizer for dry particles. *Ann Occup Hyg*; 50:843-850.

Rylander R, Haglind P, Lundholm M (1985) Endotoxin in cotton dust and respiratory function decrement among cotton workers in an experimental cardroom. *Am Rev Respir Dis*; 131:209-213.

Schlünssen V, Madsen AM, Skov S, Sigsgaard T (2010) Does the use of biomasss affect respiratory symptoms or lung function among male Danish Heat- and Power plant workers? *Occup Environ Med*; doi:10.1136/oem.2009.054403

Sebastian A, Madsen AM, Mårtensson L, Pomorska D, Larsson L (2006) Assessment of microbial exposure risks from handling of biofuel wood chips and straw - effect of outdoor storage. *Ann Agric Environ Med*; 13:139-145.

Smid T (1993) *Exposure To Organic Dust and Respiratory Disorders an epidemiological study in the animal feed industry*: CIP gegevens Koinklijke Bibliotheek Den Haag. ISBN .

Smid T, Heederik D, Houba R, Quanjer PH (1992) Dust- and endotoxin-related respiratory effects in the animal feed industry. *Am Rev Respir Dis*; 146:1474-1479.

Summers MD, Blunk SL, Jenkins BM (2003) How Straw Decomposes: Implications for Straw Bale Construction. *Ecological Building Network*-6pp.

Svedberg U, Petrini C, Johanson G (2009) Oxygen depletion and formation of toxic gases following sea transportation of logs and wood chips. *Ann Occup Hyg*; 53:779-787.

Thörnqvist T, Lundström H (1982) Health hazards caused by fungi in stored wood chips. *Forest products journal*; 32:29-32.

Timm M, Madsen AM, Hansen JV, Moesby L, Hansen EW (2009) Assessment of the total inflammatory potential of bioaerosols by using a granulocyte assay. *Appl Environ Microbiol*; 75:7655-7662.